# Lecture Notes in Applied and Computational Mechanics

## Volume 14

Series Editors

Prof. Dr.-Ing. Friedrich Pfeiffer
Prof. Dr.-Ing. Peter Wriggers

Springer

*Berlin*
*Heidelberg*
*New York*
*Hong Kong*
*London*
*Milan*
*Paris*
*Tokyo*

Engineering ONLINE LIBRARY

springeronline.com

# Novel Approaches in Civil Engineering

Michel Frémond
Franco Maceri (Editors)

Springer

Professor Dr. Michel Frémond
Laboratoire Lagrange
Laboratoire Central des Ponts et Chaussées
75732 Paris cedex 15
France

e-mail: fremond@lcpc.fr

Professor Dr. Franco Maceri
Laboratorio Lagrange
Università di Roma
Via di Tor Vergata 110
00133 Roma
Italy

e-mail: maceri@lagrange.uniroma2.it

With 209 Figures

Cataloging-in-Publication Data applied for
Bibliographic information published by Die Deutsche Bibliothek
Die Deutsche Bibliothek lists this publication in the Deutsche Nationalbibliografie; detailed bibliographic data is available in the Internet at <http://dnb.de>

ISBN 3-540-41836-9 Springer Verlag Berlin Heidelberg New York



Springer-Verlag is a part of Springer Science+Business Media

springeronline.com


Printed in Germany



Cover design: design & production GmbH, Heidelberg
Typesetting: Digital data supplied by editors

Printed on acid-free paper    62/3020Rw -5 4 3 2 1 0

# Preface

Ever since the ancient times, construction went along with growth and development of the human society: if not for other reasons, and in spite of the fact that the conception of the engineer as distinct from the architect or the master builder is relatively modern, civil engineering can be considered as old as civilization. In all ages, civil engineers were able to innovate and devise new ingenious schemes and techniques to cope with the new demands from the society: houses and water, for survival; roads, for mobility, a trait of freedom; public buildings, for life in associative form; and so on. In the 18th century, the term *ingénieur* was coined in France to denote a civil servant with a scientifically based training. This was a true turning point, from a conception of engineering - especially, civil - as a system of arts and techniques to one of engineering as the sagacious application of scientific principles. The successive, great evolution of civil engineering, as well as the accompanying impressive progress in developping new materials and structures, has always been strictly related to the posing, understanding and efficiently solving of a number of significant mathematical problems: more and more over the centuries, the central role of modelling and computation has been recognised. In this general framework, the contribution of Italian and French scientists has been crucial, structural elasticity being just a good example among many.

This book collects some research results of French and Italian engineers, mechanicians and mathematicians, all associated within the Lagrange Laboratory. Various novel approaches to problems of current interest in civil engineering are demonstrated. The topics covered range from dynamic and seismic problems to the analysis of long-span structures and ancient buildings; from studies on functionally-graded or composite materials to frictional or adhesive contact problems, including collisions; from granular materials to geotechnics.

The Lagrange Laboratory has benefitted from the support of many individuals and organizations. Special thanks are due to prof. Marie T. Janot-Giorgetti, scientific counsellor at the French Embassy in Italy, to prof. Lucio Bianco, president of CNR in Italy, and to prof. Catherine Brechignac, general director of CNRS in France. Finally, we wish to express our gratitude to Messrs. Salvatore Leonardi, Mario Bolognari and Giuseppe Biondo, majors of the cities of Messina, Taormina and Castelmola, in Sicily.

Archimedes' island hosted the Lagrange Laboratory's meeting from which this book originated. We believe that the contents offer a good perspective of the scientific approach of the French and Italian schools to modern civil engineering.

Rome,
June 30, 2003

*Michel Frémond*
*Franco Maceri*

# Contents

# An introduction to Unilateral Dynamics

Jean Jacques Moreau

Laboratoire de Mécanique et Génie Civil,
cc 048, Université Montpellier II,
F-34095 Montpellier Cedex 5, France

**Abstract.** The paper is devoted to mechanical systems with a finite number of degrees of freedom. After showing how inequality requirements in evolution problems can be handled through differential inclusions, one introduces dynamics by an elementary example of unilateral mechanical constraint. Then a general setting is constructed for multibody multicontact systems. The description of unilateral interaction at each possible contact point is formalized, with account of possible friction. This generates the numerical time-stepping policy called Contact Dynamics. The treatment of collisions or other frictional catastrophes in this framework leads to measure-differential inclusions, an essential tool in nonsmooth dynamics. The energy balance of nonsmooth evolutions is discussed. Two illustrations of the proposed numerical methods are presented. The former concerns the mechanisms of collapse of a bridge arch under local forcing. In the latter, the construction of a conical pile of grains is simulated, in order to investigate stresses in the bulk and the distribution of pressure on ground.

## 1 Purpose

### 1.1 Computation in multibody dynamics

The numerical dynamics of collections of bodies treated as perfectly indeformable, subject to the constraints of non-interpenetrability, with friction taken into account in the event of contact, currently is an active domain of research. Applications include the dynamics of machines, in particular robots, the dynamics of masonry works submitted to transient actions (earthquakes, gusts of wind or impacts), animated computer graphics and numerical simulation in granular mechanics. In all these domains, assuming the perfect indeformability of each part of the system leads to efficient numerical procedures which, in many circumstances, satisfy the needs. Possibly, some variables are added in order to also account for a certain deformability of these parts [2][37], without essentially changing the computational strategies. Of course, for the handling of non-interpenetrability and friction, much may be learned from the rich literature devoted to the treatment of contact between deformable media discretized through finite elements, even if in most cases only quasistatic evolutions are addressed (see e.g. [3][12][18][20][35] or, concerning a dynamical situation [75]).

The techniques used in multibody dynamics may be classified into the three following categories.

### 1.2 Event-driven methods

The methods so qualified (abbr. ED) are practical mainly when the concerned time-interval equals the union of not too many subintervals, a priori unknown, over which the *status* of the various contacts remain unchanged, i.e. no collision which would create new contacts occurs, no contact either gets loose, nor any critical situation needing a change in the analytical expression of the Coulomb law of dry friction is met. On each of these subintervals, the same numerical techniques as in the investigation of machines with classical bilateral, possibly frictional, constraints may be used [28]. As integration proceeds, there only is to watch the evolution of some indicators. In particular, the contact forces will be calculated. If, after a certain instant, some of the values computed for these forces are found to have directions incompatible with the unilaterality of the non-interpenetrability constraints (here we neglect adhesive, i.e. gluing effects), the programme decides that the motion has to be calculated otherwise. But one should keep in mind that the contacts which get loose after the critical instant are not necessarily those for which an unfeasible contact force has just been evaluated [22]. A popular approach to such discussions consists of reducing them to *complementarity problems*, similar to what is commonly met in constrained optimization.

More embarrassing is the calculation of the new velocity in case the critical instant is that of a *collision*. Contact forces take very large values during a very short time-interval and only poor phenomenological information is usually available about such physical circumstances. Even in the simplest case, that of the collision of two otherwise free members of the system, the traditional *coefficient of restitution* is known to depend not only on the materials these bodies are made of, but also on their shapes and relative orientations at the collision locus [72]; only the case of spherical beads appears relatively comfortable [25]. Furthermore, if some of the colliding bodies belong to clusters of already contacting ones, percussional reactions should be expected at all existing contact points. Though rigid body collisions currently are an active domain of research [9][13][16][27][73], computation has to rely on pragmatic rules whose validity has to be checked in each domain of application.

### 1.3 Smoothing methods

'Nonsmoothness' is the salient feature of the problems in view. In fact, after the set of the possible positions of the investigated system has been parametrized through an element $q$ of $\mathbf{R}^n$, the geometric restriction that the non-interpenetrability constraints impose on $q$ are expressed by a set of inequalities. Hence, instead of running in a smooth submanifold as in traditional analytical dynamics, the point $q$ is confined in a region of $\mathbf{R}^n$ whose boundary is made of a lot of pieces of hypersurfaces (millions or billions of them in current applications to granular materials): this is *nonsmoothness in space*. Furthermore, collisions are expected to induce velocity jumps: this is *nonsmoothness in time*. To end, the contact forces or 'reactions' associated with the non-interpenetrability constraints are governed by highly irregular laws. These forces vanish as soon as the corresponding contacts break while, if contact holds, the commonly stipulated mechanical conditions do not express them as functions of $q$. If, in addition, dry friction is taken into account (most usually in the form of Coulomb's law) it introduces some irregular relationships between contact forces and the sliding velocities. All this may be called *nonsmoothness in law*.

In such a state of the affairs, a natural move is to replace, approximately, the nonsmooth governing relationships by some *regularized* ones. First the non-interpenetrability constraints will be replaced by some stiff repulsion laws which take effect as soon as two members of the system come close to each other. This automatically handles the possible collisions, as far as one considers them as 'elastic', while the dissipativity of collisions may be accounted for by adding some damping actions or also by using different repulsion laws in the episodes of approach and of separation. Similarly, frictional contact may be somewhat regularized through the introduction of local elastic microdeformation and of viscosity-like effects. The dynamics of the approximate system is then governed by differential equations with sufficient regularity to be handled through standard numerical techniques. The drawback is that the need of precision requires the use of very stiff appproximate laws. Hence the time-stepping procedures applied have to resort to very small step-length and possibly also have to enforce numerical stability by introducing artificial damping or artificially increasing inertia. When treating dynamical applications, the effect of such an artificial alteration of the mechanical data may blur the picture. Significant simulations of loose (collisional) flows of granular materials have been obtained in that way, but when dense collections of bodies are concerned (pieces of masonry or compact granulates) the method is mainly applied to *quasi-static evolutions* in which only a succession of equilibrium states is looked for. Dynamical computation is then used only as a way of attaining each of these equilibrium states, a strategy referred to as 'Dynamic Relaxation'.

The pioneering work of P. Cundall [19] was precisely based on the regularization strategy, today implemented in the majority of commercial pieces of software intended to handle non-interpenetrability. Because such computation techniques are close to those applied in molecular simulations, they are commonly referred to as 'Molecular Dynamics' methods (abbr. MD), specially in the domain of granular mechanics [76].

## 1.4 Contact Dynamics

This is the technique (abbr. CD) advocated by the author, to which the present paper is meant to provide an introduction. It originated from [52], where the (unilateral) contact, possibly frictional and/or collisional, between rigid bodies received a formulation in terms of elementary convex analysis which proves suitable for computation. Mathematically, the resulting evolution problems are governed in smooth cases by *differential inclusions*. General information on the latter concept may be found in [5][21], but the need of treating also non-smooth evolutions calls for its extension to that of a *measure differential inclusion*. The Sweeping Process introduced earlier by the author, with motivation in the quasi-static evolution of elastoplastic systems [48][50] seems to have provided the first occurence of measure differential inclusions in literature. This process and some related evolution problems are still today the objet of mathematical research; see references in [42]. In Sect.2 of the present paper a description of the process, of purely kinematical nature, has been estimated able to make a tutorial introduction to the handling of unilaterality in mechanics.

A directing idea of the Contact Dynamics approach is that the main object of computation is the *velocity function* $t \mapsto u \in \mathbf{R}^n$. Time-stepping algorithms essentially have to determine the evolution of this function, by applying the principles of dynamics and the specified force laws. The position function $t \mapsto q$ is only to be updated at each step through adequate integration. Possibly, some members of $u$ are 'pseudo-parameters' such as the components of the spin vector of a solid, instead of the time-derivatives of position parameters. From the geometrical standpoint, one should observe that, in analytical dynamics, the position $q$ ranges in some differential manifold, while for each $q$ the possible velocities are elements of the tangent space to the manifold at this point. The latter is a *vector space* relatively to which such concepts as linearity, convexity, etc. make sense, while no algebraic concept of this sort is generally available in the position manifold.

Contact Dynamics procedures rest on drawing the balance of momentum of the investigated mechanical system over each time-step. No estimation of the acceleration is needed and the burden of calculating mathematically the curvatures of the involved surfaces is avoided. These curvatures are simply accounted for by the fact that, from one step to the other, the normal directions to the detected contacts vary.

The method results in time-stepping schemes which, at least in what concerns the velocity function, are of the *implicit* type. For this reason, a single computation step is needed in particular to decide whether *equilibrium* in a given position is a possible motion.

In principle, the Contact Dynamics algorithms are ready at each step to treat collisions on the same footing as persistent contacts but, of course, the need of physical information about such phenomena is not overcome.

### 1.5 Contents of the paper

The contents of Sect. 2 is purely kinematical, in other words it pertains to differential and integral calculus in the variables of time and space. There is shown how inequality requirements imposed on a moving point in $\mathbf{R}^n$ may be enforced through conditions involving its velocity vector. The example of the Sweeping Process, which is governed by a differential inclusion of first order, provides an introduction to the mathematical aspects of Unilateral Dynamics.

In Sect. 3 is developed the elementary example of the dynamics of a punctual particle unilaterally confined by a frictionless material boundary with prescribed motion. The traditional aim of rational mechanics, that of eliminating the unknown reactions of frictionless constraints, is attained in the form of a differential inclusion. In this simple mechanical example, the decisive step leading to the Contact Dynamics method is introduced. It consists in replacing the familiar formulation of frictionless contact by an apparently more complicated relationship involving velocities. The advantage of this transformation is illustrated by its application to a time-stepping scheme for the numerical approximation of the motion.

The framework for treating multibody multicontact systems is constructed in Sect. 4. This starts, as usual, with the parametrization of the system positions by means of an element $q$ of $\mathbf{R}^n$, after what the geometrical conditions of non-interpenetration of the system members and their possible confinement by external obstacles are assumed expressed by a finite set of inequalities. A contact corresponds to one of these inequalities being satisfied as equality. At every instant such that the derivative $dq/dt = u \in \mathbf{R}^n$ exists, every possible contact involves a vector $\mathcal{U}$ of local relative velocity of the contacting objects which, generically, is expressed as an affine function of $u$. As far as the contact actions consist of a simple force $\mathcal{R}$, calculating the element $r$ of $\mathbf{R}^n$ made of its generalized components is only the matter of matrix transposition. If the traditional assumption of differentiability of $t \mapsto u$ is made, the equations of the system dynamics may be written down through Lagrange's formalism or any other method of stereodynamics.

Exploiting these equations requires some information about the possible contact forces. Such an information receives a codified form in Sect. 5 under the name of a *contact law*, generically a relationship between $\mathcal{R}$ and $\mathcal{U}$, usually depending on the actual configuration attained by the system. Here again, the special case of frictionless contact is taken as an introduction. It allows one to stress the importance of the concept of a 'prospective' contact law. Roughly speaking such a law, rather than connecting the actual values of $\mathcal{R}$ and $\mathcal{U}$, pertains to their limits on the right of the concerned instant. The Coulomb law of dry friction receives an adaptation to such a formalism.

Then comes in Sect. 6 the application of the preceding to the construction of a time-stepping numerical scheme for the numerical approximation of solutions, first developed in the traditional framework of smooth motions, i.e. the unknown function $t \mapsto u$ is assumed locally absolutely continuous. This scheme is of the implicit type with regard to $u$. Due to the form given to contact laws and to their discretized forms, this time-stepping procedure automatically secures the preservation of the non-interpenetration conditions and manages the possibility of contact breaking.

In contrast, the event of the sudden introduction of new contacts, i.e. collisions, leads in Sect. 7 to giving up the framework of smooth dynamics, since $u$ is expected to exhibit jumps. Henceforth, the function $u$ is assumed to be of (locally) bounded variation on the considered time-interval. With such a function, there classically is associated an $\mathbf{R}^n$-valued measure $du$ on the time-interval, which may be called the differential measure of $u$. Smooth dynamics is retrieved as the special case where $du$ possesses a density function with regard to the Lebesgue measure $dt$.

The measure $du$ presents an atom at every instant of discontinuity of $u$, but this functional framework may also accomodate other sorts of frictional catastrophes, as referred to in 7.1. Instead of a classical differential equation, dynamics is now governed by a measure-differential equation. Contact actions are not necessarily represented as time-dependent forces but as impulsion measures. Their integrals on a time-interval constitute impulsions, a familiar concept in traditional dynamics.

Since the time-stepping numerical procedure of Sect. 6 consists of drawing the balance of impulsions on each time-step, it remains formally applicable to non-smooth evolutions. There does not seem objectionable to apply the precedingly identified contact laws so long as the measure $du$ is diffuse, i.e. $u$ continuous. The case of a collision occuring at some instant $t_c$ is more embarrassing since each contact-impulsion throughout the system is liable to possess an atom at point $t_c$. The weight of such an atom is nothing but the vector traditionally called a percussion, say $\mathcal{P}$. In 7.5 an efficient computation trick is introduced, by which $\mathcal{P}$ is connected, through Coulomb's law in the prospective form, with some weighted mean of the pre- and post-collision values of the local velocity $\mathcal{U}$. The resulting calculation is definitely richer than the traditional laws of restitution in that all contacts present in the system at instant $t_c$ are collectively taken into account, but its phenomenological validity has to be tested in each application context.

Nonsmooth dynamics does not allow one to draw energy balances as precise as those traditionally available for smooth evolutions. The reason is that some rules of the differential and integral calculus have to be replaced by a calculus of differential measures, shortly presented in Sect. 8. The thermodynamic correctness of collision models requires dissipativity, a property which is not always confirmed by energy balances, as drawn there.

As final illustrations, two examples of the numerical application of the foregoing are presented.

In Sect. 9 a two-dimensional model of a stone bridge is used to show how the collapse mechanism under some localized forcing depends on the value admitted for the interblock friction coefficient.

Section 10 is devoted to the numerical simulation of the construction of a conical pile of grains. Eager controversy took place in recent years about the experimental finding of a local minimum of ground pressure at the vertical of the pile apex. The numerical exploration of stress in a numerical simulation conducted by the CD method appears to remove any paradoxical features from this subject.

## 2 The differential handling of inequality conditions

### 2.1 Expressing viability in terms of velocity

Let us understand by a *moving point* a mapping $q$ of a (time) interval $I$ into $\mathbf{R}^n$. We want $q(t)$ to comply for every $t$ in $I$ with some inequality

$$f(t, q(t)) \leq 0, \tag{1}$$

where $f : I \times \mathbf{R}^n \to \mathbf{R}$ denotes a $\mathcal{C}^1$ real function. In other words, $q(t)$ is required to belong for every $t$ to $\Phi(t) := \{x \in \mathbf{R}^n \mid f(t, x) \leq 0\}$.

It is assumed that, for $t$ in $I$ and $x$ in $\mathbf{R}^n$, the *gradient* $\nabla f(t, x) := (\partial f / \partial x^1, \ldots, \partial f / \partial x^n)$ is a nonzero $n$-vector.

Let some $t$ be such that the *right-side derivative* $q'^{+}(t)$, called the *right-side velocity* of the moving point, exists. Through the chain rule, the real function $\tau \mapsto f(\tau, q(\tau))$ is found to possess at $\tau = t$ a right-side derivative equal to $f'_t(t, q(t)) + q'^{+}(t).\nabla f(t, q(t))$. This derivative should be $\leq 0$ if (1), assumed to hold everywhere in $I$, is satisfied at $t$ as *equality*. In contrast, if inequality holds strictly at $t$, no sign condition comes to restrain right-side derivatives.

For $t$ in $I$ and $x$ in $\mathbf{R}^n$, put

$$\Gamma(t, x) := \begin{cases} \{v \in \mathbf{R}^n \mid f'_t(t, x) + v.\nabla f(t, x) \leq 0\} & \text{if} \quad f(t, x) \geq 0 \\ \mathbf{R}^n \quad \text{otherwise.} \end{cases}$$

so that the above observation means $q'^{+}(t) \in \Gamma(t, q(t))$.

What follows may be viewed as providing a *converse*.

Suppose that the interval $I$, nonnecessarily compact, contains its origin $t_0$ and that $q$ is *locally absolutely continuous* on $I$. Equivalently, the (two-side) derivative $dq/dt$ exists almost everywhere in $I$ and equals a function $u : I \to \mathbf{R}^n$, the *velocity function*, which is locally integrable with regard to the Lebesgue measure on $I$; notation : $u \in \mathcal{L}^1_{\text{loc}}(I; \mathbf{R}^n)$, meaning that $u$ is Lebesgue-integrable over every compact subinterval of $I$. And, for every $t$ in $I$, one has

$$q(t) = q(t_0) + \int_{t_0}^{t} u(s)\, ds. \tag{2}$$

The following is established in [58]:

**Viability Lemma**. *Assume that $q$ is locally absolutely continuous on $I$ and that the inclusion*

$$\frac{dq}{dt} \in \Gamma(t, q(t)) \tag{3}$$

*holds for almost every $t$ in $I$. If inequality* (1) *is verified at the initial instant $t_0$, it is verified for every subsequent $t$.*

The term *viability* is used in system theory and control, in particular when economic systems are concerned, to express that the trajectory of a process should remain in a specified set (one may refer to [4], a book actually developed in a too sophisticated topological context for such an elementary statement as the above Lemma to find place in it).

## 2.2 Selectors

A condition of the form (3) is called a *differential inclusion* [5][21]. Since the right-hand side is set-valued, uniqueness is a priori not expected to hold for the solutions (if any) to initial value problems. By a *selector* of the 'multifunction' $(t, x) \mapsto \Gamma(t, x)$, one means a single-valued function, say $(t, x) \mapsto \gamma(t, x)$, such that $\gamma(t, x) \in \Gamma(t, x)$ for every $t$ and $x$. Then

$$\frac{dq}{dt} = \gamma(t, q(t)) \tag{4}$$

is a differential equation whose (locally absolutely continuous) solutions, if any, consequent to some initial condition verifying (1), meet the assumptions of the Viability Lemma, making $q(t)$ belong to $\Phi(t)$ for every subsequent $t$.

A basic example is provided by taking as $\gamma$ the 'lazy selector' of $\Gamma$, i.e. by defining $\gamma(t, x)$ as *the element of $\Gamma(t, x)$ with minimal Euclidean norm.* When $f(t, x) < 0$, i.e. when $x$ belongs to the interior of $\Phi(t)$, the set $\Gamma(t, x)$ consists of the whole of $\mathbf{R}^n$, so that $\gamma(t, x)$ equals $\mathbf{0}$, the zero of $\mathbf{R}^n$. Otherwise, $\Gamma(t, x)$ equals a half-space which contains $\mathbf{0}$ if $f'_t(t, x) \leq 0$, in which case $\gamma = \mathbf{0}$ again. If $f'_t(t, x) > 0$, one finds $\gamma = -(f'_t/\|\nabla f\|^2)\nabla f$, a vector oriented in the direction of decreasing $f$, normal at $x$ to the hypersurface $f(t, .) =$ Const drawn through this point.

For such a choice of $\gamma$, a solution to (4) consequent to some initial position $q(t_0)$ in $\Phi(t_0)$ may be described as follows. The point $q(t)$ belongs for every $t$ to the moving region $\Phi(t)$. As long as it lies in the interior of $\Phi(t)$, $q$ stays at rest. It is only when the boundary of $\Phi(t)$, i.e. the hypersurface with equation $f(t, .) = 0$, moves inward and reaches $q$ that the point takes on a velocity in inward normal direction, so as to go on belonging to $\Phi(t)$. The magnitude of the velocity vector equals the 'normal speed' of the hypersurface.

We have proposed to call *Sweeping Process* the above kinematical association of point motions to the given motion of a set (in $\mathbf{R}^n$ or in a real Hilbert space). See [42][59] for references and recent developments of the subject.

## 2.3 Characterizing the process by a differential inclusion

If, at time $t$, a point $x$ lies on the hypersurface $f(t, .) = 0$, the vector $\nabla f(t, x)$ (we have assumed it nonzero) normal to this hypersurface is directed outward of the region $\Phi(t)$. The half-line emanating from the origin of $\mathbf{R}^n$, generated by $\nabla f(t,x)$, is said to constitute the (outward) *normal cone* to $\Phi(t)$ at point $x$; notation: $\mathrm{N}_{\Phi(t)}(x)$. The definition of a normal cone for less regular sets than $\Phi$ is a vast subject; some other cases will be met in the sequel. For $x$ in the interior of $\Phi(t)$, it proves consistent to view $\mathrm{N}_{\Phi(t)}(x)$ as reduced to the zero of $\mathbf{R}^n$, while the cone shall be defined as empty if $x \notin \Phi(t)$. By discussing the various cases occuring in the calculation of the lazy selector, one sees that if $\gamma(t,x)$ equals this selector, every solution $q$ to the differential equation (4) verifies, for almost every $t$, the differential inclusion

$$-\frac{dq}{dt} \in \mathrm{N}_{\Phi(t)}(q(t)). \tag{5}$$

Unexpectedly *the converse is true*, i.e.(5) in spite of its multivalued right-hand side actually is *equivalent* to the differential equation (4), as far as locally absolutely continuous solution are concerned.

In fact let $q : I \to \mathbf{R}^n$, be such a solution to (5). For almost every $t$, the two-side derivative $q' = dq/dt$ exists, so that the right-hand member is non-empty and therefore $q(t) \in \Phi(t)$; the same is true for every $t$, by continuity. For $t$ such that $q(t)$ happens to lie in the interior of $\Phi(t)$, (5) implies $q' = \mathbf{0}$, which makes that (4) is also satisfied. Otherwise, suppose that $q(t)$ belongs to the boundary, i.e. the function $\tau \mapsto f(\tau, q(\tau))$ vanishes at $\tau = t$. Then the right-derivative $f_t'(t,q(t)) + q'^{+}(t).\nabla f(t,q(t))$, if it exists, is $\leq 0$ while, symmetrically, the left-derivative is $\geq 0$. Therefore $q'(t)$, when it exists, satisfies $f_t'(t,q(t)) + q'(t).\nabla f(t,q(t)) = 0$, i.e. it belongs to the boundary of the half-space $\Gamma(t,q(t))$. Furthermore, (5) entails that $q'(t)$ is directed along the inward normal to the half-space. All this elementarily characterizes $q'(t)$ as the proximal point to $\mathbf{0}$ in $\Gamma(t,q(t))$, namely $\gamma(t,q(t))$.

It was under the formulation (5) that the Sweeping Process was primitively introduced [47][50], with $\Phi(t)$ denoting a nonempty closed convex subset of a real Hilbert space $H$. The motivation then was in the quasi-static evolution of elastoplastic systems [48][49]. The convexity assumption allows one to establish the existence of solutions under rather mild conditions concerning the evolution of $\Phi(t)$, even discontinuous. Another consequence of this convexity is that the multifunction $x \mapsto \mathrm{N}_{\Phi(t)}(x)$ is *monotone* in the following sense (see e.g. [8]) : whichever are $x_1, x_2$ in $H$, $y_1$ in $\mathrm{N}_{\Phi(t)}(x_1)$, $y_2$ in $\mathrm{N}_{\Phi(t)}(x_2)$, one has $(x_1 - x_2).(y_1 - y_2) \geq 0$, with the dot denoting the scalar product of $H$. By elementary calculation, this inequality entails that, if $t \mapsto q_1(t)$ and $t \mapsto q_2(t)$ are two solutions to (5), the Hilbert distance $\|q_1 - q_2\|$ is a non-increasing function of $t$. From this *non expansion* property, it follows that at most one solution to (5) can agree with some initial position $q(t_0)$.

Another source of interest of the formulation (5) is to render evident that the successive positions of the point $q$ are connected with those of the given region $\Phi$ in a *rate-independent* way. In fact, because the right-hand member is a cone, the differential inclusion is found invariant under any non-decreasing differentiable change of variable.

## 2.4 Implicit versus explicit time-stepping

Coming to the numerical approximation of solutions through time-stepping schemes, let us denote by $[t_{\rm i}, t_{\rm f}]$, with length $h$, a time-step ('i' as in *initial*, 'f' as in *final*). From an estimate $q_{\rm i}$ of $q(t_{\rm i})$, obtained as the result of the antecedent time-step, computation has to deliver an estimate $q_{\rm f}$ of $q(t_{\rm f})$.

The formulation (4) naturally leads to take $u_{\rm i} = \gamma(t_{\rm i}, q_{\rm i})$ as an estimate of the velocity throughout the time-step, generating the prediction $q_{\rm f} = q_{\rm i} + hu_{\rm i}$. This is a computation scheme of the *explicit* type.

If (5) is discretized by viewing $(q_{\rm f} - q_{\rm i})/h$ as a representative of the velocity, a strategy of the explicit type would not allow one to express $q_{\rm f}$, since the right-hand member is multivalued. In contrast, the *implicit* strategy consists in invoking the value that this right-hand member would take at the unknown point, so one has to solve

$$q_{\rm i} - q_{\rm f} \in {\rm N}_{\Phi(t_{\rm f})}(q_{\rm f}) \tag{6}$$

(the positive factor $h$ has been dropped since ${\rm N}_{\Phi(t_{\rm f})}$ is a cone). This qualifies $q_{\rm f}$ as an *orthogonal projection* of $q_{\rm i}$ onto $\Phi(t_{\rm f})$. In the case where $\Phi(t_{\rm f})$ is convex, the projection is unique and this characterizes $q_{\rm f}$ as the nearest point to $q_{\rm i}$ in $\Phi(t_{\rm f})$. In particular $q_{\rm f} = q_{\rm i}$ when $q_{\rm i}$ happens to belong to $\Phi(t_{\rm f})$. We have proposed to call this procedure the *catching-up algorithm* [50].

## 2.5 Complementarity

From the description made of the Sweeping Process in 2.2 it is clear that velocity may be discontinuous. As for the explicit prediction $q_{\rm f} = q_{\rm i} + hu_{\rm i}$, it only requires of $u_{\rm i}$ to be the derivative of the function $q$ *on the right* of $t_{\rm i}$. The following observation makes an introduction to analogous, but more complicated, situations we are to meet in Dynamics.

Let $q$, associated with $u$ through (2), verify (5) almost everywhere in $I$. Let $t_1 \in I$ and assume that the function $u$ possesses a limit on the right of $t_1$, say $u_1^+$; in view of (2) this limit also provides the right derivative $\dot{q}^+(t_1)$. As a stronger assertion than (4), which pertained to bilateral derivative and was only declared to hold almost everywhere, let us prove that $u_1^+ = \gamma(t_1, q(t_1))$.

Trivially, both members of this equality are $\mathbf{0}$ if $f_1 := f(t_1, q(t_1)) < 0$. Otherwise, i.e. when $f_1 = 0$, it was seen that

$$\dot{f}_1^+ = f_t'(t_1, q(t_1)) + u_1^+ . \nabla f(t_1, q(t_1)) \leq 0. \tag{7}$$

Inclusion (5) means the existence of a function $t \mapsto \lambda(t) \leq 0$ such that $u(t) = \lambda(t) \nabla f(t, q(t))$. Since $\nabla f$ is continuous and nonzero, the assumed existence of $u_1^+$ secures that of the right-limit $\lambda_1^+$ and

$$u_1^+ = \lambda_1^+ \nabla f(t_1, q(t_1)). \tag{8}$$

If $\dot{f}_1^+ < 0$, instant $t_1$ is followed by an interval throughout which $f < 0$. This has been observed to imply $u = 0$, so that $\lambda$ vanishes on this interval and consequently also its right-limit $\lambda_1^+$. Summing up, one has

$$\dot{f}_1^+ \leq 0, \quad \lambda_1^+ \leq 0, \quad \dot{f}_1^+ \lambda_1^+ = 0, \tag{9}$$

a system of *complementarity conditions*. This is a popular formalism in many research domains where inequality requirements are faced. Solving (8) (9), with $\dot{f}_1^+$ defined as in (7), constitutes a *linear complementarity problem*. Through arguments from Convex Analysis, such problems are shown to be equivalent to finding the critical points of some quadratic functions over polyhedral convex sets. In the present setting, where inequality $\dot{q}_1^+ \leq 0$ simply expresses that $u_1^+$ belongs to the half-space $\Gamma(t_1, q(t_1))$, one readily checks that the above system of conditions characterizes $u_1^+$ as the minimizing point of the function $x \mapsto \|x\|^2/2$ over $\Gamma(t_1, q(t_1))$, namely $\gamma(t_1, q(t_1))$ as announced.

### 2.6 A hydromechanical illustration

Assume that $n = 2$ and that $t, x^1, x^2$ are Cartesian coordinates in physical space, with the $t$-axis vertical and directed downward. Picture the region $f(t, x^1, x^2) \leq 0$ as an underground cavity and the curve $x^1 = q^1(t), x^2 = q^2(t)$ as a stationary waterstream dripping down into it.

The differential equation (4), on account of the diverse circumstances met in the definition of the lazy selector, expresses that: i) any part of this stream which happens to be detached from the cavity wall is rectilinear and vertical; ii) when water runs on the wall, it follows a *line of steepest descent* (this agrees with hydrodynamics under the simplifying assumption that inertia effects are negligible with regard to gravity and to liquid/wall friction); iii) the dependence of $\gamma$ on the sign of $f'_t$ makes that the stream can run only on a part of the wall exposed upward: when it reaches the rim of a possible overhang, water gets loose and falls vertically down as described in i).

In this example, under the complication typically added by unilaterality, the comparison of (4) and (5) merely reflects the classical equivalence between the two standard properties of the lines of steepest descent in a surface: at each point on such a line i) the slope is maximal; ii) the direction is orthogonal to the level curve of the surface.

## 3 Frictionless confinement of a particle

### 3.1 Primary formulation

Notations are the same as in Sect.2, with $n = 3$. The element $q := (q^1, q^2, q^3)$ now consists of the *orthonormal* coordinates of a material point $Q$, with mass $m$, moving under the action of a given force field $(t, x) \mapsto X(t, x)$ and constrained in the region $\Phi(t)$ by the impenetrability of its boundary, assumed to be realized as a material surface with imposed motion. This is the occasion of recalling that, in Mechanics, the description of a *constraint* never reduces – as it does, for instance, in Optimization or in Abstract System Theory – to imposing a geometrical restriction on positions. Some information is needed about the mechanical process through which this restriction is enforced. For instance, using some *servomechanism* in order to secure (1) could result in quite different motions than those obtained under the present assumption of confinement by the *contact action* of a given material boundary.

On a time interval $I$ throughout which the motion is smooth enough for the *velocity function* $u$ in (2) to be locally absolutely continuous (this precludes collisions, events to which we shall come back later), the motion of the particle $Q$ obeys, almost everywhere in $I$,

$$m\frac{du}{dt} = X(t, q(t)) + r(t), \tag{10}$$

where the force $r := (r^1, r^2, r^3)$ denotes the unknown *reaction* possibly exerted by the confining boundary. In this Section, the confinement process is assumed to comply at every instant with the following model:

- this is a *contact* process

$$f(t, q) < 0 \quad \Rightarrow \quad r = 0, \tag{11}$$

- the possible contact is *frictionless*

$$f(t, q) = 0 \quad \Rightarrow \quad \exists\lambda \in \mathbf{R} : r = \lambda \nabla f(t, q), \tag{12}$$

- without adhesion

$$\lambda \leq 0. \tag{13}$$

If compared with the definition given in 2.3 for the normal cone $\mathrm{N}_{\Phi(t)}(x)$ at a point $x$, the above system of three conditions is found equivalent to

$$-r \in \mathrm{N}_{\Phi(t)}(q). \tag{14}$$

Therefore, the traditional aim of Analytical Dynamics, namely the elimination of the unknown reactions of the so-called perfect constraints is attained by rapproaching conditions (10) and (14)

$$X(t, q(t)) - m\frac{du}{dt} \in \mathrm{N}_{\Phi(t)}(q(t)), \tag{15}$$

an integro-differential inclusion, as the unknown functions $q$ and $u$ are essentially connected by (2). Formally, this inclusion implies $q(t) \in \Phi(t)$ for almost every $t$ in $I$, since otherwise the right-hand member would be empty. By continuity $q(t) \in \Phi(t)$ holds for every $t$.

### 3.2 The 'Contact Dynamics' approach

From the theoretical standpoint as well as in the elaboration of approximation schemes, the differential inclusion (15) proves difficult to handle (see however [64], under the assumption of convexity for the function $f$). In what concerns the position function $q$, the problem at hand is of differential order 2, as expected in a dynamical context, while in the purely kinematical setting of Sect. 2, $q$ was the unknown of a differential problem of order 1. However, some of the arguments used then will be transposed in what follows.

*Let $\Gamma$ be defined as in 2.1. Provided the initial position $q(t_0)$ lie in $\Phi(t_0)$, the solutions of (15) are the same as those of*

$$X(t, q(t)) - m\frac{du}{dt} \in \mathrm{N}_{\Gamma(t,q(t))}(u(t)). \tag{16}$$

In fact (16), assumed to hold for almost every $t$ in $I$, implies $u(t) \in \Gamma(t, q(t))$. Since $q(t_0) \in \Phi(t_0)$, the Viability Lemma secures $q(t) \in \Phi(t)$ for every $t$ in $I$. The definition of $\Gamma$ then makes that the normal cone $\mathrm{N}_{\Gamma(t,q(t))}(u)$ is contained in $\mathrm{N}_{\Phi(t)}(q(t))$ whatever is $u$, so that (15) is a fortiori satisfied.

Conversely, let $q$ be a solution to (15), hence $q(t) \in \Phi(t)$ for every $t$ in $I$. When $f(t, q(t)) < 0$, the set $\mathrm{N}_{\Phi(t)}(q(t))$ reduces to the singleton $\{\mathbf{0}\}$ and the same is true for the right-hand member of (16). In contrast, for $t$ such that $f(t, q(t)) = 0$, the set $\Gamma(t, q(t))$ is a half-space and the argument used in 2.3 proves that the element $u(t) = u^+(t) = u^-(t)$ belongs to its boundary plane. Consequently, $\mathrm{N}_{\Gamma(t,q(t))}(u)$ consists of the cone generated by $\nabla f(t, q(t))$ hence equals $\mathrm{N}_{\Phi(t)}(q(t))$.

### 3.3 First example of a *CD* numerical scheme

As before, let us denote by $[t_\mathrm{i}, t_\mathrm{f}]$, with length $h$, a time-step. From the approximate values $q_\mathrm{i}, u_\mathrm{i}$ obtained for $q$ and $u$ at $t_\mathrm{i}$, one has to calculate $q_\mathrm{f}, u_\mathrm{f}$, pertaining to $t_\mathrm{f}$. The given force field $X$ is assumed to depend smoothly on its arguments, so that one chooses to approximate it throughout the time-step by the value it takes at $t_\mathrm{m} := t_\mathrm{i} + h/2$ and $q_\mathrm{m} := q_\mathrm{i} + hu_\mathrm{i}/2$. It is also at the point $(t_\mathrm{m}, q_\mathrm{m})$ that $f$ is calculated, in order to decide whether boundary contact is in effect or not and to determine the set $\Gamma$ accordingly. Depending on the sign of $f(t_\mathrm{m}, q_\mathrm{m})$, the latter equals the whole of $\mathbf{R}^3$ or a half-space with $\nabla f(t_\mathrm{m}, q_\mathrm{m})$ as normal vector. Inclusion (16) is thus discretized in the form

$$X(t_\mathrm{m}, q_\mathrm{m}) - \frac{m}{h}(u_\mathrm{f} - u_\mathrm{i}) \in \mathrm{N}_{\Gamma(t_\mathrm{m},q_\mathrm{m})}(u_\mathrm{f}), \tag{17}$$

i.e. in view of the right-hand member being a cone,

$$u_\mathrm{i} + \frac{h}{m}X(t_\mathrm{m}, q_\mathrm{m}) - u_\mathrm{f} \in \mathrm{N}_{\Gamma(t_\mathrm{m},q_\mathrm{m})}(u_\mathrm{f}).$$

This classically characterizes $u_{\mathrm{f}}$ as the *proximal point* to the known element $u_{\mathrm{i}} + hX(t_{\mathrm{m}}, q_{\mathrm{m}})/m$ in $\Gamma(t_{\mathrm{m}}, q_{\mathrm{m}})$. One finishes the calculation with

$$q_{\mathrm{f}} = q_{\mathrm{m}} + \frac{h}{2}\, u_{\mathrm{f}}.$$

This algorithm automatically handles the possible breaking of contact : this happens if $u_{\mathrm{i}} + hX(t_{\mathrm{m}}, q_{\mathrm{m}})/m$ falls into the interior of $\Gamma(t_{\mathrm{m}}, q_{\mathrm{m}})$.

*Remark 1.*– Provided $q(t_0) \in \Phi(t_0)$, inclusion (16) has been precedingly observed to secure $f(t, q(t)) \leq 0$ at every consequent $t$. But, in the above time-stepping procedure, it is only $u_{\mathrm{f}}$ which, at each step, is constructed as an element of $\Gamma(t_{\mathrm{m}}, q_{\mathrm{m}})$, so that the Viability Lemma is just involved through time-discretization. One thus may fear that some violation of the inequality would build up from step to step. Actually, if the step-length is not too large, numerical experiments show some self-corrective effect which, on the contrary, tends to reduce violations. This effect seems related to $\Phi(t)$ possessing a nonempty interior. In contrast, in industrial softwares devoted to the dynamics of machines, some *bilateral* constraints are commonly introduced, leaving a set of feasible positions with empty interior. Their treatment in terms of velocities [28] then requires corrective procedures to prevent cumulative errors.

*Remark 2.*– The calculation of $u_{\mathrm{f}}$ from $u_{\mathrm{i}}$ is based on mechanical elements evaluated at the mid-position $q_{\mathrm{m}}$; in turn, $u_{\mathrm{f}}$ is used to calculate from $q_{\mathrm{m}}$ the final position of the current step and, from there, the mid-position of the subsequent time-step. This interleaving makes the above time-stepping procedure resemble the policy called 'leapfrog' in Molecular Dynamics simulations. In order to figure out how it improves precision, compared with an ordinary Euler explicit time-stepping scheme, one may apply it to calculating the parabolic unconstrained motion of a particle in a uniform gravity field. If $h$ is constant, the values found for $q$ at the successive steps coincide with the exact solution, while the Euler scheme generates cumulative errors. Of course, when an algorithm of the above sort is implemented with constant $h$, one may calculate each $q_{\mathrm{m}}$ from the antecedent one by a single incrementation. The proper output of the computation, namely $q_{\mathrm{f}}$, may not be needed at each step.

*Remark 3.*– In contrast with the approach of the motion of a point in a surface through the traditional methods of dynamics, the above time-stepping procedure does not require calculating the curvature of the boundary. This curvature is implicitely accounted for by the fact that the direction of $\nabla f$ evolves from one step to the other. The question of existence of solutions to (15) or (16) is not addressed in this paper; one naturally expects that $f$ has to be twice dfferentiable.

# 4 Multicontact systems

## 4.1 Parametrization

Let the possible configurations of a body collection be parametrized (at least locally) through generalized coordinates, say $q := (q^1, \dots, q^n)$. For the sake of reducing the number $n$, this parametrization may be constructed with account of possible permanent, frictionless linkages imposed on the members of the system. After that, the constraints of *non-interpenetrability* are additionally considered. The geometric restriction consequently imparted on the system positions is assumed expressed by a finite set of inequalities

$$f_\alpha(t, q) \leq 0, \quad \alpha \in \{1, \dots, \kappa\}, \tag{18}$$

where $f_1, \dots, f_\kappa$ are given functions. Through the presence of $t$ as an argument of $f_\alpha$, provision is made for the inequality to describe the confinement of a member of the system by some external obstacle or boundary with prescribed motion. Under this parametrization, a motion of the system consists of a mapping $t \in I \mapsto q(t) \in \mathbf{R}^n$ and, as in the foregoing, this mapping is assumed locally absolutely continuous, i.e. there exists a locally integrable velocity function $u : I \to \mathbf{R}^n$ from which $q$ may be retrieved in the form (2).

As an example, one may consider a pair of members of the system whose positions in a chosen reference frame are well located as soon as the value of the element $q$ of $\mathbf{R}^n$ is known (together with the time $t$ in case of a time-dependent parametrization). Then, one may take as $f_\alpha$ the expression, as a function of $(t, q)$, of some measure of the *overlap* of the two bodies. This overlap should be understood as a directed quantity, so that it becomes negative in case the bodies lie apart from each other. The convention applied in (18) of characterizing the permitted configurations through the $\leq 0$ inequality, comes from Convex Optimization theory, where such a sign convention offers technical advantages. No convexity hypothesis is made here concerning the functions $f_\alpha$: since such an assumption would not be preserved under a change of parametrization, it cannot in general have any mechanical meaning. If one prefers to deal with the $\geq 0$ symbol, there only is to consider, instead of the overlap the opposite quantity, usually called the *gap* between the considered bodies.

The above formalism is not limited to collections of strictly rigid bodies, since $q$ may also include parameters accounting for a finite-freedom approximation of deformability. Such additional parameters possibly arise from some modal representation of the deformation dynamics or from the Finite Element discretization of deformable parts [37].

## 4.2 Contact kinematics

Suppose that inequality $f_\alpha \leq 0$ expresses the local non-interpenetration of some pair of members of the system, say $\mathcal{B}$ and $\mathcal{B}'$, so that equality $f_\alpha =$

0 corresponds to these bodies touching each other at some point of space denoted by $M_\alpha$. This we shall assume here to be an isolated contact point, but other contacts, associated with different values of $\alpha$, may also be in effect between the same bodies at the same instant. For every imagined motion $t \mapsto q(t)$ bringing the system through the considered contacting position for some $t$, with a definite value of $u = dq/dt \in \mathbf{R}^n$, the velocity vectors $\mathcal{V}_\alpha$ and $\mathcal{V}'_\alpha$, relative to the chosen reference frame, of the respective particles of $\mathcal{B}$ and $\mathcal{B}'$ passing at point $M_\alpha$ let themselves be expressed as affine functions of $u$. The same is thus true for the *relative velocity* $\mathcal{U}_\alpha = \mathcal{V}_\alpha - \mathcal{V}'_\alpha$ of body $\mathcal{B}$ with respect to body $\mathcal{B}'$ at this point, say

$$\mathcal{U}_\alpha = G_\alpha\, u + \mathcal{W}_\alpha, \tag{19}$$

where $G_\alpha : \mathbf{R}^n \to \mathbf{E}^3$ (the space of the vectors of physical space) denotes a linear mapping, depending on $t$ and $q$. No attention is paid at this stage to the imagined motion preserving contact or not. The term $\mathcal{W}_\alpha \in \mathbf{E}^3$, a known function of $t$ and $q$, vanishes in the familiar case of a time-independent parametrization.

Similar formula holds if inequality $f_\alpha \leq 0$ expresses the confinement of a member $\mathcal{B}$ of the system by some external material boundary with prescribed motion. Assume that equality $f_\alpha = 0$ corresponds to contact taking place at some point, here again denoted by $M_\alpha$. The local velocity, at this point, of body $\mathcal{B}$ with respect to the boundary has the same form as $\mathcal{U}_\alpha$ in (19), where $\mathcal{W}_\alpha$ now reflects the known velocity of the boundary (for a time-independent parametrization, $\mathcal{W}_\alpha$ equals the negative of this velocity vector).

At the contact point $M_\alpha$, we assume that a *common tangent plane* to the respective surfaces of the concerned bodies has been defined. This does not require of both surfaces to be smooth; for instance, contact may take place between a smooth body and some corner point or sharp asperity of the other. Let $\mathbf{n}^\alpha$ denote the unit vector normal to this plane, *directed toward* $\mathcal{B}$. In computation, as well as in existential studies, it proves useful that the definition of the above elements would be conventionally extended to a neighbourhood of the concerned value of $(t, q)$ in $\mathbf{R} \times \mathbf{R}^n$. This allows one to express as a function of $(t, q)$ the *normal gap*, say $g_\alpha(t, q)$, between $\mathcal{B}$ and $\mathcal{B}'$, *counted as negative in the case of overlap.* Classically, the derivative of the function $t \mapsto g_\alpha(t, q(t))$ is found equal to $\mathcal{U}_\alpha.\mathbf{n}^\alpha$, the normal component of the relative velocity of the contacting bodies at point $M_\alpha$.

Sometimes, in Computation literature, the second time-derivative of the gap is improperly referred to as the 'normal relative acceleration'. Actually, since the material particles involved in the definition of $\mathcal{U}_\alpha$ are not the same from an instant to the other, this second derivative has in general nothing to do with the relative acceleration vector. As an example, one may consider a body of circular or spherical shape: a variety of rotations may be imparted to such a body without altering its overall location, so yielding the same gap while the normal acceleration is changed.

*Remark.*– The representation of non-interpenetration through of a finite set of inequalities is operative in most practical situations. However, non-interpenetration cannot be described in that way in the neighbourhood of a configuration where two sharp asperities or corners come into contact by their points. This is evidenced by the fact that, in this case, the set of the values of the local right-velocity $\mathcal{U}_\alpha^+$ which are compatible with non-interpenetration is not a convex cone anymore. In contexts where the probability of such an event cannot be treated as negligible, numerical techniques have to resort to adequate procedures (possibly involving $\mathcal{U}_\alpha$) for the identification of a mechanically plausible contact plane.

## 4.3 Contact forces

Assume that the contact actions that body $\mathcal{B}$ experiences at point $M_\alpha$ from body $\mathcal{B}'$ are described as a simple force $\mathcal{R}^\alpha$ (there would be no conceptual difficulty in adding to this description some local torque, accounting for a resistance to rolling). Then $\mathcal{B}'$ experiences from $\mathcal{B}$ the force $-\mathcal{R}^\alpha$. The standard machinery of Analytical Dynamics needs a representation of this pair of forces, in regard to the chosen parametrization, through its *covariant components* (or 'generalized components'), namely the element $r^\alpha$ of $\mathbf{R}^n$ expressed as

$$r^\alpha = G_\alpha^* \mathcal{R}^\alpha, \tag{20}$$

with $G_\alpha^* : \mathbf{E}^3 \to \mathbf{R}^n$ denoting the *transpose* of $G_\alpha$.

*The convention of implicit summation will never be applied to Greek indices.*

If inequality $f_\alpha \leq 0$, expresses the confinement of a member $\mathcal{B}$ of the system by some external obstacle with prescribed motion, (19) still holds with $\mathcal{U}_\alpha$ denoting the local velocity of $\mathcal{B}$ relative to this obstacle. Then it is found that $r^\alpha$ in (20) consist of the covariant components of the force $\mathcal{R}^\alpha$ alone, acting on $\mathcal{B}$ at the contact point. Its counterpart, exerted by $\mathcal{B}$ upon the obstacle, is not in this case a force experienced by the system. Incidentally observe that the term $\mathcal{W}_\alpha$ does not appear in (20).

## 4.4 The equation of Dynamics

As before, the context here is that of standard dynamics, involving the second derivative of the function $t \mapsto q(t)$. Therefore the velocity function $t \mapsto u(t)$ is required to be locally absolutely continuous. Using Lagrange's technique or any other tools from classical solid dynamics, one obtains a differential equation, to be read as an equality of elements of $\mathbf{R}^n$

$$A(t,q)\frac{du}{dt} = F(t,q,u) + \sum_\alpha r^\alpha, \tag{21}$$

where $A$ denotes the $n \times n$ *inertia matrix*. The expression $F$ comprises certain standard terms (commonly referred to as 'centrifugal' and 'gyroscopic') and also the covariant components of some applied forces, supposed given as functions of time, the position of the system and its velocity. The elements $r^\alpha, \alpha \in \{1, 2, \ldots, \kappa\}$, are made of the covariant components of the respective contact forces, as expressed in (20).

The same formalism remains more generally valid with $q$ related to some velocity function $u$ by other kinematical relations than (2). For instance, when dealing with 3-dimensional rigid bodies, it is usual to attach to each of them a frame of principal axes of inertia emanating from its center of mass. Then one may choose to enter, among the constituents of the $\mathbf{R}^n$-valued function $u$, the three components relative to these axes of the *spin vector* of the rigid body, instead of the time-derivatives of some directional parameters. This offers the considerable advantage of generating a contribution in the matrix $A$ which is diagonal and constant with regard to $t$ and $q$. Retrieving from these spin components the evolution of some directional parameters of the concerned body is only the matter of integrating adequate kinematical formulas. Correlatively, if forces are applied to the rigid body, the total moments of these forces about the same axes should be entered as covariant components into the corresponding lines of the right-hand side of (21).

Since contact forces vanish when contact is not in effect, the summation in the right-hand side of (21) may be restricted to the values of $\alpha$ belonging to

$$J(t, q) := \{\alpha \in \{1, \ldots, \kappa\} \mid f_\alpha(t, q) \geq 0\}. \tag{22}$$

The geometric conditions (18) of non-interpenetrability, joined to the differential equation of dynamics (21), clearly are not enough for determining the motion consequent to initial data. Some phenomenological information should be added, concerning the contact forces. Since the contact phenomenon takes place in physical space, this information is expected to involve the vectors $\mathcal{R}^\alpha$, $\mathcal{U}_\alpha$, as well as the values of $t$ and $q$ specifying the actual configuration of the system. Hence, for every $\alpha$ labelling a possible contact, a relationship of the form

$$\mathrm{law}_\alpha(t, q, \mathcal{U}_\alpha, \mathcal{R}^\alpha) = \mathrm{true}, \tag{23}$$

called a *contact law*, should be available.

## 5 Contact laws

### 5.1 Frictionless contact

Under the present notations we are to meet the same circumstances as in Sect. 3. If the contact labelled $\alpha$ concerns two bodies denoted by $\mathcal{B}$ and $\mathcal{B}'$, with common normal unit $\mathbf{n}^\alpha$ directed toward $\mathcal{B}$, the assumptions of no-friction an no-adhesion mean $\exists \rho_\alpha \geq 0 : \mathcal{R}^\alpha = \rho_\alpha \mathbf{n}^\alpha$. It has been agreed in the foregoing to extend the definition of $\mathbf{n}^\alpha$, at least in a neighbourhood of the concerned values of $t$ and $q$, to cases where $g_\alpha$, the *normal gap*, takes nonzero values and to state $\mathcal{R}^\alpha$=0 if $g_\alpha > 0$. Define

$$\mathcal{K}_\alpha(t,q) := \begin{cases} \{\mathcal{V} \in \mathbf{E}^3 \mid \mathcal{V}.\mathbf{n}^\alpha \geq 0\} & \text{if} \quad g_\alpha(t,q) \leq 0 \\ \mathbf{E}^3 \quad \text{otherwise.} \end{cases}$$

This is the set of the values of the local *right-velocity* of $\mathcal{B}$ relatively to $\mathcal{B}'$ (the latter may be a member of the system or an external obstacle with prescribed motion) which are compatible with non-interpenetration. In the first line, $\mathcal{K}_\alpha$ equals a half-space, hence the *normal cone* $\mathrm{N}_{\mathcal{K}_\alpha}$, evaluated at the origin $\mathbf{0}$ of $\mathbf{E}^3$, equals the half-line generated in this vector space by $-\mathbf{n}^\alpha$. Otherwise $\mathcal{K}_\alpha = \mathbf{E}^3$, so that the cone $\mathrm{N}_{\mathcal{K}_\alpha}(\mathbf{0})$ reduces to the set $\{\mathbf{0}\}$.

Therefore, at time $t$, the no-friction and no-gluing assumptions (including the case of no-contact) are equivalent to assert

$$-\mathcal{R}^\alpha \in \mathrm{N}_{\mathcal{K}_\alpha(t,q)}(\mathbf{0}). \tag{24}$$

The move made in Sect.3 of replacing inclusion (15) by inclusion (16) admits as a counterpart here the replacement of inclusion (24) by

$$-\mathcal{R}^\alpha \in \mathrm{N}_{\mathcal{K}_\alpha(t,q)}(\mathcal{U}_\alpha), \tag{25}$$

In fact, in this context where $u$ is continuous, the same argument as in 2.3 shows that $\mathcal{U}_\alpha$ belongs to the boundary plane of the half-space $\mathcal{K}_\alpha$, hence $\mathrm{N}_{\mathcal{K}_\alpha}(\mathcal{U}_\alpha) = \mathrm{N}_{\mathcal{K}_\alpha}(\mathbf{0})$, while in case of no-contact $\mathrm{N}_{\mathcal{K}_\alpha}(\mathcal{U}_\alpha) = \{\mathbf{0}\}$ whatever is $\mathcal{U}_\alpha$. In short, (25) contains all the stipulations implied when a contact is declared frictionless.

But, *in addition*, (25) entails $\mathcal{U}_\alpha \in \mathcal{K}_\alpha(t,q)$, since otherwise $\mathrm{N}_{\mathcal{K}_\alpha}(\mathcal{U}_\alpha)$ would be empty. If $g_\alpha(t,q) > 0$ this actually imparts no restriction on $\mathcal{U}_\alpha$ while if $g_\alpha(t,q) \leq 0$, i.e. $\alpha \in J(t,q)$, this implies $\mathcal{U}_\alpha.\mathbf{n}^\alpha \geq 0$. It has been precedingly recalled that $\mathcal{U}_\alpha.\mathbf{n}^\alpha$ equals the derivative of the function $t \mapsto g_\alpha(t,q(t))$. This allows one to invoke the Viability Lemma (2.1), with $f = -g_\alpha$, in order to prove that the assumption of (25) being verified for almost every $t$ in $I$ entails : *non-interpenetration holds for every* $t > t_0$, provided it holds at $t_0$.

The latter statement applies more generally to any contact law which, among other phenomenological stipulations, secures the following

- in all cases $\mathcal{U}_\alpha \in \mathcal{K}_\alpha$,
- if $\mathcal{U}_\alpha \in$ interior $\mathcal{K}_\alpha$, then $\mathcal{R}^\alpha = 0$.

In other words, one has the implications

$$g_\alpha(t,q) \leq 0 \Rightarrow \mathbf{n}^\alpha.\mathcal{U}_\alpha \geq 0 \tag{26}$$

$$\mathbf{n}^\alpha.\mathcal{U}_\alpha > 0 \Rightarrow \mathcal{R}^\alpha = 0. \tag{27}$$

We propose to say that a package of information, concerning the possible contact labelled $\alpha$, if it possesses these two properties, is a *contact law of prospective type* (or, in the terminology precedingly used by the author [58][59], a *complete* contact law). The underlying idea is that such a law does not properly govern the values of $\mathcal{U}_\alpha$ and $\mathcal{R}^\alpha$ at the actual instant, but their limits on the right of this instant, assumed to exist. In fact, if $\mathcal{U}_\alpha.\mathbf{n}^\alpha > 0$, the concerned instant is followed by a contactless time-interval. Since $\mathcal{R}^\alpha$ vanishes over this interval, the same is true for its right-limit.

### 5.2 Multicontact frictionless Dynamics

We are now to see how the observations made in Sect.3 let themselves be transposed into the present setting. Even the simple case then considered, of a particle confined by a frictionless material boundary, may exhibit multicontact features if the boundary consists of several parts with smooth equations, each of them enforcing an inequality of the form (18). If two of these smooth surfaces meet to form an edge, the particle, when lying on this edge, experiences contact forces from both parts, the resultant of which may take any value in the convex cone generated by the respective inward normals. Therefore, the writing in (14) remains valid provided $\mathrm{N}_{\Phi(t)}(q)$ is defined as the convex cone generated by the outward normals to the surfaces which form the edge.

When coming to general multicontact systems, one has to connect normality in the linear Euclidean space $\mathbf{E}^3$ of the vectors of physical space with normality in the space $\mathbf{R}^n$ of the abstract components. The following relationship is found to hold [52] between the element $\nabla f_\alpha$ of $\mathbf{R}^n$ and the normal unit vector $\mathbf{n}^\alpha$ at point $M_\alpha$ to the contacting bodies, directed toward $\mathcal{B}$

$$\exists \lambda_\alpha \geq 0 \quad \text{such that} \quad G^*_\alpha \, \mathbf{n}^\alpha = -\lambda_\alpha \, \nabla f_\alpha . \tag{28}$$

The proof of this rests on a unilateral version of the algebraic theorem of Lagrange multipliers, known in Convex Analysis as *Farkas' lemma* [69].

In all the sequel, we shall assume that the mapping $G_\alpha$ is *surjective* of $\mathbf{R}^n$ onto $\mathbf{E}^3$; equivalently, its transpose $G^*_\alpha$ is *injective* of $\mathbf{E}^3$ into $\mathbf{R}^n$. Then $\lambda_\alpha$ in (28) is nonzero. Some special positions of a multibody system may give rise to 'wedging' effects which contradict this assumption.

In view of (19) and (20) this allows one to replace the laws of frictionless contact, either (24) or (25), by equivalent relationships involving only the abstract components $u$ and $r^\alpha$ instead of $\mathcal{U}_\alpha$ and $\mathcal{R}^\alpha$. Under the definition (22) of $J$, put

$$W(t,q) := \{v \in \mathbf{R}^n \,|\, \forall \alpha \in J(t,q) \, : \, \frac{\partial f_\alpha}{\partial t} + v.\nabla f_\alpha \leq 0\}, \tag{29}$$

a polyhedral closed convex set. One then finds [59] that a value $r$ of the sum $\sum_\alpha r^\alpha$ is compatible with the contact law (25) holding for all $\alpha$, if and only if $-r \in \mathrm{N}_{W(t,q)}(u)$. Consequently, the elimination of frictionless reactions from the dynamical equation (21) is achieved in the writing

$$F(t,q,u) - A(t,q)\,\frac{du}{dt} \in \mathrm{N}_{W(t,q)}(u). \tag{30}$$

This allows one to derive a time-stepping scheme for the numerical approximation of solutions quite similar to the one presented in 3.3. The occurrence of the matrix $A$ in (30) at the place occupied in (16) by the scalar factor $m$ does not constitute an essential complication. It only means that, instead of the standard Euclidean metric of $\mathbf{R}^3$, one is using in $\mathbf{R}^n$ the Euclidean metric defined by the positive definite matrix $A$.

### 5.3 Handling inequality conditions in terms of acceleration

The differential inclusion (30) formally resembles (5) which has been found to characterize the Sweeping Process. In fact, in the special case where $F = 0$ and $A = \mathbf{1}$, (30) makes the function $t \mapsto u$ appear as a solution to the Sweeping Process by the moving (closed, convex) set $t \mapsto W(t, q(t))$. Of course, the latter is not given, since it depends on $q$ which itself is connected to $u$ by (2), but one may infer from this analogy that, similarly to the equivalence of (5) to (4), the inclusion (30) could be replaced by a differential *equation* whith right-hand member defined by a minimization property. This is the object of the forthcoming.

A time-stepping scheme of the *implicit* type, rests on predicting the velocity whithout resorting to any expression of the acceleration. In contrast, what follows is aimed at determining the *right-acceleration* $\dot{u}^+$, which is needed when a scheme of the *explicit* type is being planned, as well as an Event Driven calculation [1][65].

Let $t_1$ denote an instant preceded by some time-interval throughout which the motion, with absolutely continuous $u$, satisfies (30). Assume that $u$ remains continuous at instant $t_1$, i.e. *no collision occurs.* Non-interpenetration entails that, for every $\alpha$ in $J(t_1, q(t_1))$, the right-derivative of the function $t \mapsto f_\alpha(t, q(t))$ at $t_1$ is $\le 0$, while the left-derivative is $\ge 0$. Hence this function has zero derivative of order one at $t_1$.

Let us assume in this Subsection that the functions $f_\alpha$ are $C^2$. The investigation of the motion by explicit time-stepping or by an ED policy rests on the assumption that $t_1$ is followed by a nonzero interval throughout which (30) is verified again. It just may happen that some contacts break at $t_1$, inducing a change of $J(t,q)$. If $u$ possesses a derivative on the right of $t_1$, the function $t \mapsto f_\alpha(t, q(t))$ possesses a second derivative on the right of $t_1$, expressed through the chain rule with a certain $a_\alpha$, in the form $a_\alpha(t_1, q_1, u_1) + \dot{u}_1^+ . \nabla f_\alpha(t_1, q_1)$. Since the function is zero at $t_1$, as well as its first derivative, non-interpenetration requires of this second derivative to be $\leq 0$. If it is strictly negative, $f_\alpha(t, q(t))$ becomes strictly negative on a subsequent interval, making the corresponding quantities $r^\alpha = \mu_\alpha \nabla f_\alpha$ vanish on this interval, as well as their limits for $t \downarrow t_1$, assumed to exist. One thus obtains a set of complementarity conditions

$$a_\alpha + \dot{u}_1^+ . \nabla f_\alpha \leq 0, \quad \mu_\alpha \leq 0, \quad (a_\alpha + \dot{u}_1^+ . \nabla f_\alpha)\, \mu_\alpha = 0.$$

By joining them to the equation of dynamics

$$A\,\dot{u}_1^+ - F = \sum_\alpha \mu_\alpha \nabla f_\alpha$$

one reduces the determination of $\dot{u}_1^+$ and of the multipliers $\mu_\alpha$ to a *Linear Complementarity Problem* in standard form.

Because the matrix $A$ is positive definite, such a problem is classically equivalent to minimizing a convex quadratic function in a closed convex polyhedral subset of $\mathbf{R}^n$. It has been shown in [45][46] that this extremal characterization of the acceleration may be viewed as extending Gauss' *Principle of the Least Deviation* to mechanical systems subject to unilateral frictionless constraints. Some *dual* minimization property is also found to characterize the contact forces.

### 5.4 Coulomb friction

The presence of dry friction, governed by the law of Coulomb, at the possible contact with label $\alpha$, is expressed by a relationship of the form (23). Traditionally, the law of Coulomb is only invoked for persistent contact, but when devising numerical schemes, making it meaningful for $\mathcal{U}_\alpha . \mathbf{n}^\alpha \geq 0$ and securing that a contact law of the prospective type is so stated, is just the matter of writing the code adequately.

Apart from the numerical success, the consistency of the concept of prospective type is illustrated by the following feature.

Dropping the label $\alpha$ for brevity, one defines the (non adhesive) Coulomb friction at some contact point by giving the *Coulomb cone* $\mathcal{C}$, a closed convex conical region of $\mathbf{E}^3$ to which the contact force $\mathcal{R}$ exerted by $\mathcal{B}'$ upon $\mathcal{B}$ should belong in any circumstance. In the standard case, $\mathcal{C}$ is rotationally symmetric about the normal vector $\mathbf{n}$ and contains it, but more general situations, accounting for anisotropic friction, are possible. The law consists in a relationship between the force $\mathcal{R}$ and the local velocity $\mathcal{U}$ of $\mathcal{B}$ relative to $\mathcal{B}'$ which resembles a plasticity law in that the values of $\mathcal{R}$ lying in the interior of $\mathcal{C}$ are compatible with $\mathcal{U} = 0$ only. But the 'flow rule' which characterizes the values of $\mathcal{U}$ compatible with a value of $\mathcal{R}$ lying on the boundary of $\mathcal{C}$ does not involve the normality of $\mathcal{U}$ to this boundary. In short, this is not an 'associated' flow rule (except in the case of zero friction coefficient).

The concept of *bipotential* has been introduced by De Saxcé and Feng [23] as a tool for handling non-associated force/velocity (or stress/strain-rate) laws, from the theoretical standpoint, as well as in numerical techniques. Here is the aspect this concept takes in the case of Coulomb friction.

As usual, one decomposes the vectors $\mathcal{R}$ and $\mathcal{U}$ into their normal and tangential components. Let $\mathbf{T}$ denote the subspace of the vectors in $\mathbf{E}^3$ which are orthogonal to $\mathbf{n}$.

$$\mathcal{R} = \mathcal{R}_{\mathrm{T}} + \mathcal{R}_{\mathrm{N}}\,\mathbf{n}, \quad \mathcal{R}_{\mathrm{T}} \in \mathbf{T}, \quad \mathcal{R}_{\mathrm{N}} \in \mathbf{R},$$

$$\mathcal{U} = \mathcal{U}_{\mathrm{T}} + \mathcal{U}_{\mathrm{N}}\,\mathbf{n}, \quad \mathcal{U}_{\mathrm{T}} \in \mathbf{T}, \quad \mathcal{U}_{\mathrm{N}} \in \mathbf{R}.$$

Classically, if the Coulomb law is restricted to situations where $\mathcal{R}_{\mathrm{N}}$ is known, say $\mathcal{R}_{\mathrm{N}} = 1$, the law turns out to reduce to a relationship between $\mathcal{R}_{\mathrm{T}}$ and $\mathcal{U}_{\mathrm{T}}$ of the associated type. Let $D_1 := \{\mathcal{R}_{\mathrm{T}} \in \mathbf{T} \mid \mathcal{R}_{\mathrm{T}} + \mathbf{n} \in \mathcal{C}\}$ (the 'unit section' of the cone $\mathcal{C}$) and define in $\mathbf{T}$ the real function (the 'dissipation function' of the said restricted law)

$$\mathcal{T} \in \mathbf{T} \mapsto \varphi_1(\mathcal{T}) := \sup\{\mathcal{S}.\mathcal{T} \mid \mathcal{S} \in -D_1\}.$$

In the traditional case of isotropic friction with coefficient $\gamma$, one simply has $\varphi_1(\mathcal{T}) = \gamma\|\mathcal{T}\|$.

Similarly to 5.1, define

$$\mathcal{K}(t,q) := \begin{cases} \{\mathcal{V} \in \mathbf{E}^3 \mid \mathcal{V}.\mathbf{n} \geq 0\} & \text{in case of contact or overlap} \\ \mathbf{E}^3 & \text{in case of no-contact.} \end{cases}$$

The Coulomb cone depends on $t$ and $q$; put $\mathcal{C} = \{\mathbf{0}\}$ in case of no-contact. Using arguments from Convex Analysis, one establishes that the relation between the elements $\mathcal{U}$ and $\mathcal{R}$ of $\mathbf{E}^3$ consisting of the system of conditions

$$\mathcal{U} \in \mathcal{K},\ \mathcal{R} \in \mathcal{C},\ -\mathcal{U}.\mathcal{R} = \varphi_1(\mathcal{U}_{\mathrm{T}})\mathcal{R}_{\mathrm{N}} \tag{31}$$

is a contact law of the prospective type which, in the standard situation, reduces to the law of Coulomb.

Furthermore, one may prove

$$\forall \mathcal{V} \text{ in } \mathcal{K}, \quad \forall \mathcal{S} \text{ in } \mathcal{C}: \quad \mathcal{V}.\mathcal{S} + \varphi_1(\mathcal{V}_{\mathrm{T}})\,\mathcal{S}_{\mathrm{N}} \geq 0$$

so that (31) expresses that the real function $(\mathcal{V}, \mathcal{S}) \mapsto \mathcal{V}.\mathcal{S} + \varphi_1(\mathcal{V}_{\mathrm{T}})\,\mathcal{S}_{\mathrm{N}}$, separately convex with regard to $\mathcal{V}$ and $\mathcal{S}$, attains at point $(\mathcal{U}, \mathcal{R})$ its minimal value relative to the product set $\mathcal{K} \times \mathcal{C}$ and that *this minimal value is zero.*

# 6 Time-stepping

## 6.1 Numerical dynamics of multicontact systems

As before, denote by $[t_{\mathrm{i}}, t_{\mathrm{f}}]$, with length $h$, a time-step. From the approximate values $q_{\mathrm{i}}, u_{\mathrm{i}}$ obtained for $q$ and $u$ at $t_{\mathrm{i}}$ as the result of antecedent computation, one has to calculate $q_{\mathrm{f}}, u_{\mathrm{f}}$, corresponding to $t_{\mathrm{f}}$.

The inertia matrix $A(t,q)$ is assumed to depend smoothly on $t$ and $q$, so that one chooses to approximate it throughout the time-step by the value $A_{\mathrm{m}}$ it takes at the 'midpoint' $t_{\mathrm{m}} := t_{\mathrm{i}} + h/2$, $q_{\mathrm{m}} := q_{\mathrm{i}} + hu_{\mathrm{i}}/2$. Similarly, the term $F(t,q,u)$ in (21) is assumed to vary slowly enough for allowing one to replace it by the value $F_{\mathrm{m}}$ it takes at $t = t_{\mathrm{m}}$, $q = q_{\mathrm{m}}$, while the value $u_{\mathrm{i}}$ of $u$ is used. Thereby it is understood that the force fields or pull-back actions which, among other terms, are compounded into the function $F$, do not depend on $q$ in too steep a way. If such is not the case, a trick to overcome this difficulty is presented in [36]; it results in replacing $A$ by an artificial inertia matrix in the construction of which the possible stiff elasticity coefficients are entered.

It is also at the point $(t_{\mathrm{m}}, q_{\mathrm{m}})$ that the functions $f_\alpha$ are calculated, in order to decide which contacts are to be treated as effective. The set $J(t, q(t))$ is thus estimated to equal $J_{\mathrm{m}} := J(t_{\mathrm{m}}, q_{\mathrm{m}})$ throughout the time-step.

The dynamical equation (21) is then discretized in the form

$$A_{\mathrm{m}}(u_{\mathrm{f}} - u_{\mathrm{i}}) = hF_{\mathrm{m}} + \sum_{\alpha \in J_{\mathrm{m}}} p^\alpha, \tag{32}$$

where the element $p^\alpha$ of $\mathbf{R}^n$ is made of the covariant components of the *impulsion* at contact $\alpha$, i.e. the integral $\mathcal{P}^\alpha$ over $[t_{\mathrm{i}}, t_{\mathrm{f}}]$ of the contact force $\mathcal{R}^\alpha$. Throughout the time-step, one ascribes to the linear mapping $G^*_\alpha$ of (20) its value computed at $(t_{\mathrm{m}}, q_{\mathrm{m}})$, hence

$$p^\alpha = G^*_\alpha \mathcal{P}^\alpha. \tag{33}$$

One does the same with the linear mapping $G_\alpha$ and with the rheonomic term $\mathcal{W}_\alpha$ of (19). Since a discretization scheme of the implicit type is being planned, the *final* velocity $u_{\mathrm{f}}$ is invoked, so as to calculate

$$\mathcal{U}_{\alpha,\mathrm{f}} := G_\alpha u_{\mathrm{f}} + \mathcal{W}_\alpha. \tag{34}$$

In the exact problem, the contact law with label $\alpha$ should connect $\mathcal{U}_\alpha$ to the contact force $\mathcal{R}^\alpha$ at every instant. Our approximation scheme of the implicit type consists in connecting, through the same relationship, the final value $\mathcal{U}_{\alpha,\mathrm{f}}$ of $\mathcal{U}_\alpha$ to the average of this force, namely $\mathcal{P}^\alpha/h$. In usual situations such as the no-friction case described by (25) or also the case of Coulomb friction, the contact law happens to be *positively homogeneous* with degree zero relatively to the contact force, so the discretized law reads

$$\mathrm{law}_\alpha(t_{\mathrm{m}}, q_{\mathrm{m}}, \mathcal{U}_{\alpha,\mathrm{f}}, \mathcal{P}^\alpha) = \mathrm{true}. \tag{35}$$

### 6.2 Nonlinear Gauss-Seidel iterations

Solving the system of conditions (32) to (35) is the hard part of the computation. From (32) and (33), one obtains

$$u_{\rm f} = u_{\rm r} + A_{\rm m}^{-1} \sum_{\beta \in J_{\rm m}} G_\beta^* \mathcal{P}^\beta, \tag{36}$$

where $u_{\rm r} := u_{\rm i} + hA_{\rm m}^{-1} F_{\rm m}$ may be called the *relaxed velocity.* Here is an iteration technique à la Gauss-Seidel which amounts to treating a succession of single-contact problems.

Let an estimated solution $(u_{\rm f}^{\rm esti}, \mathcal{P}_{\rm esti}^\beta)$, $\beta$ running through $J_{\rm m}$, be obtained with (36) satisfied. One expects to obtain a corrected estimate, say $(u_{\rm f}^{\rm corr}, \mathcal{P}_{\rm corr}^\beta)$, by selecting a label $\alpha$ and altering only $\mathcal{P}^\alpha$, i.e. by making $\mathcal{P}_{\rm corr}^\beta = \mathcal{P}_{\rm esti}^\beta$ for $\beta \neq \alpha$. The new estimate is astrained to verify (36); equivalently, since the old estimate satisfies the same,

$$u_{\rm f}^{\rm corr} = u_{\rm f}^{\rm esti} + A_{\rm m}^{-1} G_\alpha^* (\mathcal{P}_{\rm corr}^\alpha - \mathcal{P}_{\rm esti}^\alpha) \tag{37}$$

and to satisfy the discretized contact law (35). By applying $G_\alpha$ to both members of (37), one gives to (35) the form

$$\mathrm{law}_\alpha(t_{\rm m}, q_{\rm m}, G_\alpha\, u_{\rm f}^{\rm esti} + \mathcal{W}_\alpha + H_\alpha(\mathcal{P}_{\rm corr}^\alpha - \mathcal{P}_{\rm esti}^\alpha), \mathcal{P}_{\rm corr}^\alpha) = \mathrm{true} \tag{38}$$

where $H_\alpha := G_\alpha A_{\rm m}^{-1} G_\alpha^*$ is a symmetric positive definite $3 \times 3$ matrix.

Solving (38) in the unknown $\mathcal{P}_{\rm corr}^\alpha$ is easy in some significant cases [52], such as two-dimensional Coulomb contact or the case where $H_\alpha$ is axissymmetric about $\mathbf{n}^\alpha$. Generally, some iterative procedures may be applied, in which the normal and tangential components of the contact force are alternatively treated as known. Anyway, when a solver has been devised, the above procedure of correcting successive estimates is iterated, with $\alpha$ ranging cyclically in $J_{\rm m}$. The decision of stopping iterations may be made on observing the magnitude of the vector $\mathcal{P}_{\rm corr}^\alpha - \mathcal{P}_{\rm esti}^\alpha$; this actually is equivalent to checking the precision at which each pair $(u_{\rm f}^{\rm esti}, \mathcal{P}_{\rm esti}^\alpha)$ satisfies the corresponding contact law.

Observe that, provided such a numerical convergence check is made, the linear operator $H_\alpha$ in (38) may be replaced by any other mapping of $\mathbf{E}^3$ into itself with zero limit at the origin, with the possible advantage of making resolution easier. This replacement is also used in tricks for accelerating convergence.

The mathematical convergence of algorithms of this sort, in the case of Coulomb contact, as well as the very existence of solutions to the problems addressed, has only been established in special situations [44][40]; uniqueness cannot be expected in general.

If all the contact laws invoked are of the prospective type, one sees that, for each $\alpha$, the final relative velocity $\mathcal{U}_{\alpha\rm f}$ belongs to $\mathcal{K}_\alpha(t_{\rm m}, q_{\rm m})$. Similarly to 3.3, Remark 1, this ensures non-interpenetration with a welcome self-corrective effect if the step-length is not too large. Contact laws of the prospective type also manage automatically the possible breaking of contacts, without resorting to any analysis of complementarity conditions.

To start iterations, one needs an estimate $(u_{\mathrm{f}}^{\mathrm{esti}}, \mathcal{P}^{\beta}_{\mathrm{esti}})$ verifying (36). One may take $u_{\mathrm{f}}^{\mathrm{esti}} = u_r$ and all $\mathcal{P}^{\beta}_{\mathrm{esti}}$ equal to zero. In cases where the set $J_{\mathrm{m}}$ of the active contacts does not change much from one step to the other, it could be much more efficient to take as first guess for the $\mathcal{P}^{\beta}_{\mathrm{esti}}$ the values calculated at the antecedent time-step for the contacts already present and to adjust $u_{\mathrm{f}}^{\mathrm{esti}}$ by means of (36).

Anyway, iterations do not require the handling of large matrices since, in the representation of each $G_\alpha$ for a rigid body, only a $3 \times 6$ submatrix is nonzero ($2 \times 3$ in the case of a two-dimensional multibody system).

Also observe that, due to the computation being of the implicit type relative to velocity, a single computation step is needed to check whether a given position of the system is that of a possible equilibrium. One just have to launch iterations with $u_{\mathrm{i}} = 0$ ; if $u_{\mathrm{f}}$ is found zero, the corresponding set of computed values for the contact forces is compatible with equilibrium. In this sort of application, the inertia matrix $A$ only provides the backbone of the computation and its value may be chosen so as to ease convergence.

## 7 Nonsmooth dynamics and collisions

### 7.1 Discontinuous velocity functions

The preceding Sections were developed in the context of usual 'smooth' dynamics, in which the velocity function $u$ is assumed locally absolutely continuous on the time interval $I$. The properly 'unilateral' feature was only the possibility offered to contacts of breaking at any instant.

In contrast, the sudden occurrence at some instant $t_c$ of a new contact, i.e. a *collision* is expected to generate a velocity jump. Even without such an event, the presence of dry friction at a contact point has long been known, in some cases, to forbid the existence of a smooth solution beyond some catastrophic instant, on the left of which the contact force, as well as the derivative $u'$, may become unbounded [24]. This is a dynamic analogue to the *locking* situations familiarly met in the statics of frictional systems. Around year 1900, such a lack of smooth solution for an apparently well set problem seemed inadmissible to P. Painlevé and induced him to question the very concept of a contact force. Hence the inadequate denomination of 'Painlevé's paradox' which has been given to the observation. Today, one is accustomed to see models reaching the limits of their validity domain. Such is the case for smooth dynamics at a catastrophic instant of any sort, so that one is induced to enlarge the function space where $u$ is looked for [52][71].

In order to develop *Nonsmooth Dynamics*, the widely accepted extension of the functional framework is that of $\mathbf{R}^n$-valued functions with *bounded variation*. More precisely, since the time interval $I$ has not been supposed compact, we consider the linear space of the functions with *locally* bounded variation, i.e. they have bounded variation over every compact subinterval ; notation $\mathrm{lbv}(I, \mathbf{R}^n)$. Any $u$ in this space is locally Lebesgue-integrable, so that (2) remains meaningful. The classical property of a function with bounded variation to possess a derivative almost everywhere has no relevance in the present situation, for such a pointwise defined derivative of $u$ (generating values for the accelerations of the system elements in physical space) does not allow one to retrieve the function through integration. In fact, the possible discontinuity points of $u$, which make at most a countable, hence Lebesgue-negligible, subset of $I$ may be responsible in an essential way for the evolution of this function. For instance, a step-function possesses almost everywhere a derivative equal to zero without having to be a constant.

The role of the function $u'$ is now played by an $\mathbf{R}^n$-valued *measure* on $I$, called the Stieltjes measure or *differential measure* of $u$, that we shall denote by $du$. In the smooth case, where $u$ is locally absolutely continuous, one has $du = u'\,dt$, with $dt$ denoting the Lebesgue measure (this is nothing but the differential measure of the real function $t \mapsto t$).

Classically, a function $u$ in $\mathrm{lbv}(I, \mathbf{R}^n)$ possesses a left- and a right-limit at every point $t$ of $I$, say $u^-(t)$ and $u^+(t)$ (with the convention $u^-(t_0) = u(t_0)$ and symmetrically at the possible other end of $I$) and one has for every compact subinterval $[\sigma, \tau]$ of $I$

$$\int_{[\sigma,\tau]} du = u^+(\tau) - u^-(\sigma).$$

In particular, the integral of $du$ over the singleton $\{\sigma\}$ equals the possible *jump* of $u$ at point $\sigma$.

### 7.2 A measure-differential equation

Since one cannot rely on the second derivative of the function $t \mapsto q(t)$ anymore, the accelerations of the elements of the system in physical space are also missing, as well as forces. The latter have to be replaced by $\mathbf{E}^3$-valued measures, the integrals of which over every compact subinterval of $I$ constitute *impulsions*. In smooth dynamics, the impulsion measures admit, relatively to $dt$, some *density functions* whose values, at a given instant, are properly forces, so that impulsions, in accordance with the traditional definition, equal their time-integrals.

In particular, instead of a contact force $\mathcal{R}^\alpha$, there will more generally be invoked a *contact impulsion*, an $\mathbf{E}^3$-valued measure on the time interval $I$. For the homogeneity of notations, let us denote by $d\mathcal{S}^\alpha$ this measure, without having to pay attention to any function $\mathcal{S}^\alpha$ admitting it as differential measure. The *covariant components* of $d\mathcal{S}_\alpha$, relative to the parametrization in use, are $\mathbf{R}^n$-valued measure on $I$, say $ds_\alpha$, which, similarly to (20), are defined as

$$ds^\alpha = G^*_\alpha \, d\mathcal{S}^\alpha . \tag{39}$$

This expression makes sense if the linear mapping $G^*_\alpha$ depends continuously on $t$ and $q$, an assumption usually verified in multibody dynamics.

The differential equation of dynamics (21) has then to be replaced by

$$A(t,q)\, du = F(t,q,u)\, dt + \sum_\alpha ds^\alpha , \tag{40}$$

an equality of $\mathbf{R}^n$-valued measures on $I$, called a *measure-differential equation.*

Concerning the connexion of this writing with a general formulation of classical dynamics, refer to [54]. It may be asked why (40) is given the restricted form of an equality of measures, instead of an equality of distributions of arbitrary order on the interval $I$. Our answer, in the present context, is that nonsmoothness comes from the contact actions which, due to unilaterality are subject to inequality requirements (even in the more general case of adhesive contact [26][68]). This is a basic fact that distributions verifying inequalities are necessarily measures.

Such an equality of measures may equivalently be exploited as an equality of functions, thanks to the following observation. Given a finite collection of $\mathbf{R}$-valued or $\mathbf{R}^n$-valued measures, such are $dt$, $du$ and $ds^\alpha$ above, there exists (non uniquely) a nonnegative real measure on $I$, say $d\mu$, relatively to which these measures respectively admit density functions $t'_\mu \in \mathcal{L}^1_{\mathrm{loc}}(I, d\mu; \mathbf{R})$, $u'_\mu \in \mathcal{L}^1_{\mathrm{loc}}(I, d\mu; \mathbf{R}^n)$, $(s^\alpha)'_\mu \in \mathcal{L}^1_{\mathrm{loc}}(I, d\mu; \mathbf{R}^n)$. Then (40) is equivalent to the equality of $\mathbf{R}^n$-valued functions

$$A(t,q)\, u'_\mu = F(t,q,u)\, t'_\mu + \sum_\alpha (s^\alpha)'_\mu , \tag{41}$$

holding $d\mu$-almost everywhere in $I$. Replacing the *base* measure $d\mu$ by another admissible one amounts to multiply densities by positive real functions [53].

There remains to precise how contact laws may be entered into this formalism. As observed before, in the case of Coulomb friction as well as in the no-friction case, contact laws in the form (23) are conic, i.e. positively homogeneous of degree zero with regard to the argument $\mathcal{R}^\alpha$. If the density of contact impulsion $(\mathcal{S}^\alpha)'_\mu$ is introduced at the place of $\mathcal{R}^\alpha$ in the contact law, the relationship so stated between the contact impulsion measure and the local velocity function $\mathcal{U}_\alpha$ is thus indifferent to the change of the base measure. Such a policy does not seem questionable as long as the measure $d\mu$ and, consequently, the measure $d\mathcal{S}^\alpha$ are *diffuse* on the considered time-interval. In this case $u$ is continuous, so that the local velocity $\mathcal{U}_\alpha$ is well defined for every $t$.

### 7.3 Collisions

In contrast, assume that, at some instant $t_c$, a velocity jump occurs as the result of a collision.

The measure $du$ then presents an *atom* at the point $t_c$ of $I$ and the same should be expected from the other measures involved in the calculation. Necessarily the base measure $d\mu$ in use presents an atom at $t_c$ and $d\mu$ may be adjusted in such a way that this atom has unit mass, i.e. the said atom is a Dirac measure located at $t_c$. With $d\mu$ so specified, the value $\mathcal{P}^\alpha = (\mathcal{S}^\alpha)'_\mu(t_c)$ of the density of contact impulsion for the contact labelled $\alpha$, is nothing else than the vector called *percussion* in the traditional formalization of shocks. Consequently, the measure equation (40) readily applies to the standard question of determining the post-collision velocity $u^+(t_c)$, knowing the pre-collision value $u^-(t_c)$, *provided some information is available about contact impulsions.*

An information of this sort is implicitely included in the various pragmatic approaches proposed in the form of *collision laws*, since the very times of Descartes or Newton. These laws usually involve empirical parameters meant to account for the 'bounciness' of the collision (coefficients of restitution) and for the possible role of friction during the process. It is today recognized that such parameters can only be identified in narrowly delimited situations [72] such as the collision of two, otherwise free, spherical beads. The coefficients possibly determined for a pair of free bodies are not valid anymore if any of them is subject to additional constraints [32]. It should be kept in mind that if one of the bodies involved in a collision is part of a cluster of contacting objects, nonzero contact percussions should be expected at all contacts in the cluster.

Viewing a collision as a strictly instantaneous process is a schematic model, pertaining to the dynamics of strictly indeformable bodies. In contrast, if some slight deformability of the colliding bodies is evoked, a nonzero time-interval, say $[t_c, t_c + \theta]$, should be ascribed to the interaction. Large values of the contact forces are expected, the time-integral of which over this interval makes the contact percussion vector, but at this scale smooth dynamics remains applicable.

Studies of the latter sort are usually developed through a *multiple scaling* approach : a micro-time is introduced, in order to investigate the variation of the velocity function, while the variation of the position $q$ is neglected. The direction of the sliding velocity $\mathcal{U}_\alpha$ at some of the involved contacts may be found to exhibit large variations as a function of the micro-time, preventing the identification of any representative value which could, phenomenologically, be connected with the total impulsion $\mathcal{P}^\alpha$. The situation is better if $\mathcal{U}_\alpha$ remains zero : because the Coulomb cone, closed and convex, is constant with regard to the micro-time (see however the discussion in 8.1), the condition of $\mathcal{R}^\alpha$ belonging to it commutes with the integration invoked in defining $\mathcal{P}^\alpha$.

Deeper investigation of collision processes has also been conducted, by taking into account the whole deformation of the interacting bodies, either analytically [43] or numerically thanks to the finite element representation of deformability [76]. This also comes to confirm that collision laws can only roughly approximate the complexity of the process.

### 7.4 Frictionless inelastic collision

In the time-stepping scheme of Sect. 6, contact forces were involved only through their total impulsions on each time-interval, so that it remains applicable to problems formulated in the nonsmooth framework. This numerical procedure handles collisions in an automatic way. It only happens that, in the course of computation, a time-step exhibits some contacts not detected at the preceding step. Since the velocity $u_{\rm f}$ has to comply with some non-interpenetration conditions from which $u_{\rm i}$ is exempt, the contact impulsions $p^\alpha$ involved in (32) are expected to take values of larger magnitude than in the case of constant $J_{\rm m}$. But calculation remains the same and delivers kinematically admissible post-collision velocities. In short, because the contact laws (35) have been assumed *positively homogeneous* with degree zero in their last arguments, the algorithm treats collisions on the same footing as permanent contacts.

There remains to precise which collision model the algorithm turns out to approximate. Assume for simplicity that a single contact has come to increment $J_{\rm m}$, i.e. there occured a single impact at some unknown instant $t_{\rm c}$ of $[t_{\rm i}, t_{\rm f}]$. The endpoint values $u_{\rm i}$ and $u_{\rm f}$ are viewed as approximating $u^-(t_{\rm c})$ and $u^+(t_{\rm c})$ respectively.

We first have a look at the calculation to be made in the special frictionless setting of 5.2. The polyhedral set $W$ defined in (29), namely the set of the values of $u^+$ compatible with all non-interpenetration conditions, including the new one, is approximated by

$$W_{\rm m} := \{v \in \mathbf{R}^n \,|\, \forall\alpha \in J_{\rm m} \;:\; \frac{\partial f_\alpha}{\partial t}(t_{\rm m}, q_{\rm m}) + v.\nabla f_\alpha(t_{\rm m}, q_{\rm m}) \leq 0\}. \tag{42}$$

As in 6.2, one constructs the relaxed velocity $u_{\rm r} := u_{\rm i} + hA_{\rm m}^{-1}F_{\rm m}$. If the positive definite matrix $A_{\rm m}$ is used to define in $\mathbf{R}^n$ a Euclidean metric, the core of the computation simply consists in constructing $u_{\rm f}$ as *the nearest point to* $u_{\rm r}$ *in* $W_{\rm m}$. By imagining the step-length $h$ arbitrarily close to zero, one sees that the calculation approximates the following *collision law* :

*The post-collision velocity* $u^+(t_{\rm c})$ *equals, in the sense of the kinetic metric* $A(t_{\rm c}, q(t_{\rm c}))$, *the nearest point to* $u^-(t_{\rm c})$ *in* $W(t_{\rm c}, q(t_{\rm c}))$.

Let us refer now to the contact locus, in the notations of 5.1. The above calculation amounts to admit that, for each $\alpha$, the contact impulsion $\mathcal{P}^\alpha$ is connected, through the law of frictionless contact (25), with the *post-collision* velocity $\mathcal{U}_\alpha^+$. Due to this law being of prospective type, the implication $\mathcal{P}^\alpha \neq \mathbf{0} \Rightarrow \mathcal{U}_\alpha^+.\mathbf{n}^\alpha = 0$ holds, meaning that, if the contact labelled $\alpha$ takes an effective part in the nonsmooth process, it exhibits the feature traditionally formulated by saying that *the Newton restitution coefficient is zero.* But, depending on the circumstances met in constructing the projection, one may also have $\mathcal{P}^\alpha = \mathbf{0}$, with $\mathcal{U}_\alpha^+ . \mathbf{n}^\alpha$ only astrained to be nonnegative.

The present model (it was introduced in [51] under the name of Standard Inelastic Shock) is somewhat more realistic than the traditional Newton statement, in that *all the contacts present at instant* $t_c$ *are treated collectively.* This produces plausible results for instance in the case, popular in the Earthquake Engineering literature, of a rectangular block *rocking* on a horizontal plane: if the traditional formulation, with zero restitution coefficients, was applied to all contacts, no rocking could be found. In contrast, the above collision law allows one to discuss the outcome with regard to the aspect ratio of the block. A slender block presents a succession of oscillations, progressively damped by the inelasticity of collisions, while, below some critical aspect ratio, a single episode of motion leads to permanent contact.

## 7.5 A three-parameter collision law

In the same line, there remains to show how non-zero friction and non-zero restitution can be entered into a computationally efficient collision model. At an instant where the system experiences a velocity jump we propose, for every contact $\alpha$, to relate through the Coulomb law of friction (stated in prospective form) with friction coefficient $\gamma_\alpha$, the contact percussion $\mathcal{P}^\alpha$ with some *average value* $\mathcal{U}_\alpha^a$ of the local velocity. This artificial value is defined as a weighted mean of the (known) pre-collision velocity $\mathcal{U}_\alpha^-$ and of the (unknown) post-collision velocities $\mathcal{U}_\alpha^+$. Introducing as before the normal and tangential components of the concerned vectors, put

$$\mathcal{U}_{\alpha\mathrm{N}}^a = \frac{\rho_\alpha}{1+\rho_\alpha}\mathcal{U}_{\alpha\mathrm{N}}^- + \frac{1}{1+\rho_\alpha}\mathcal{U}_{\alpha\mathrm{N}}^+ \tag{43}$$

$$\mathcal{U}_{\alpha\mathrm{T}}^a = \frac{\tau_\alpha}{1+\tau_\alpha}\mathcal{U}_{\alpha\mathrm{T}}^- + \frac{1}{1+\tau_\alpha}\mathcal{U}_{\alpha\mathrm{T}}^+ . \tag{44}$$

The empirical parameters $\rho_\alpha$ and $\tau_\alpha$ will be called *the normal coefficient of restitution* and *the tangential coefficient of restitution* at the contact labelled $\alpha$, denominations justified by what follows.

Since the invoked contact law is of the prospective type, implications (26) and (27) show that $\mathcal{P}^\alpha$ can be nonzero only if $\mathcal{U}_{\alpha\mathrm{N}}^a = 0$, i.e. $\mathcal{U}_{\alpha\mathrm{N}}^+ = -\rho_\alpha \mathcal{U}_{\alpha\mathrm{N}}^-$, which formally is Newton's restitution law. But the present formulation is richer than stipulating a normal restitution law separately for each contact ; it also allows $\mathcal{P}^\alpha = 0$, in which case only the inequality $\mathcal{U}_{\alpha\mathrm{N}}^a \geq 0$ happens to be asserted. It is the global calculation, involving all the contacts together through the equation of dynamics, which decides between these two alternatives.

Similarly, the global calculation, if friction is large enough, may end in the zero sliding case of Coulomb's law at contact $\alpha$. Then $\mathcal{U}^+_{\alpha\mathrm{T}} = -\tau_\alpha \mathcal{U}^-_{\alpha\mathrm{T}}$, which is a *law of tangential restitution.*

The construction of the average velocity may readily be incorporated into the time-stepping computation scheme of Sect.6, so that the algorithm is ready to face possible collisions at every instant. At the price of a few extra lines of code at each time-step before launching the Gauss-Seidel iterations, no computation cost at all is added to the iterations themselves

Of course the same collision model may be used in conjonction with other solvers [38][39].

The time-stepping scheme also yields plausible results when some of the frictional catastrophes referred to in 7.1 are met [52].

It should however be admitted that relating the percussions $\mathcal{P}^\alpha$ to the respective average velocities $\mathcal{U}^a_\alpha$ is nothing but a pragmatic trick, generating a collision law of the sort which precisely was criticized in 7.3. Before relying on it in a specific context, one has to check its practical value against any available experimental measurement.

In the simplest case of all, that of the collision of two otherwise free spherical beads, the resulting law turns out to be mathematically equivalent to another three-parameter collision model anteriorly proposed on the basis of quite different arguments [76] and which, experimentally, has been found in fairly good agreement with reality [25].

The various circumstances met when a ball bounces against a fixed plane are also convincingly reproduced, in particular the exotic behaviour of a 'superball' [55][56].

The rocking of a slender block supported by an oscillating table has provided another occasion of testing the model. Some sensitive features of the rocking regimes have been correctly predicted by CD computation [66].

It is in the domain of the numerical simulation of granular materials that the computational simplicity of the procedure proves invaluable. In fact, one may then have several ten thousands of contacts to investigate at each time-step.

Some satisfactory comparisons of computation with experiments are presented in [55][56], pertaining in particular to the circulatory currents observed in shaken granular samples and their connection with the segregation of grain sizes [41].

The study of the flow of grains over a rough incline has also benefited from joining experimental investigation with CD numerical simulations [14].

The efficiency of the method in the dynamics of granular materials mainly stems from that, once a time-discretization has been chosen, all the collisions which have been detected as occurring on a given time-step are treated together. This of course entails some trade-offs which have to be assessed. It does not seem to matter that the ordering of collisions, which mechanically should be successive, is only internal to the algorithm. In fact the problems in view are physically *undeterministic*, since a slight change in the initial conditions is liable to generate after a short time a completely different sequence of positions and collisions. A more critical issue is that each contact point is only counted once in the analysis. An accumulation of bounces, such as those of a ping-pong ball coming to rest, is thus viewed as a single collision as soon as the successive bounces are all comprised in the time-step. This could result in underestimating the total energy loss arising from inelastic collisions. This source of error may be checked by repeating computation with step-length reduced to half : if no appreciable change is found in the energy versus time curve, one may conclude that no harm was done.

It has been stressed in the foregoing that, in compact assemblies, the bodies involved in a collision may be part of clusters of already contacting objects, between which percussions are likely to be induced. The described procedure of global dynamical calculation does take this into account. Physically however, the transmission of impulses in clusters should involve elasticities in a way analogous to sound propagation, an effect which cannot be analyzed through the present model of strictly rigid objects.

Anticipating on the question of energy balance, to be studied in the next Section, one may finally rise another objection. Like other popular collision laws, the trick of average velocities offers no security against the possible violation of the laws of Thermodynamics. In fact, for colliding objects of not too usual shapes, the calculated outcome may entail some energy creation. This is a physically inacceptable conclusion, except of course for artificially boosted collisions as in electric billiard games. A systematic thermodynamic formalism has been proposed by M. Frémond [27], securing that the empirical collision laws one may insert in it are dissipative [15][16].

# 8 Energy balance and calculus for differential measures

## 8.1 Departing from smooth dynamics

Drawing the energy balance of a dynamical evolution is essentially relevant to the case where the external obstacles or boundaries are fixed in the (Galilean) reference frame in use. In the smooth dynamical context of Sect.4 it was alluded to the possibility of constructing the parametrization $(q)$ with account of linkages or permanent hinges through which members of the system would be connected with some external supports. Also these external supports have to be assumed fixed, in order to obtain a 'scleronomic', i.e. time-independent parametrization. This makes the inertia matrix $A(t, q)$ constant with regard to its first argument and gives to the kinetic energy of the system an expression which, for every $q$, is a positive definite quadratic form in $u$, namely $\mathcal{E}_{\mathrm{k}} = A_{ij} u^i u^j / 2$. In this framework of smooth dynamics, one classically establishes that the time-derivative of $\mathcal{E}_{\mathrm{k}}$ equals the total power of the forces experienced by the system, including the contact forces between members and the possible action of external supports. If contacts are frictionless as well as the (fixed) possible external hinges the corresponding terms in the power vanish. In the familiar case where, additionally, the other forces in presence derive from a potential energy, the integration of derivatives yields the well known property of *energy conservation.*

In contrast, the elementary example of a particle hitting 'inelastically' a fixed frictionless plane shows that, in nonsmooth situations, *the absence of friction does not secure energy conservation anymore.* Mathematically, the reason is that the formula for the time-derivative of a product of (locally absolutely continuous) functions, used in establishing the energy balance in the smooth case, has to be replaced by less precise relationships to be presented in 8.2 below. The physical background needs some comments.

The discussion sketched in 7.3, in which a nonzero time-interval $[t_c, t_c+\theta]$ was, at least qualitatively, assigned to the process, makes one understand that the small deformations that the contacting bodies undergo during the collision may generate values of the local velocity vector $\mathcal{U}_\alpha$ disagreeing with condition $\mathcal{U}_\alpha.\mathbf{n}^\alpha = 0$, while the vector $\mathbf{n}^\alpha$ itself may exhibit transient variations. The instant power of contact forces can then be nonzero, yielding for their total work over the time-interval a nonzero, commonly negative, value. The same remark applies to the external linkages asserted to connect the system with fixed supports : even frictionless, they may transmit work to the outside world. This induces one to question, when nonsmooth circumstances are met, the legitimity of involving these linkages in the construction of the parametrization : the fact that such linkages are perfect in the sense of the traditional smooth dynamics does not secure that the percussions they impart in a collisional situation should have zero generalized components.

## 8.2 Nonsmooth differential calculus

Some rules of a calculus for l.b.v. vector functions of a real variable $t$ are established in [53]. In particular, one finds expressions for the differential measures of functions constructed from elements of l.b.v. through multilinear operations. For instance one has

$$d(u^i u^j) = u^{i+} du^j + u^{j-} du^i \tag{45}$$

and various similar formulas where left- and right-velocities are differently combined. If the scalar product in $\mathbf{R}^n$ is denoted by a dot, this yields

$$d\|u\|^2 = (u^+ + u^-).\, du. \tag{46}$$

Furthermore, thanks to the quadratic form $u \mapsto u.u = \|u\|^2$ being nonnegative, one establishes the following inequality, in the sense of the ordering of real measures on $I$

$$2u^-.du \leq d\|u\|^2 \leq 2u^+.du. \tag{47}$$

This holds true, more generally, in any Euclidean or real Hilbert linear space.

By integrating the concerned measures on the singleton $\{t_c\}$, one may derive from these rules of extended differential calculus a formula for the jump of kinetic energy at the instant of a possible collision

$$\mathcal{E}_k^+ - \mathcal{E}_k^- = \frac{1}{2}(u^+ + u^-).A(u^+ - u^-). \tag{48}$$

This result, sometimes called Kelvin's theorem, actually is of purely algebraic nature and one may recognize in it the elementary identity $(a+b).(a-b) = \|a\|^2 - \|b\|^2$ applied to the Euclidean metric associated with the inertia matrix $A = A(q(t_c))$. If (40) is used here to express $A(u^+ - u^-)$ (recall that the measure $dt$ has zero integral on a singleton), this yields

$$\mathcal{E}_k^+ - \mathcal{E}_k^- = \frac{1}{2}(u^+ + u^-).A(u^+ - u^-) = \frac{1}{2}(u^+ + u^-).\sum_\alpha p^\alpha. \tag{49}$$

where $p^\alpha = G_\alpha^* \mathcal{P}^\alpha$ is the $n$-vector made of the generalized component of the percussion $\mathcal{P}^\alpha$ occurring at the contact labelled $\alpha$. The right-hand member equivalently writes down as

$$\frac{1}{2}\sum_\alpha (u^+ + u^-).G_\alpha^* \mathcal{P}^\alpha = \sum_\alpha \frac{1}{2}(G_\alpha u^+ + G_\alpha u^-).\mathcal{P}^\alpha$$
$$= \sum_\alpha \frac{1}{2}(\mathcal{U}_\alpha^- + \mathcal{U}_\alpha^+).\mathcal{P}^\alpha. \tag{50}$$

Therefrom the jump of $\mathcal{E}_k$ is expressed as a sum of terms corresponding to the respective contacts effective in physical space. Mathematically, it may be convenient to view each of these terms as defining the *energy* or *work* of the corresponding percussion. But one should keep in mind that nothing in this analysis can justify the conception that such terms could, each for its part, express a separate energy balance for some physical phenomenon occurring at the corresponding contact locus.

As an illustration, one may come back to the frictionless collisions presented in 7.4. Since each percussion $\mathcal{P}^\alpha$ was then assumed connected with the corresponding $\mathcal{U}_\alpha^+$ through the law of frictionless contact (25), one has $\mathcal{U}_\alpha^+.\mathcal{P}^\alpha = 0$. From the right-side inequality in (47), one concludes that such collisions essentially involve some energy loss. In contrast, one might stipulate that the same law of frictionless contact should hold between $\mathcal{P}^\alpha$ and the arithmetic mean $(\mathcal{U}_\alpha^- + \mathcal{U}_\alpha^+)/2$. Then (50) would entail energy preservation. Note that if some of the contacts involved take place between a member of the system and some external obstacle, the latter must be fixed.

When invoked in the space $\mathbf{R}^n$ of the parameters, the norm and scalar product should be understood in the sense of the kinetic metric. More generally than in the above investigation of an instant phenomenon, the energy balance of the possibly nonsmooth motion of the system over a time interval may be stated as an equality of measures, in which the evolution of the matrix $A$ is taken into account [52][53].

## 9 Behaviour of a masonry structure

The dynamics of block assemblies has been submitted to CD computation, in order to simulate their response to seismic actions [34]. In this domain of application, it may be necessary to take also into account a slight deformability of the blocks and the presence of mortar joints [2].

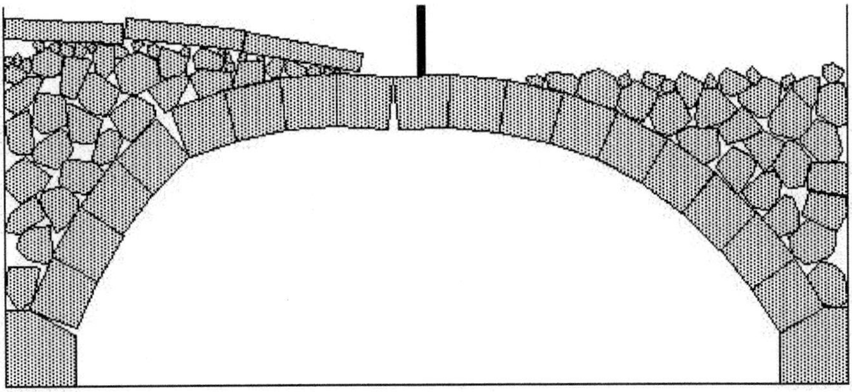

**Fig. 1.** Friction coefficient 0.6

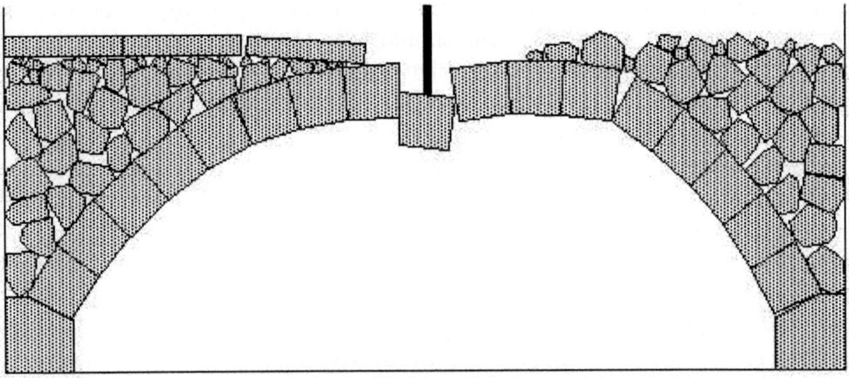

**Fig. 2.** Friction coefficient 0.3

Numerical simulation is applied here to a two-dimensional model of stone bridge in order to display its collapse under some localized forcing. The purpose is pedagogical, with a view to make clear that the ruin is not the result of the crushing strength of stone beeing exceded, but has to be discussed in terms of geometry and of the unilaterality of non-interpenetrability constraints [29][30]. If mortar joints are present between stones, their strength is neglected and the contact described as obeying the Coulomb law of friction. Coulomb himself, when investigating this issue, observed that, in practical instances, the friction coefficient was large enough for no sliding to occur between archstones so that the evolution toward collapse consisted of the opening of some joints through the relative rotation of the corresponding archstones about *hinge points*.

The structure, initially in equilibrium under its own weight, is submitted to the action of an external object with imposed motion, forcing the central archstone down. Computation of the subsequent motion is performed twice, with the respective values of 0.6 and 0.3 for the friction coefficient at all contacts.

# 10 Stress distribution in a conical grain pile

## 10.1 Constructing a pile

Numerical simulation is a popular mode of investigation in granular mechanics. In particular, dry cohesionless granular materials are naturally modelled as collections of bodies which interact by contacts affected with Coulomb friction. Provided the model is sufficiently validated by comparing some measurements, feasible in laboratory on physical experiments, with the corresponding numerical results, computer simulation may bring precious informations about intimate mechanisms and quantities otherwise inaccessible.

The following observation has, in recent years, started a lot of speculation and controversy [11][70]. If a conical pile is created by pouring grains from a point source onto a rough rigid horizontal ground, some experimentalists have been surprised to find that the distribution of pressure at various ground contact points was not proportional to the height of the material above and even that some local minimum of ground pressure could be present at the vertical of the apex.

The distribution of stress in a granular material not only depends on the forces actually exerted on it, but strongly also on the way the granular mass has been prepared. Numerical simulation thus has to reproduce the preparation process.

Figure 3 displays a view of the following numerical experiment [61], performed with the Contact Dynamics algorithm of Sect. 6. A pile of about 14000 spherical grains is constructed by depositing grains, one by one, at the contact of already existing ones. Grains diameters are distributed at random, uniformly from 0.25 cm to 0.5 cm. The fixed horizontal ground roughness is simulated by a random pavement of grains with the same distribution of

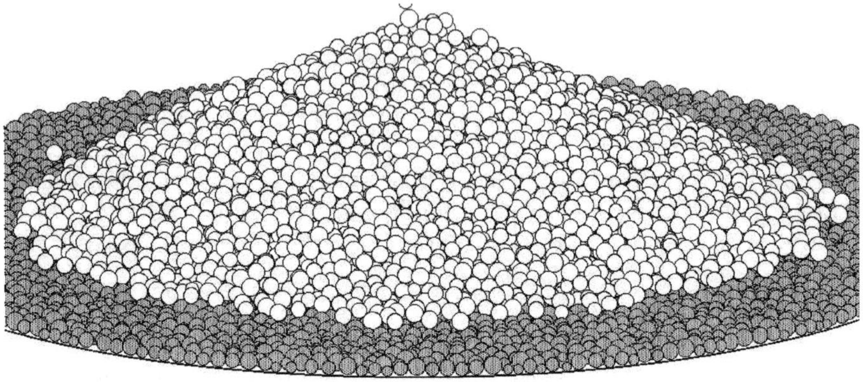

**Fig. 3.** View of the pile

sizes. Friction coefficient : 0.4 everywhere. Restitution coefficients : 0. Gravity : 981 cm/s$^2$. The vertical of a grain center at the time of its deposition is chosen at random in the neighbourhood of the $Oz$ axis at a maximal distance of 0.8 cm. Deposition frequency : 200 per second.

A large part of the deposited grains run down the pile slope before stopping, sometimes triggering avalanches. The velocity of running grains may reach 24 cm/s. For sufficient precision in calculating these motions, including the avalanches, the step length is fixed at $2 \times 10^{-4}$ s. The duration of the simulated phenomenon equals 72 s, so that 360 000 steps have been needed. In the final stages the number of contact points is about 33 000. The Gauss-Seidel procedure remains practical on a microcomputer (but a matter of weeks) for a system of such a size because the largest part of the pile stays in quasi-equilibrium : the contact impulsions calculated in the antecedent time-step for the contacts already effective are used as first guess in iterations.

Figure 4 shows, in four stages of the pile building, a slice 2 cm thick containing $Oz$. The grains deposited between $t = 15.5$ and $t = 18.7$ and which, at the considered stage, have their centers in this slice are represented in black. One observes that a noticable proportion of the deposited grains do not run out on the slopes, but accumulate in the central region so as to cause some plastic deformation of the existing granular mass. This deformation may be assessed by visualizing the distribution of these black grains at the successive stages. Let us call a *fossile layer* such a collection of grains observed at a certain date and deposited during a specified anterior period.

### 10.2 Stress distribution

The programme allows one to choose a line segment in a plane drawn through $Oz$. By rotation about $Oz$, this segment generates a curved strip in the shape of a truncated cone or of a cylinder, across which force transmission will be analyzed. To this end, the contacts occurring in grain pairs with centers

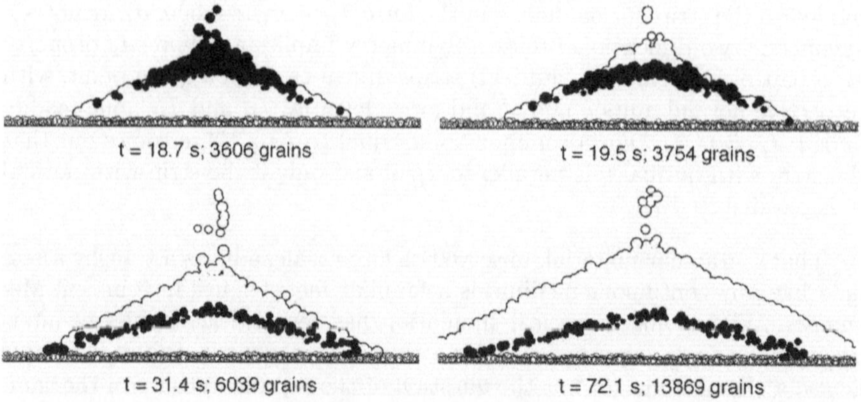


**Fig. 4.** Fossilelayers

separated by the conical or cylindrical surfaces are reviewed. A contact force $\mathcal{R}$ is retained in the list if the corresponding contact point $M$ has a projection onto the separating surface which belongs exactly to the strip. The meridian half-plane $\Pi$ containing $M$ is used, in order to decompose $\mathcal{R}$ into a component $\mathcal{R}_z$ in the $Oz$ direction and a component $\mathcal{R}_r$ in the axifugal direction (the component of $\mathcal{R}$ orthogonal to $\Pi$ is not investigated). The sum of all $\mathcal{R}_z$ in the list, as well as the sum of all $\mathcal{R}_r$, are divided by the area of the strip. This yields the components in $\Pi$ of a vector $\mathcal{T}_\Pi$ which may be viewed as the *average density of meridian force* transmitted across the strip.

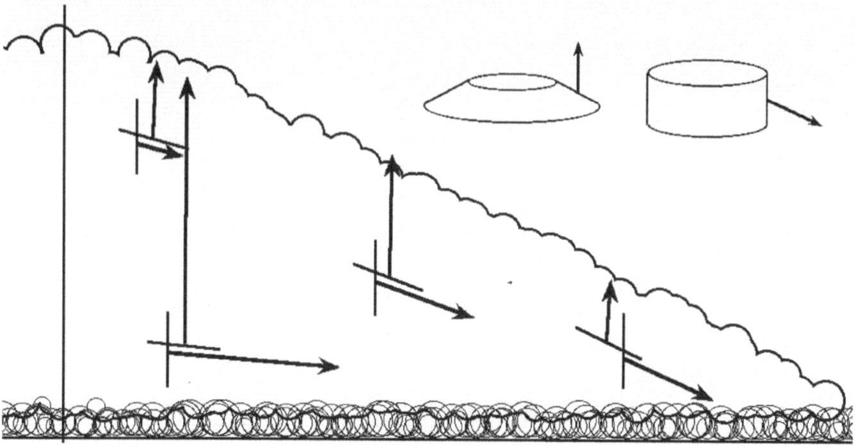

**Fig. 5.** Reciprocal cuts

In a classical continuous material, possessing a Cauchy stress tensor field, the similar construction would deliver an average density of meridian force related to the strip normal unit $\nu$ in the form $\mathcal{T}_\Pi = \sigma_\Pi \nu$, where $\sigma_\Pi$ denotes a *symmetric* two-dimensional tensor. Symmetry implies a *reciprocity* property for a pair of (conical or cylindrical) strips drawn through a given point, with respective normal units $\nu$ and $\nu'$ and force densities $\mathcal{T}_\Pi$ and $\mathcal{T}'_\Pi$ : one readily finds $\nu.\mathcal{T}'_\Pi = \nu'.\mathcal{T}_\Pi$. For a common value equal to zero, there comes out that the strip with normal $\nu$ is parallel to $\mathcal{T}'_\Pi$ if and only if the strip with normal $\nu'$ is parallel to $\mathcal{T}_\Pi$.

That a granular material, observed at large scale, admits a Cauchy stress field like any continuous medium is a familiar fact, for instance in Soil Mechanics. When some numerical simulation has revealed the values of intergranular forces, the assessement of the average stress over a delimited sample is also a classical matter but the question of the representativity of the sample, in regard to its size, needs investigation. In fact, photoelastic experiments as well as numerical simulations reveal that the transmission of forces in a granular mass in equilibrium exhibits a certain inhomogeneity. Loads happen to be carried by privileged grains forming *force chains.* A sample has naturally to exceed the size of these chains in order to yield a representative value of stress. For an investigation of this subject through Contact Dynamics simulations, see [57][60].

The purpose of Fig. 5 is to check that, in the present numerical experiment, the numbers of grains is large enough for the above reciprocity property to hold reasonably true, in spite of sampling aleas (a two-dimensional example, with irregular polygonal grains is presented in [60]).

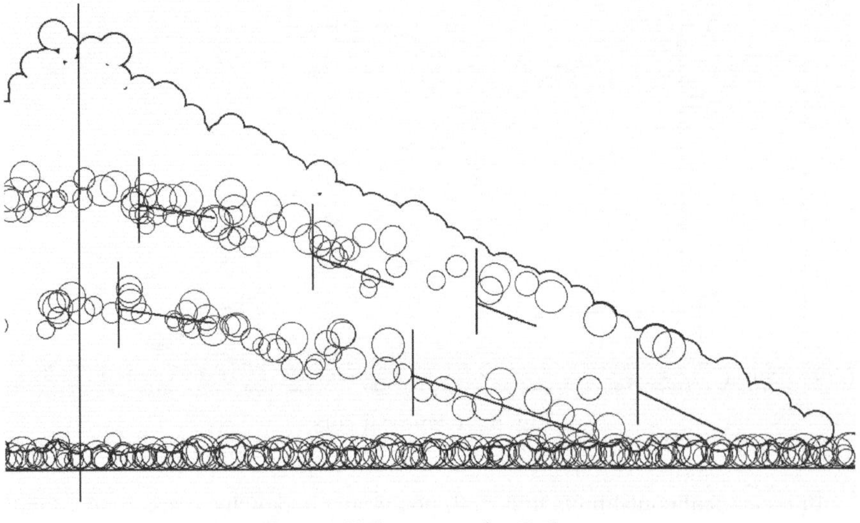

**Fig. 6.** Reference to fossile layers

Figure 6 shows the same 2 cm thick slice as before, with two fossile layers outlined. This suggests the following approximate law :

*In each point of the pile, the average density of meridian force $\mathcal{T}_\Pi$ corresponding to a cylindrical strip with axis $Oz$ has the same direction as the meridian section of a fossile layer containing this point.*

This distribution of meridian forces expresses an *arching* effect, present even in the central region of the pile.

In the peripheral bank, which has been constructed by the deposition of avalanching grains, the fossile layers are nearly parallel to the free surface. Through the reciprocity property, this is equivalent to saying that, on a conical cut parallel to the free surface, the average density of transmitted force is vertical. This feature looks natural since the bank is made of layers successively deposited by avalanches. When such a layer stops, the supporting forces it experiences from the material beneath should equilibrate its weight, and this verticality of transmitted forces is expected to persist after other layers possessing the same property have been superimposed.

### 10.3 Ground pressure

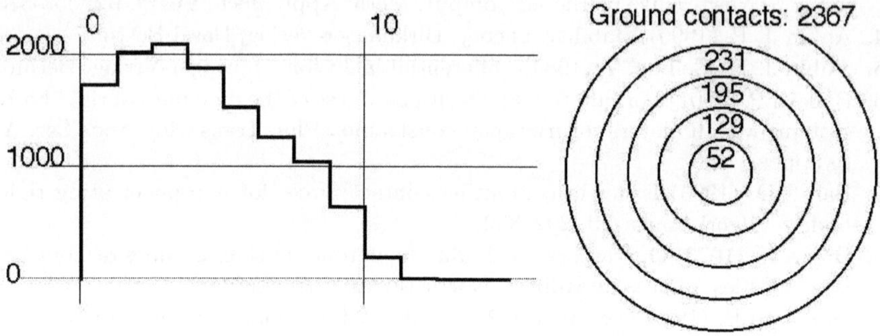


**Fig. 7.** Average ground pressure in annular regions

The central core of the pile clearly has part of its weight supported by the arching effect of the peripheral bank and this explains that in some experiments, a local minimum of ground pressure has been found at the vertical of the pile apex. Figure 7 displays the results obtained in the present numerical simulation. The supporting ground surface has been divided into annular regions with axis $Oz$ and uniform width. For all the contact points between the pile and the rough gound surface wich project in one of the annular regions, the total vertical component of the contact forces is divided by the area of the region, yielding the corresponding average pressure. The graph shows the respective values of this pressure (in deciPascal) plotted against the distance to $Oz$. Regions of small radii are naturally more subject to sampling aleas than the larger ones, but a central local minimum is in evidence. The pressure at center is smaller than the hydrostatic pressure which would be found in a liquid column of the same height and the same average volume mass, namely 3320 dPa corresponding to the height of 5.5 cm and a volume mass of 0.615 (in computation, the volume mass of the material the grains are made of has been taken equal to 1).

Anyway, the statics of a pile of dry grains has nothing to do with hydrostatics. This is evidenced by the fact that the free surface of the pile at rest is not horizontal. The merit of the reciprocity property disclosed in the foregoing is to establish a mathematical connexion between the arching effect and the inclination of the free surface.

## References

1. Abadie M. (2000) Dynamic simulation of rigid bodies : modelling of frictional contact. In: Brogliato B. (Ed.) Impacts in Mechanical Systems. Analysis and Modelling. Springer, Berlin Heidelberg, 61–144
2. Acary V. (2001) Contribution à la modélisation mécanique et numérique des édifices maçonnés, Thesis, Université Aix-Marseille II
3. Alart P, Curnier A. (1991) A mixed formulation for frictional contact problems prone to Newton like methods. Comput. Meth. Appl. Mech. Engrg. **92**, 353-375
4. Aubin J. P. (1990) Viability Theory. Birkhäuser-Verlag, Basel Boston
5. Aubin J. P., Cellina A. (1984) Differential inclusions. Springer-Verlag, Berlin
6. Ballard P. (2001) Formulation and well-posedness of the dynamics of rigid body systems with unilateral or frictional constraints. Phil. Trans. Roy. Soc., Ser. A, to appear
7. Baraff D. (1993) Issues in computing contact forces for non-penetrating rigid bodies. Algorithmica **10**, 292–352
8. Brézis H. (1973) Opérateurs maximaux monotones et semi-groupes de contraction dans les espaces de Hilbert. North-Holland, Amsterdam
9. Brogliato B. (1999) Nonsmooth Mechanics, 2d. edition. Springer, London
10. Brogliato B., ten Dam A. A., Paoli L., Génot F., Abadie M. (2001) Numerical simulation of finite dimensional multibody nonsmooth mechanical systems. ASME Applied Mechanics Reviews, to appear.
11. Cates M. E., Wittmer J. P., Bouchaud J.-P., Claudin P. (1998) Development of stresses in cohesionless poured sand. Phil. Trans. Roy. Soc. London, Ser. A, **356**, 2535–2560
12. Chabrand P., Dubois F., Raous M. (1998) Various numerical methods for solving unilateral contact problems with friction. Mathl. Comput. Modelling **28**, 97–108
13. Chatterjee A., Ruina A. (1998) Two interpretations of rigidity in rigid-body collisions. ASME J. Appl. Mech. **65**, 894–900
14. Chevoir F., Prochnow M., Jenkins J., Mills P. (2001) Dense granular flows down an inclined plane. Powders and Grains 2001 (Sendai, Japan). Balkema, Rotterdam Brookfield, to appear
15. Cholet C. (1998) Chocs de solides rigides. Thesis, Univ. Paris 6

16. Cholet C., Dimnet E., Chevoir F., Moucheront P., Frémond M. (1997) Experimental study of collisions of angular particles. In: Behringer R., Jenkins J. (Eds.) Powders and grains 97. Balkema, Rotterdam Brookfield, 543–546
17. Cottle R. W., Pang J.-S., Stone R. E. (1992) The Linear Complementarity Problem. Academic Press, Boston, etc.
18. Christensen P. W., Klarbring A., Pang J. S., Strömberg N. (1998) Formulation and comparison of algorithms for frictional contact problems. Int. J. Num. Meth. Engng. **42**, 145–173
19. Cundall P. A. (1971) A computer model for simulating progressive large scale movements of blocky rock systems, Proceedings of the Symposium of the International Society of Rock Mechanics, Vol.1, Nancy, France, 132–150
20. Curnier A. (1984) A theory of friction. Int. J. Solids Struct. **20**, 637–647
21. Deimling K. (1992) Multivalued Differential Equations. De Gruyter, Berlin
22. Delassus E. (1917) Mémoire sur la théorie des liaisons finies unilatérales. Ann. Sci. Ecole Norm. Sup. **34**, 95–179
23. De Saxcé G., Feng Z. Q. (1991) New inequation and functional for contact with friction. J. Mech. of Struct. and Machines **19**, 301–325
24. Génot F., Brogliato B. (1999) New results on Painlevé paradoxes. European Journal of Mechanics, A/Solids, **18**, 653–677
25. Foerster S., Louge M., Chang H., Allia K. (1994) Measurements of the collision properties of small spheres. Phys. Fluids **6**, 1108-1115.
26. Frémond M. (1987) Adhérence des solides. J. Mécanique Théorique et Appliquée **6**, 383–407
27. Frémond M. (1995) Rigid body collisions. Physics Letters A **204**, 33–41
28. Haug, E. J. (1989) Computer Aided Kinematics and Dynamics, vol. 1. Allyn and Bacon, Boston
29. Heyman J. (1966) The stone skeleton. Int. J. Solids Structures **2**, 249–279
30. Heyman J. (1995) The stone skeleton, structural engineering of masonry architecture. Cambridge University Press, Cambridge U. K.
31. Isac G. (1992) Complementarity Problems. Lecture Notes in Math., Vol. **1528**. Springer-Verlag, Berlin
32. Ivanov A. P. (1999) On the problem of constrained collision. In: Pfeiffer F., Glocker Ch. (Eds.) Unilateral Multibody Contacts. Kluwer, Dordrecht Boston London, 107–116
33. Jean M., Moreau J. J. (1987) Dynamics in the presence of unilateral contact and dry friction. In: Del Piero G., Maceri F. (Eds.) Unilateral Problems in Structural Analysis 2, CISM Courses and Lectures Vol. **304**. Springer-Verlag, Wien, New York, 151–196
34. Jean M., Moreau J. J. (1992) Unilaterality and dry friction in the dynamics of rigid body collections. In: Curnier A. (Ed.) Contact Mechanics International Symposium. Presses Polytechn. et Universit. Romandes, Lausanne, 31–48.
35. Jean M., Touzot G. (1988) Implementation of unilateral contact and dry friction in computer codes dealing with large deformations problems. J. Mécanique Théorique et Appliquée **7** (suppl. Nr. 1), 145–160
36. Jean M. (1995) Frictional contact in collections of rigid or deformable bodies: numerical simulation of geomaterials. In: Selvadurai A.P.S., Boulon M.J. (Eds.) Mechanics of Geomaterial Interfaces. Elsevier, Amsterdam, 453–486
37. Jean M. (1999) The Non Smooth Contact Dynamics method. In: Martins J. A. C., Klarbring A. (Eds.) Computational Modeling of Contact and Friction, special issue of Computer Meth. in Appl. Mech. and Engng. **177**, 235–257

38. Johansson L. (1999) A linear complementarity algorithm for rigid body impact with friction. European Journal of Mechanics, A/Solids, **18**, 703–717
39. Johansson L., Klarbring A. (2000) Study of frictional impact using a nonsmooth equation solver. ASME J. Appl. Mech. **67**, 267–273
40. Jourdan, F., Alart, P., Jean, M. (1998) A Gauss-Seidel-like algorithm to solve frictional contact problems. Computer Meth. Appl. Mech. Engng. **155**, 31–47
41. Knight J. B., Jaeger H. M., Nagel S. R. (1993) Vibration-induced size separation in granular media : the convection connection. Phys. Rev. Lett. **70**, 3728–3731
42. Kunze M., Monteiro Marques M. D. P. (2000) An introduction to Moreau's sweeping process. In: Brogliato B. (Ed.) Impacts in Mechanical Systems. Analysis and Modelling. Springer, Berlin Heidelberg, 1–60
43. Maw N., Barber J. R., Fawcett J. N. (1977) The rebound of elastic bodies in oblique impact. Mech. Res. Comm. **4**, 17–21
44. Monteiro Marques M. D. P. (1993) Differential Inclusions in Nonsmooth Mechanical Problems: Shocks and Dry Friction. Birkhäuser, Basel, Boston, Berlin
45. Moreau J. J. (1963) Les liaisons unilatérales et le principe de Gauss. Comptes Rendus Acad. Sci. Paris **256**, 871–874
46. Moreau J. J. (1966) Quadratic programming in mechanics: dynamics of one-sided constraints. SIAM J. Control **4**, 153–158.
47. Moreau J. J. (1973) Problème d'évolution associé à un convexe mobile d'un espace hilbertien. Comptes Rendus Acad. Sci. Paris **276**, 791–794
48. Moreau J. J. (1974) On unilateral constraints, friction and plasticity. In Capriz G., Stampacchia G. (Eds.) New variational techniques in mathematical physics, C.I.M.E. II ciclo 1973. Edizioni Cremonese, Roma,173–322.
49. Moreau J. J. (1976) Application of convex analysis to the treatment of elastoplastic systems. In: Germain P., Nayroles B. (Eds.) Applications of functional analysis to problems of mechanics, Lecture Notes in Mathematics **503**, Springer-Verlag, Berlin, Heidelberg, New York, 56-89.
50. Moreau J. J. (1977) Evolution problem associated with a moving convex set in a Hilbert space. J. Diff. Equ. **26**, 347–374
51. Moreau J. J. (1985) Standard inelastic shocks and the dynamics of unilateral constraints. In: Del Piero G., Maceri F. (Eds.) Unilateral Problems in Structural Analysis, CISM Courses and Lectures Vol. **288**. Springer-Verlag, Wien New York, 173–221
52. Moreau J. J. (1988) Unilateral contact and dry friction in finite freedom dynamics. In: Moreau J. J., Panagiotopoulos P.D. (Eds.) Nonsmooth Mechanics and Applications, CISM Courses and Lectures, Vol. **302**. Springer-Verlag, Wien New York, 1–82
53. Moreau J. J. (1988) Bounded variation in time. In: Moreau J. J. , Panagiotopoulos P. D., Strang G. (Eds.) Topics in Nonsmooth Mechanics. Birkhäuser, Basel Boston Berlin, 1–74
54. Moreau J. J. (1989) An expression of classical dynamics, Ann. Inst. H. Poincaré Anal. Non Linéaire. **6** (suppl.), 1-48. Volume also available as: Attouch H., Aubin J.-P., Clarke F., Ekeland I. (Eds.) Analyse Non Linéaire, Gauthier-Villars, Paris
55. Moreau J. J. (1994) Some numerical methods in multibody dynamics : application to granular materials. Eur. J. Mech., A/Solids **13**, n°4 - suppl., 93–114
56. Moreau J. J. (1995) Numerical experiments in granular dynamics: vibration-induced size segregation. In: Raous M. et al. (Eds.) Contact Mechanics. Plenum Press, New York, 347–358

57. Moreau J. J. (1997) Numerical investigation of shear zones in granular materials. In: Grassberger P., Wolf D. (Eds.) Proc. HLRZ-Workshop on Friction, Arching, Contact Dynamics. World Scientific, Singapore, 233–247
58. Moreau, J. J. (1999) Some basics of unilateral dynamics. In: Pfeiffer F., Glocker Ch. (Eds.) Unilateral Multibody Contacts. Kluwer, Dordrecht Boston London, 1–14
59. Moreau J. J. (1999) Numerical aspects of the sweeping process. In: Martins J. A. C., Klarbring A. (Eds.) Computational Modeling of Contact and Friction, special issue of Computer Meth. in Appl. Mech. and Engng. **177**, 329–349
60. Moreau, J. J. (1999) Évolutions en présence de liaisons unilatérales : notions de base. In: Guédra-Degeorges D., Ladevèze P., Raous M. (Eds.) Actes du 4e Colloque National en Calcul des Structures. Teknea, Toulouse, 25–40.
61. Moreau, J. J. (1999) Application des algorithmes "Contact Dynamics" aux milieux granulaires, Actes du 14e Congrès Français de Mécanique, Toulouse (CD ROM, ISBN 2-84088-040-7)
62. Moreau J. J. (2000) Contact et frottement en dynamique des systèmes de corps rigides. Rev. Europ. des Eléments Finis **9**, 9-28
63. Nouguier C., Bohatier C., Moreau J. J., Radjai F. (2000) Force fluctuations in a pushed granular material. Granular matter **2**, 171–178
64. Paoli L., Schatzman M. (1993) Mouvements à un nombre fini de degrés de liberté avec contraintes unilatérales: cas avec perte d'énergie. Math. Modelling and Num. Anal. **27**, 673–717
65. Pfeiffer F., Glocker Ch. (1996) Multibody Dynamics with Unilateral Contacts. John Wiley and Sons, New York
66. Raous M. (1993) Experimental analysis of the rocking of a rigid block. 3rd Pan American Congress of Applied Mechanics (PACAM III), Sao Paulo, Brazil
67. Raous M. (1999) Quasistatic Signorini problem with Coulomb friction and coupling to adhesion. In: Wriggers P., Panagiotopoulos P. (Eds) New developments in contact problems, CISM Courses and Lectures, Vol. **384**, Springer Verlag, Wien New York, 101–178
68. Raous M., Cangémi L., Cocu M. (1999) A consistent model coupling adhesion, friction and unilateral contact. Computer Meth. in Appl. Mech. and Engng. **177**, 383–399
69. Rockafellar R. T. (1970) Convex Analysis. Princeton Univ. Press, Princeton
70. Savage S. B. (1998) Modeling and granular material boundary value problems. In: Herrmann H. J. et al. (Eds.) Physics of dry ganular media. Kluwer, Dordrecht Boston London, 25–96
71. Stewart D. E. (1998) Convergence of a time-stepping scheme for rigid body dynamics and resolution of Painlevé's problem. Arch. Rational Mech. Anal. **145**, 215–260
72. Stoianovici S.P., Hurmuzlu Y. (1996) A critical study of the concepts of rigid body collision theory. J. Appl. Mech. **63**, 307–316
73. Stronge W. J. (2000) Contact problems for elasto-plastic impact in multi-body systems. In: Brogliato B. (Ed.) Impacts in Mechanical Systems. Analysis and Modelling. Springer, Berlin Heidelberg, 189–234
74. Trinkle J., Pang J.-S, Sudarski S., Lo G. (1997) On dynamic multirigid-body-contact problems with Coulomb friction. Zeitschr. Angew. Math. Mech. **77**, 267–279

75. Vola D., Pratt E., Jean M., Raous M. (1998) Consistent time discretization for a dynamical frictional contact problem and complementarity techniques. Rev. Europ. des Eléments Finis **7**, 149–162
76. Walton O. R. (1993) Numerical simulation of inelastic, frictional particle-particle interactions. In: Roco M. C. (Ed.) Particulate two-phase flow. Butterworth-Heinemann, Boston, 884-910

# Large Structures Behaviour: the Past and the Future

Mario Como, Antonio Grimaldi

Dipartimento di Ingegneria Civile
Università di Roma "Tor Vergata"
00133 Roma, Italia

## 1 Introduction

The most relevant structures of the past centuries are the masonry structures, for instance large masonry arch bridges and monumental buildings.
Today and in the future, cable supported bridges are the most challenging solutions to solve long span crossings
The aim of the paper is first to examine the structural behaviour of some important past masonry works, as the Pantheon, the S. Peter dome and the Colosseum, and then to analyse the behaviour of cable stayed and suspension bridges, characterizing the actual and future development of large structures.
For masonry arches and domes it is possible to emphasize that the geometrical shape and the mass distributions are the main factors that determine the static behaviour and the ultimate strength with respect to live load and to time deterioration action.
In the masonry arches and domes the dominant internal forces distribution is characterized by the compression stresses produced by the structure dead load.
For such structures infact the self-weight distribution and a clever geometrical shape define the load capacity (of the structure).
Similarly also for long span bridges the geometry and the tensile forces in the cables produced by dead loads, are the main elements influencing the bridge service behaviour with respect to the live actions due to traffic loads and wind forces.
Increasing the bridge dimension the deformability and aerodynamic stability are likely the critical aspects and the technical limits for the feasibility of very long span bridges.
The analysis and results given in the paper can help to understand the basic and common structural features of the large past and future structures.

## 2 Static and strengh of masonry constructions

The main lines of the Statics of masonry constructions, in the framework of the no tensile strength model, are here preliminarily recalled. From this assumption the real final strength capacity of the masonry structures, due only to the weight, comes out. References of this study are the Heyman works and some further developments performed at the Department of Civil Engineering of the University of Roma Tor Vergata quoted in Bibliography.

### 2.1 Some recalls of the statics of masonry structures

The key assumptions of the simple model of the masonry material:
- sliding failure cannot occur
- masonry is unable to sustain tensile stresses
- masonry has an infinite compressive strength
- elastic strains are negligible.

The masonry continuum can be represented as an assemblage of rigid particles of stone held together by compressive forces, and liable to crack as soon as tensile stresses begin to develop. The very small size of the stones compared to the dimensions of the whole structure allows us to consider a continuous body instead of a discrete system composed of a large number of particles.

Statics of masonry structures based on the assumption of rigid in compression tensionless material is confined to a "limit analysis theory" and the corresponding statical and the kinematical theorems can be formulated. So, as far as the statical theorem is concerned, the existence of any internal state of admissible stresses in equilibrium with the given loads q, is sufficient to guarantee the existence of the admissible equilibrium in a masonry body under the action of the loads. The actual state of equilibrium attained under the action of given loads remains, on the other hand, undetermined.

To these uncertainties of the theory, we can supply with the observation that the masonry structure, by means of small mechanism displacements, is able to accommodate itself to the unavoidable small settlements of foundations or to the same elastic strains. A simple arch, in fact, that suffers small unknown movements apart the abutments, accommodates itself to the slightly increased span by cracks forming "hinges" and therefore a mechanism sagging at the crown. The arch becomes a three pin statically determinate structure that, among all the possible positions of the thrust line, has the shortest span and the highest rise: the thrust thus attains its minimum value.

This simple statement, pointed out by Heyman for the first time, has a general validity even if it can become less evident as in case of structures that exhibit much more complex patterns of cracks or when settlements involve vertical displacements of columns or walls. To clarify this concept, let us consider a masonry structure in a given admissible equilibrium state under the action of the loads. We assume the existence of a thrust transmitted by the structure to its foundation or to its substructure and vice versa.

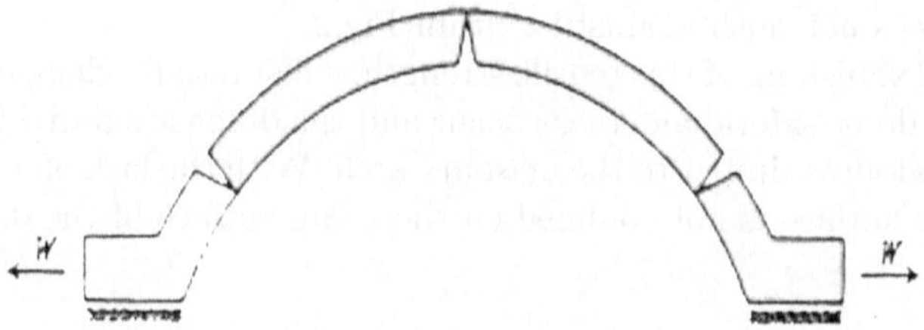


**Fig. 1.** The settled arch

In order to point out the action of this thrust upon the structure, we can remove the constraint whose action is $\mu$r. A release has been thus allowed and the masonry body is loaded by external forces represented both by the loads q and the reaction $\mu$r, this last applied at the abutments along the release direction. The equilibrium is statically admissible.
The structure now slightly cracks to accommodate itself to small settlements of its foundation or, more generally, to a deformation of its substructure. Further information can be obtained with the following argument. Let v be a mechanism corresponding to the predicted settlement and $\mu$ the corresponding kinematically admissible multiplier, in the sense that the structure attains the mechanism state v under the loads q and the reaction $\mu$r , i.e.

$$\mu \langle r, v\rangle + \langle q, v\rangle = 0 \tag{1}$$

Among all the mechanisms v, a statically admissible state of stress is attained at the actual mechanism $v_0$. The corresponding multiplier $m(v_0)$ thus reaches its maximum among all the kinematically admissible reaction multipliers $\mu$(v).
In the next, in this framework, the study of the statics of three important masonry monuments in Rome, the Pantheon, the St. Peter dome and the Colosseum, - whose behaviour in a certain sense is similar to that of a masonry dome - will be pursued.
The study of the masonry dome, particularly of the hemispherical dome, the more common for constructional grounds, shows the strong differences in behaviour between the no tensile strength model and the elastic one with tensile strength. At the same time, by comparing the theoretical previsions with the various real situations, the effects of the vanishing masonry tensile strength are also shown. Particularly, the gradual decreasing of this strength, frequently due to a slow deterioration during the time, leads the behaviour estimated by the no tensile strength model near to that of the real masonry structures.
For an hemispherical masonry dome, under its weight, - the dominating load - tensile strength involves hoop stresses and, therefore, the presence of tensile ring stresses along all the lower wide zone of the dome. In such a condition

the dome does not push against its drum(Fig.2).
The gradual vanishing of this tensile strength, substantially changes the stress state in the dome. Meridian cracks occur and the dome is subdivided in slices whose behaviour is similar to the masonry arch. With the lack of hoop stresses the pressure surface is not confined on the mean surface of the dome.

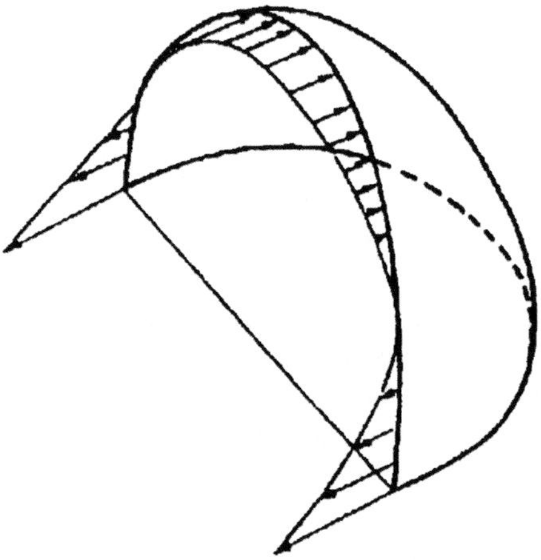

**Fig. 2.** Hoop stresses in the dome exhibiting tensile strength

At the springing, where the mean surface of the dome presents a vertical tangent, the pressure surface will form an angle with the vertical line: the thrust on the drum wall will be activated (Fig.3).

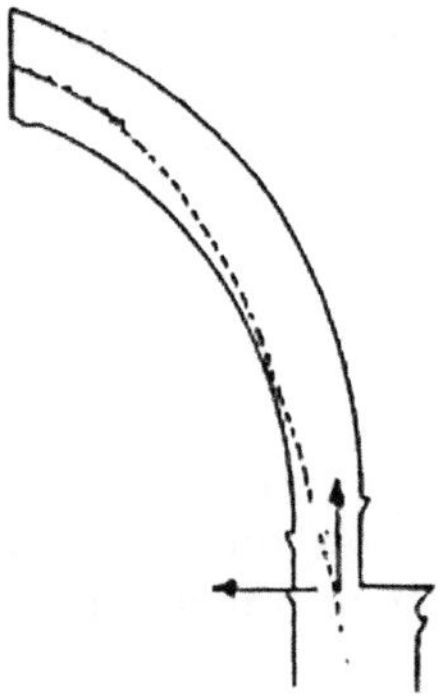

**Fig. 3.** The activation of thrust in the cracked dome

In the extreme case, the pressure line, winding through the thickness of the dome, can touch the extrados and the intrados of the dome. Circular cracks will be formed along the intrados and the extrados of the dome and a collapse mechanism can also occur, as shown in Fig. 4.

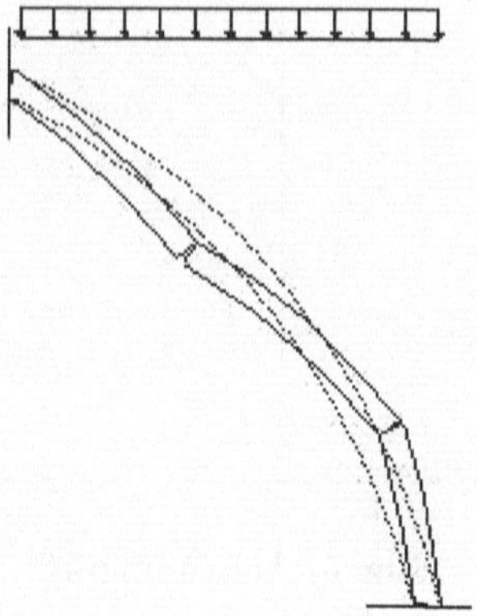

**Fig. 4.** The collapse mechanism for an hemispehrical dome under its weight

Statical efficiency of a masonry dome is based on its capacity to take up the new stress state corresponding to the activation of the thrust. A well shaped dome, with a proper meridian curve and a proper mass distribution, at the occurrence of the meridian cracking, can explicate a thrust that is well absorbed by the drum and the lower pilasters, that will sustain the loads with small eccentricities. On the contrary, a masonry dome, not well shaped and without a proper distribution of loads, can be strongly damaged when the weak tensile strength vanishes and can collapse.

### 2.2 An analysis of Pantheon dome

The Pantheon is one of the most perfect and best preserved monuments of the ancient Rome built by the Emperor Adriano, probably with the cooperation of Apollodorus, in order to replace an earlier temple built by Marcus Agrippa. The rectangular portico screens the vast hemispherical dome. Its true scale and beauty can be appreciated only from inside. The rotunda height and diameter are equal: 43.3 m. The hole at the top of the dome, the *oculus*, provides the only light. The drum of the rotunda is founded on a massive ring of concrete, 6,40 m large at the top, made of travertine fragments in layers of mortar and pozzolana, which has become rock-hard with the time. The unit weight of the foundation has been calculated equal to 2200 kg/mc.

The drum of the rotunda is a circular masonry wall of total thickness of 6 m. A series of alternately larger and smaller brick arches can be seen in the

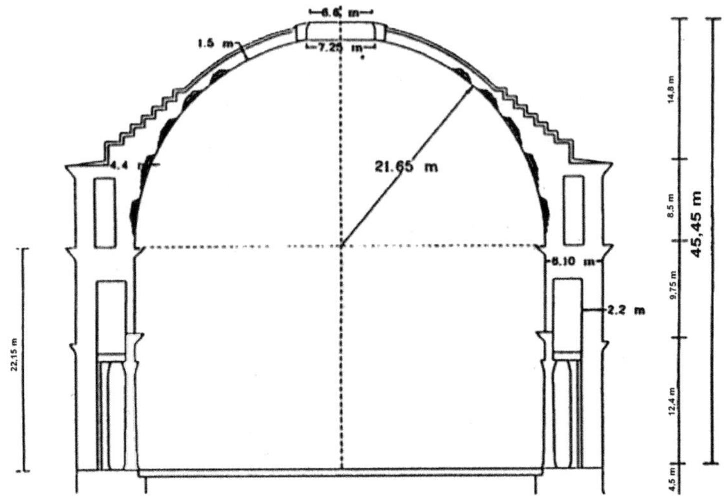


**Fig. 5.** The Pantheon

circular wall. These relieving arches span the eight large exedras that we see from the interior.

The dome is built with a reducing thickness - from 6 m to 1,5 m at the center - and with a concrete of gradually reducing specific weight from the springing to the centre. Right at the top the dome is finished with a compression ring which surrounds the opening in the centre. From investigations made in the years 1929-1934 it was determined that the cupola is without the ribs assumed in the fantastic drawings by Piranesi.

The distribution of the materials is an expression of a conscious and rational arrangement. As shown in Fig.6 at the various heights the different types of concrete are made

1) with aggregates of alternating layers of travertine fragments and lumps of *tufo* ($\gamma = 1800$ kg/mc)

2) with aggregates of alternating layers of *tufo* and fragments of tiles ($\gamma = 1700$ *kglmc)*

3) with *coementae* of predominantly broken bricks ($\gamma = 1600$ kg/mc)

4) with aggregates of alternating layers of bricks and *tufo* ($\gamma = 1500$ kg/mc)

5) with *coementae* of alternating layers of light *tufo* and volcanic slag ($\gamma = 1300$ kg/mc)

The dome was cast by pouring concrete over a temporary wooden framework containing the templets for the coffers. The workers did not lay out and concrete one horizontal ring layer until the course below it was beginning to harden so that the requirements for shutter boards and scaffolding were considerably reduced. Only the topmost of the cupola presumably required an extensive scaffolding. The concrete dome covers the cylindrical hall inside

with a simple geometrical figure, a regular hemisphere. The dome, which is decorated with coffers, springs from the top cornice. Externally the shape of the dome is more complicated, essentially for structural reasons, since the cylindrical wall above the level of the springing, which is shown on the facade by the middle cornice, is continued another stage vertically upwards. With this ring of wall around the foot of the cupola the builder inhibited side slipping and provided a counter-weight which strengthened the cupola in its lower zone, so important for stability.
Large cracks at the intrados of the dome were discovered during the works of maintenance in the years 1881-1882. Further research was pursued in the years 1932-1936. Very old cracks were observed both on the circular wall of the drum and on the cupola, and were repaired. Fig.7 shows the cracks discovered in the 1936 at the intrados of the cupola.
Probably these cracks appeared straight afterwards the completion of the construction. The inscription on the front architrave of the pronaos states that the Emperor Settimio Severo Pantheon restored the Pantheon *vetustate corruptum* in the year 202 A.D..

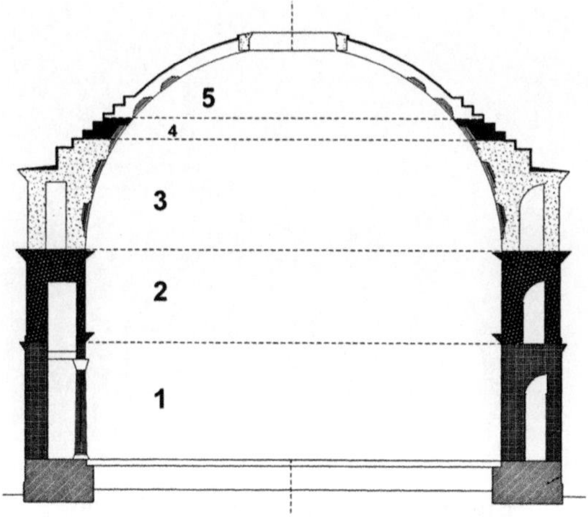


**Fig. 6.** Distribution of the materials

The minimum thrust has been evaluated considering a slice of the dome of the width of 45°. The slice was subdivided into 38 voussoirs whose weight had been appropriately evaluated, taking the position of the voussoir into account.

The weight of all the voussoirs of the considered slice is equal to 2377,7 t. The total weight W of the cupola is thus W = 2377,73X8 = 19021,8 t. The weight of the cupola, together with its superior ring of the drum, is much lighter than other smaller domes. The St. Peter dome in Rome together with

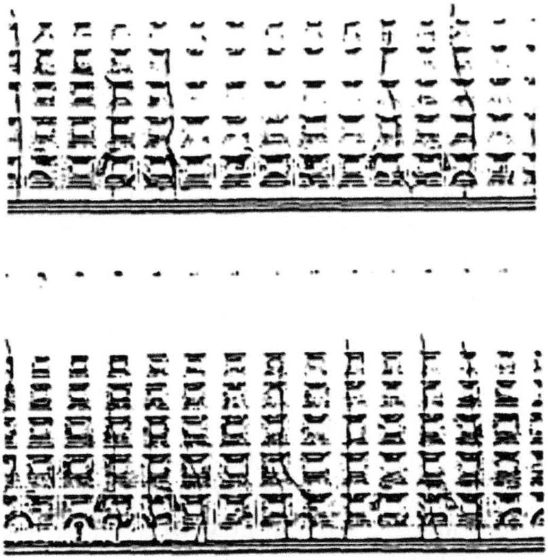

**Fig. 7.** Meridian cracks discovered in the Pantheon dome

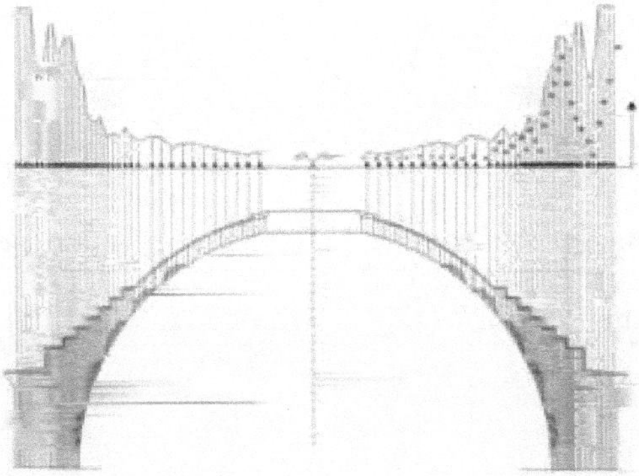

**Fig. 8.** Loads distribution on the slice

its drum weights about 40,000 t. From inspection of Fig. 8 we can recognize that the weight is strongly concentrated near the springing. This fact has important consequences on the statics of the dome.

Fig. 9 gives the pressure lines corresponding to different locations of the starting points at the crown. The minimum thrust line is the curve C3 of the figure. This curve, in the segment between the central ring and the springing, is funicular only of a part of the weights of the slice. All the weights of the more external voussoirs, in fact, intercept the pressure line under the springing.

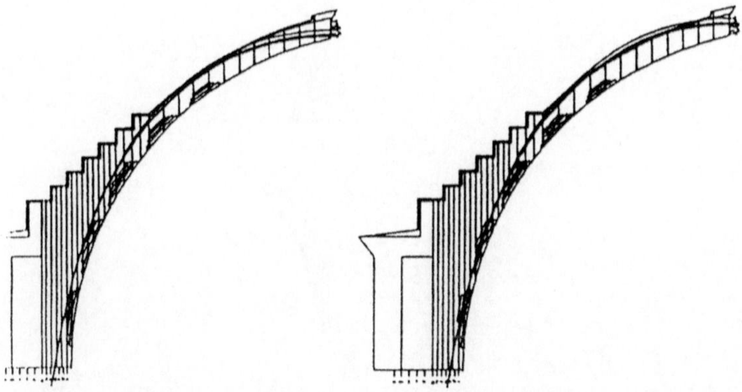

**Fig. 9.** Evaluation of the minimum thrust by statical approach

The minimum thrust "So", minimum among all the curves contained inside the thickness of the wall, is tangent at the intrados and at the extrados of the dome to imply the presence of a mechanism to support a small movement at the springing, consequence of a light increment of the span. For the considered slice the minimum thrust has been calculated in 186,22 t.

This value has been checked by using the kinematical approach. Fig. 8 shows the mechanism displacement that allows a small enlargement of the dome at its springing with arbitrary position of the hingings. With application of the work equation, the research of the position of the hingings is pursued to obtain the maximum value of the kinematical thrust multiplier.

With a mean radius of 21,05 + 3,25 m = 24,3 m, the minimum thrust for unit length of the mean internal circular ring at the springing is equal to 186,22X8/pX2X24,3 = *9,52* t/m, less twice smaller than the minimum thrust of the St. Peter dome, equal to about 20 t/m.

This so reduced thrust is transmitted to the circular drum. The weight of the part of the circular wall, that rises higher than the springing, further deviates the pressure line deep inside the wall of the rotunda. The eccentricity of the resultant of all the forces acting at the basement is very small and the compression stresses are very low.
The cracks that we can see on the external wall of the rotunda, together with all the other cracks in the dome, unlikely can be related to the thrust transmission from the dome to the drum. They, on the contrary, can be related to settlements that probably occurred just after the construction of the monument, when the concrete of the foundation ring was not yet sufficiently stiff. The soil where the Pantheon was built at the confluence of two small rivers, is alluvium above volcanic sands and settlements due to the soil consolidation certainly occurred. The settlements of the

foundations probably slowly produced cracks in the drum and in the dome, some of them still today evident. The settlement mechanism probably pro-

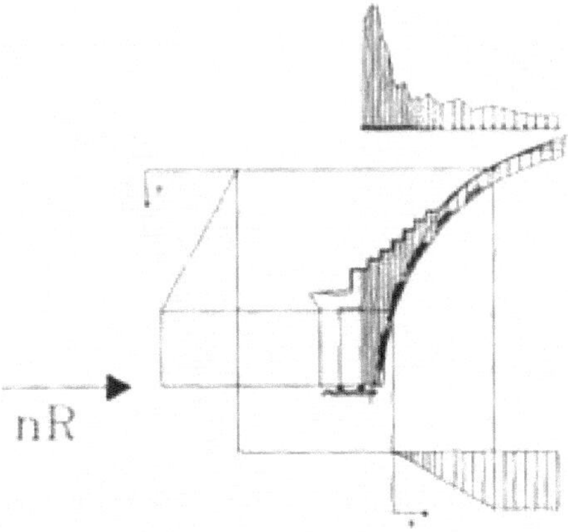


**Fig. 10.** Minimum thrust evaluated according to the kinematical approach

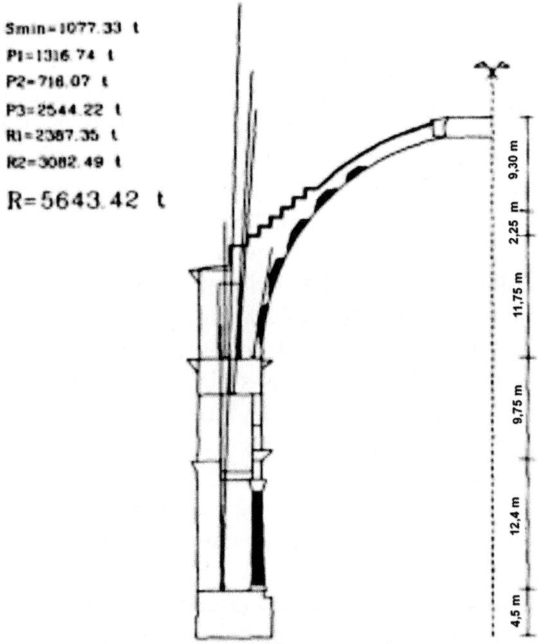


**Fig. 11.** The subsequent resultants up to the foundation along the drum

duced meridian cracks in the dome and in the circular wall activating the thrust of the hemispherical dome. Once these settlements finished, the stress state of the dome remained constant and stable in the time.
The choice and distribution of the various aggregates constituting the concrete, with decreasing density together with the strongly reducing thickness of the dome from the springing to the crown, the design of the upper ring of the drum over the springing of the dome, the lightening of the dome with the lacunarii, were all smart devices to reduce the thrust. The Pantheon dome represents the unsurpassed vertex among the construction of masonry domes.

## 2.3 A study of the old restoration works of the St. Peter Dome by Poleni and Vanvitelli

**The history of the old restoration works** At the first half of the eighteenth century S. Peter dome presented a wide cracking pattern that caused a great deal of comments in the scientific and technical world. Arches, vaults and domes were among the most interesting research problems of the structural science of that time and the discussions about the damage suffered by the big dome grew among the various Schools. Two different opinions contended for the assessment of the effective state of damage and on the remedies of restoration. According to the first opinion, supported particularly by the so called "three Mathematicians" ([6]) & Boscovich, T. le Seur and F. Jacquier, the collapse of the dome was imminent and its restoration, to undertake without delay, required important modifications to the architecture of the monument. Such a statement was not passed as correct because contradicted by facts. The dome, even if heavily damaged, assumed that was in fact standing. The other opinion, sustained by G. Poleni, was less dramatic and assumed that "the defects" of the dome could be repaired with ease.
The damage suffered by the monument was surveyed by Vanvitelli between the 1742 and 1743. The cracks in the dome were considered visible from the interior: the extrados of the dome was covered by a layer of lead that was not removed. Long meridian cracks crossed all the dome. The continuity of the rings was interrupted from the springing up to the lantern. Also one of the two iron chains, inserted in the masonry by Della Porta during the construction of the dome, was found broken. Fig. 12 sketches the long meridian cracking crossing the extrados of the dome: ring cracks on the contrary, were absent. The dome was repaired and stregthened by Poleni and Vanvitelli.

The debate between Poleni and the "three Mathematicians" didn't quiet soon. Still today it is not yet completely plain which of the two opinions was right. A wide literature has been concerned in this subject, mainly analyzing its historical aspects. Aim of this paper is to recover the problem trying to give answers to the various questions in the framework of the actual developments of the Masonry Structures Mechanics.

The Vatican dome has an ogival structure with a double calotte stiffened by sixteen large ribs. The middle surface of the double calotte can be modelled

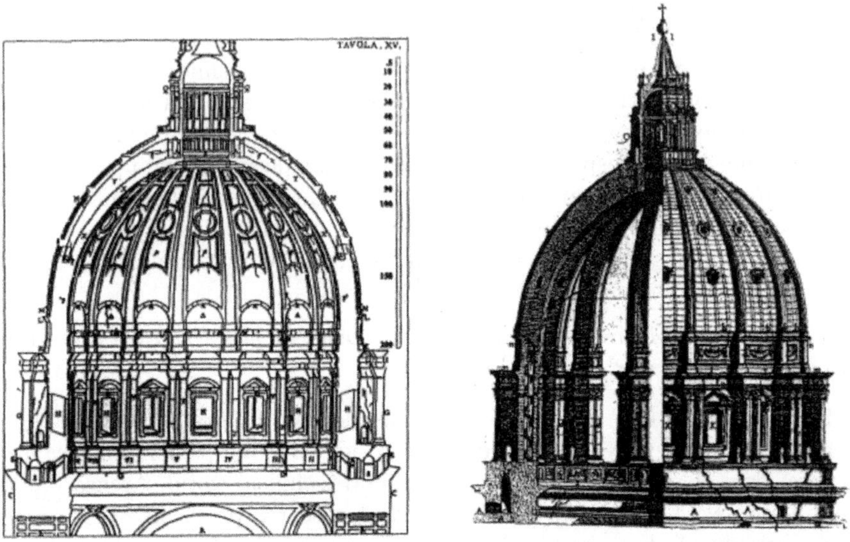


**Fig. 12.** Cracking in the St. Peter dome according to L. Vanvitelli

as spherical ogival dome. The dome was built using a large wooden scaffolding based on the drum. All the sixteen ribs were built simultaneously. The two calottes were built between the ribs. The scaffolding was taken down only after the completion of the double calotte construction. It was the year 1592, i.e. twenty eight years after the death of Michelangelo.
The thickness of the lower calotte is equal to 2,00 mt while of the higher calotte is about 1,00 mt. The ribs have a variable thickness ranging from 2,00 mt. at the base to 5,00 mt at the crown. The drum is a cylindrical wall of 3,00 mt of thickness and with an internal radius of 21,42 mt. The dome was built with bricks and *travertino* stones, assembled with common lime. The weights of the various parts of the dome, according to Poleni and to the "three Mathematicians" calculations are: lantern :1.523 tons.; calottes and ribs: 18.714 tons.; drum and attic: 17.920 tons.; buttresses: 4.980 tons.; basement: 18.695 tons, with a total weight of 61.832 tons.

The previous Fig. 2 shows a meridian section of an hemispherical linear elastic shell loaded by its weight. We can see that the hoop stresses are compressive down from the crown but then become tensile and increase rapidly in value toward the base. The thrust transmitted by compressed rings is gradually balanced by the tensile stresses acting in the lower rings.

For masonry domes the situation is different. Old masonry constructions are made with bricks or stones poorly assembled with mortar. Tensile strength is very low and cracking occurs. The hemispherical dome, as previously discussed, thus thrusts against its supports. Poleni, in evaluating the state of safety of the cracked dome, was inspired by the Hooke statement of the hanging chain and of the corresponding inverted arch.: "*ut pendet continuum flexile, sic stabil contiguum rigidum inversum*" ([21]). From a drawing of the cross section of the dome (Fig. 13), Poleni computed the weight of the sliced arch, this last obtained subdividing the dome into fifty solid sectors. Each half arch was divided in turn into 16 wedges. Then loaded a flexible string with 32 unequal weights : each weight in proportion to the corresponding wedge of the arch. In his computation also the weight of the lantern surmounting the eye of the dome, was taken into account.

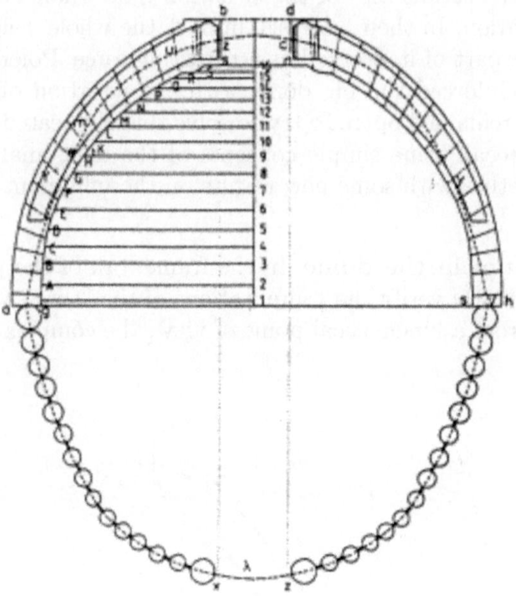

**Fig. 13.** The Poleni analysis of the dome

The chain hanged from the center of the sections of the sliced arch at the join of the dome with the attic. The length of the chain was chosen so that it passed through the center of the sections of the ring connecting the lantern with the dome. By inverting it will be seen that the shape of the chain does lie indeed within the inner and outer surfaces of the arch. Poleni concluded that the observe cracking was not critical[22]. This statement could also demonstrate that the dome divided into slices was safe and so the complete dome. It is worth to point out that Poleni, analyzing the equilibrium of a slice of the dome, passed over the analysis of the whole system composed by dome - buttress - drum. The same ingenious system of the hanging chain should be used by Poleni to analyze a lune of the dome connected to the corresponding portion of the drum, but without buttresses. In this case Poleni would recognize that the pressure line was no more contained inside the drum. Such a result should have been met for any pressure line contained in the dome. It will be shown that admissible equilibrium of the dome was in fact due only to the presence of the sixteen buttresses that strongly worked under the thrust transmitted by the drum. Smeared diagonal cracks crossed all the buttresses, as shown in the Vanvitelli drawings. The "three Mathematicians", on the contrary, were convinced that the dome was about to fail. With a kinematical procedure, they studied in fact the complete dome and pointed out the existence of an unbalance between thrust and resistance. They used the "sliding staff" mechanism (Fig 14). We will show that the "three Mathematicians", even grasping the seriousness of the situation more than Poleni carried out a wrong calculation. In their analysis, in fact, the whole weight of the dome -and not only a part of it -worked as a thrusting force. Poleni and Vanvitelli repaired and reinforced the big dome with the insertion of six iron chains but the debate remained open.To try to solve this old heated problem, firstly it is useful to recall some simple concepts of the limit analysis of masonry structures, together with some new results on the minimum thrust state.

**The equilibrium in the dome** In the framework of the previous considerations let us firstly verify the computations of the "three Mathematicians" by analysing, from a kinematical point of view, the complex system

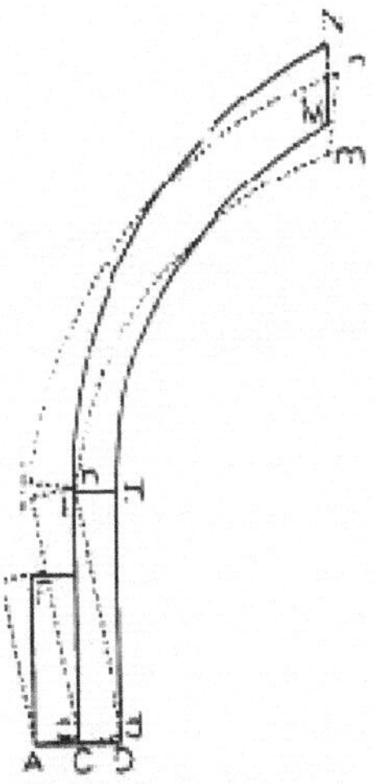


**Fig. 14.** The sliding staff mechanism of the "three Mathematicians"

composed by the dome, the drum and the buttresses. To the same slice of Poleni, we will add a portion of the attic, not included in the Poleni analysis

and the drum and the buttress. Calculation are made on the same lune of Poleni equal to 1/50 of the dome corresponding to an angle of 7°,20.
The mechanism assumed by the "three Mathematicians", as reported by the same Poleni, is represented in Fig. 15. The buttress is completely separated from the drum. The different assumption of the buttress integral with the drum is, in fact, not realistic. The contribution of the buttress is due to the raising of its weight during the mechanism deformation of the dome. The intermediate "hinge" is localized very low, near the connection of the dome with the attic. This mechanism is poorly efficient because all the central part of the dome, with the lantern, meets low sagging.

The calculation of the work made by the resisting and of the pushing forces is carried out with reference to the kinematical chain of Fig. 15, indicated as (M). In the same figure the kinematical chain corresponding to the other mechanism, indicated as (A) is reported.

Poleni analysis represents an early application of the statical theorem. Unfortunately Poleni considered only an incomplete portion of the complex system composed by dome – drum - buttress.
To improve this analysis, let us evaluate the minimum thrust transmitted by the buttresses to the drum. In this case we can obtain a safe state of stress of the whole complex system. The slice of the dome with the drum is thus in mechanism state involving outwards displacements of the buttress. The position of the intermediate hinge is unknown in this mechanism. The corresponding thrust can be thus obtained as a maximum among the all the kinetically admissible thrusts, defined by eqs. (14). We can operate considering the same mechanisms (A) and (M).

The six iron chains inserted in the dome by Poleni and Vanvitelli, considerably increased the safety level of the monument. After these restoration works no more signs of serious statical degradation of the dome occurred.

## 2.4 The Colosseum

Roman's greatest amphitheatre, commissioned by the Emperor Vespasiano in AD 72 on the marshy site of a lake, in the grounds of Nerone palace, the Domus Aurea, was built to a practical design, with its 80 arched entrances, allowing easy access to 55,000 spectators. The superimposed building has an elliptical plan with a perimeter of 577 m and an height of 48,5 m.

Outside the monument presents a threshold order of superimposed arcades and a superior cornice, the *attico*. The external structure, in fact, is composed by three ring walls connected, at the various levels, by the ambulatory vaultes. The outer ring was built in four rows, the first three had 80 arches each and the last one, the above mentioned *cornice* rectangular windows. Radial masonry walls, located transversally, in prosecution of the pilasters of the ambulatory vaultes, are the support of the *cavea*. Fig. 18 gives a sketch of the outer wall.

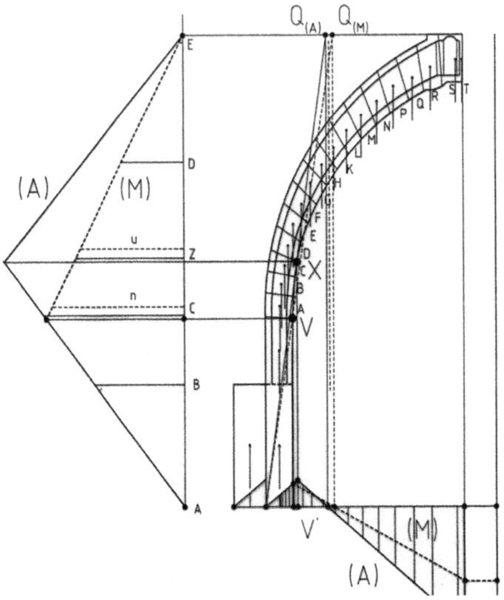


**Fig. 15.** Mechanisms (A) and (M)

The foundation is constituted by a large and massive continuous ring of concrete whose width measures more than 50 m and with a thickness of about 10 m.

The structures present different masonry composition. Fig.21 shows the composition of the three external ring walls. A peculiar aspect of the Statics of the monument is the presence of circumferential compression due to the external arches. These arches, in fact, induce horizontal thrust with radial external component to each pier. This thrust is added to the thrust produced by the ambulatory vaults. The pier are gradually corbelled from the base to the top. In this way the eccentricity of the vertical load is reduced in the piers.

The arch behaviour of the vaults is sketched in Fig.22. The ambulatory vaults, above the internal corridors, transmit loads to the peripheral arches. These arches, in turn, transmit to the pilasters the thrusts due to the weight of the arch and to the actions exerted by the ambulatory

vaults.(Thrust B and C of Fig.22). Due to the curvature of the ring walls, the thrust transmitted to the pilaster by the two adjacent peripherical arches have a radial component (Forces D and E of Fig.22). The Colosseo exhibits a large strength. Outside rotation of parts of the external wall is not allowed. Mechanisms that involve radial opposing displacements of the wall are inconsistent. The outside rotation of a portion of the outer ring, sliding against the adjacent wall, is not allowed because the friction strength explicated by the masonry circumferentially compressed.

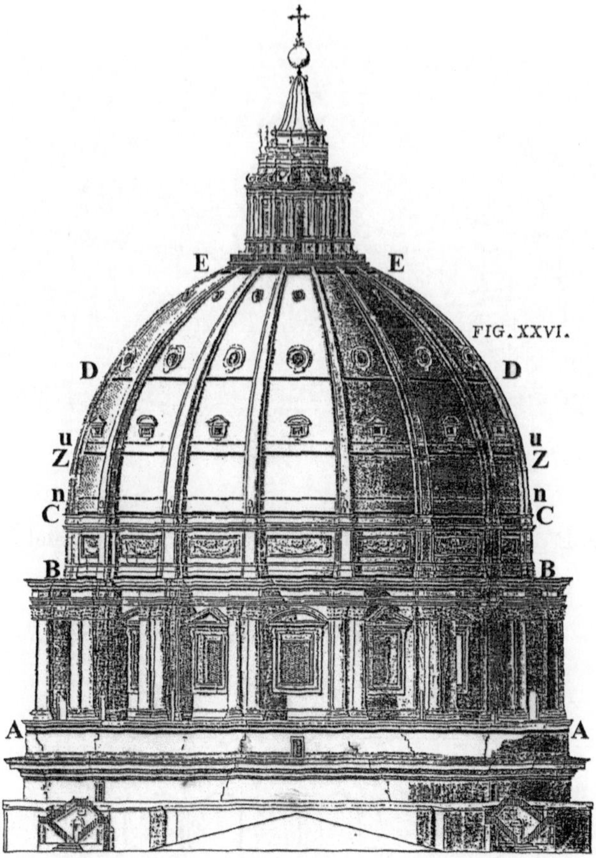


**Fig. 16.** The six iron chains inserted in the dome by Vanvitelli

Because of the particular shape of the cross section (Fig.20), only a global external rotation of the whole ring wall, involving only the outer and the intermediate ring , can occur. (Fig.23).

Under the action of the vertical loads acting along the internal corridors upon the ambulatory vaults, the whole structure can open involving a radial outside mechanism. Along this mechanism vertical loads acting upon the ambulatory vaults perform active positive work while the resisting work is dominantly made by the lifting of the weight of the walls outside rotating.(Fig.23). A calculation of the various active and resisting works gives for the collapse multiplier of the load acting upon all the circumferential vaults at the four levels the value of 1,14 7 t/mq. In this calculation the crushing of the base of the columns due to the particular round shape of the base section has been allowed.

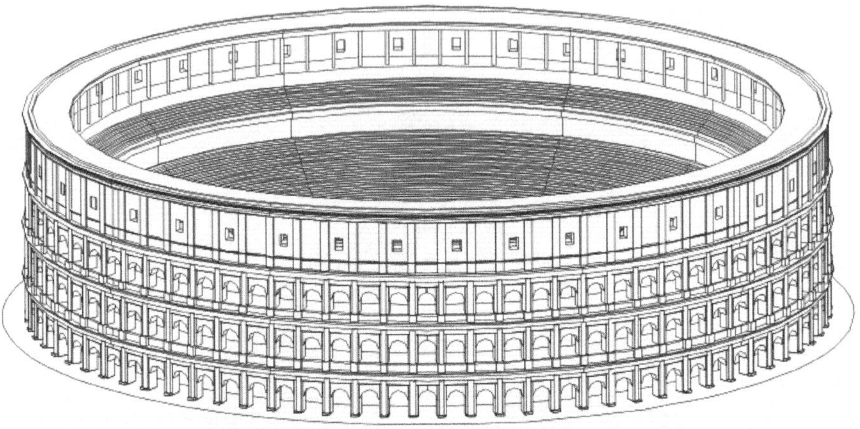

**Fig. 17.** A plastic model of the Colusseum in its original form

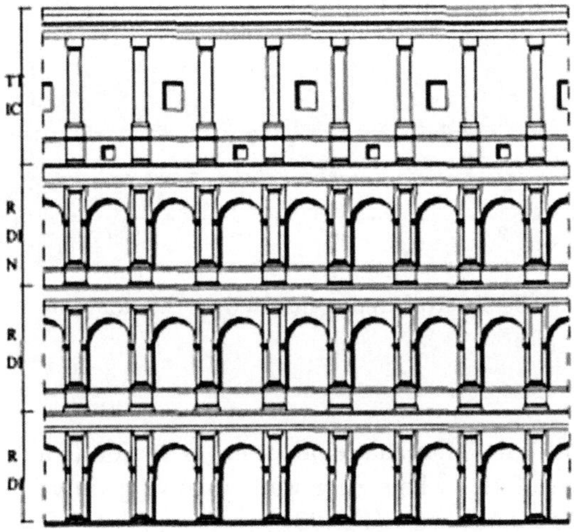

**Fig. 18.** The outer ring wall

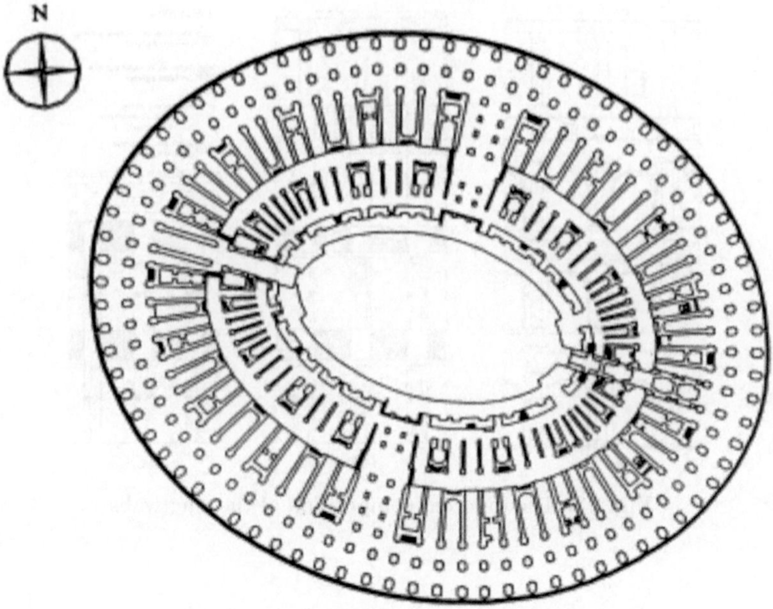


**Fig. 19.** A plan of the Colusseum

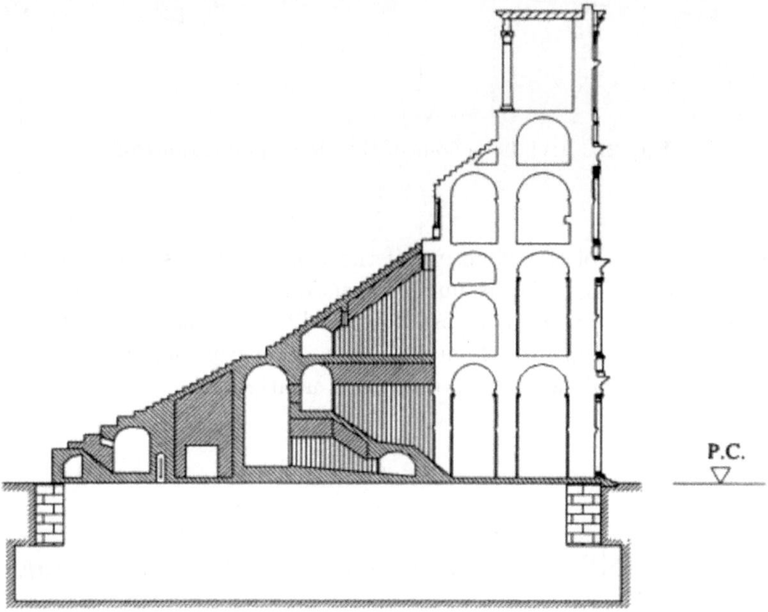


**Fig. 20.** The section of the Colusseum

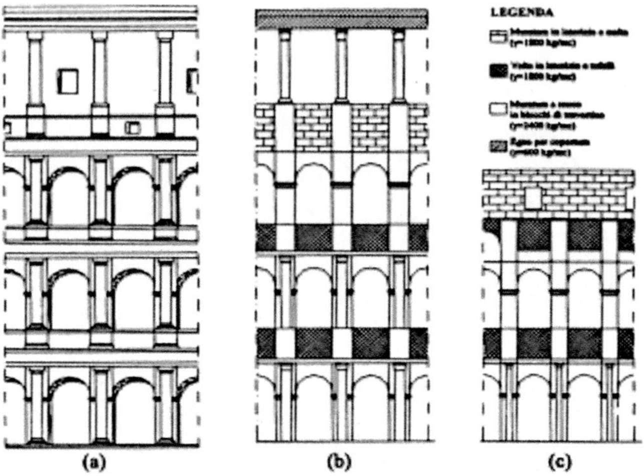


**Fig. 21.** The different composition of the ring walls

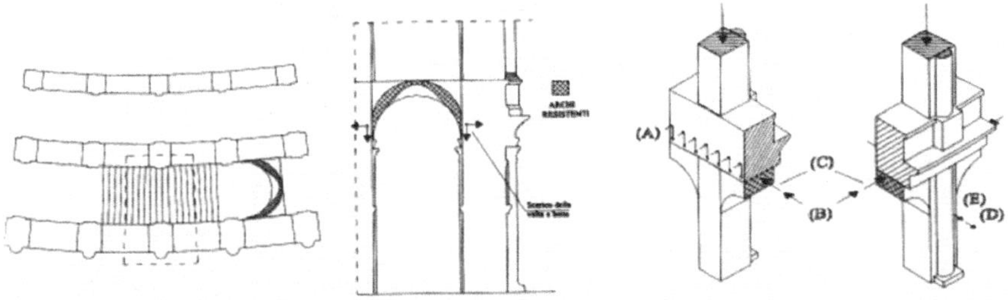


**Fig. 22.** Arch behaviour of the circumpherential vaults

The behaviour of the structure of the Colosseum is similar to that of a masonry dome which collapses under the action of central loads - for instance, the weight of the lantern - contrasted by the lifting of the weight of the lower annular bands.(Fig.25). Fig.26 and 27 show other considered mechanisms to which correspond higher values of the load multipliers.

# 3 Deformability and aerodynamic stability of long span bridges

Cable supported bridges have been of great interest in recent years, particularly with respect to the fan-shaped cable stayed scheme and the suspension

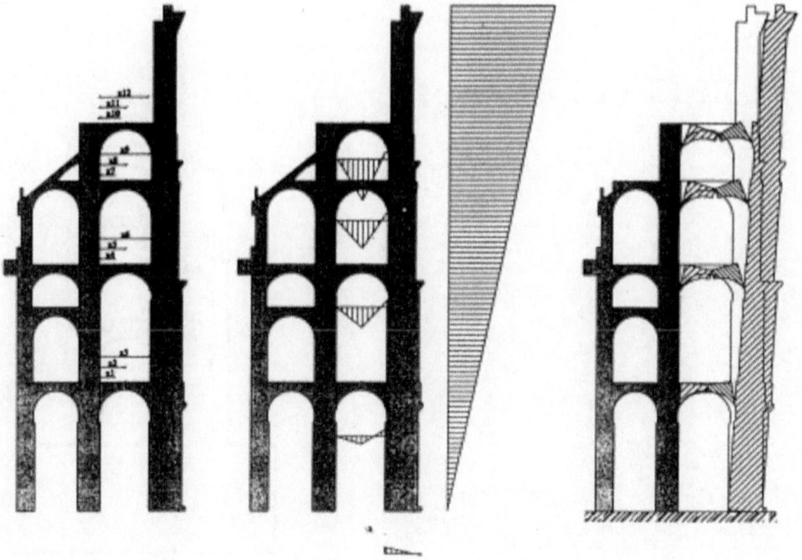

**Fig. 23.** A semiglobal mechanism involving the outside rotation of the outer wall

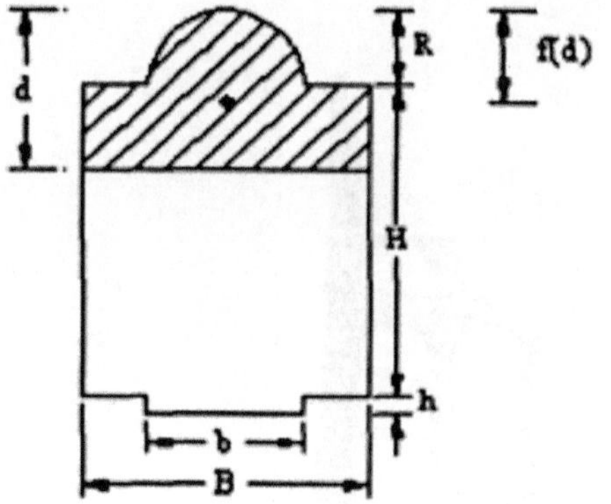


**Fig. 24.** The round shape of the base of the external pilasters

one for long spans.

For long-span bridges one of the most important problems is related to the deformability under live loads. In the case of bridges carrying both road and railway traffic, and for spans greater than 1000 m, this aspect can seriously influence the design and the feasibility of the structure. In addition, increasing the span length, the aerodynamic behaviour of the bridge becomes the feasibility key problem.

Como et al. [12] analyzed the static behaviour of long span cable stayed

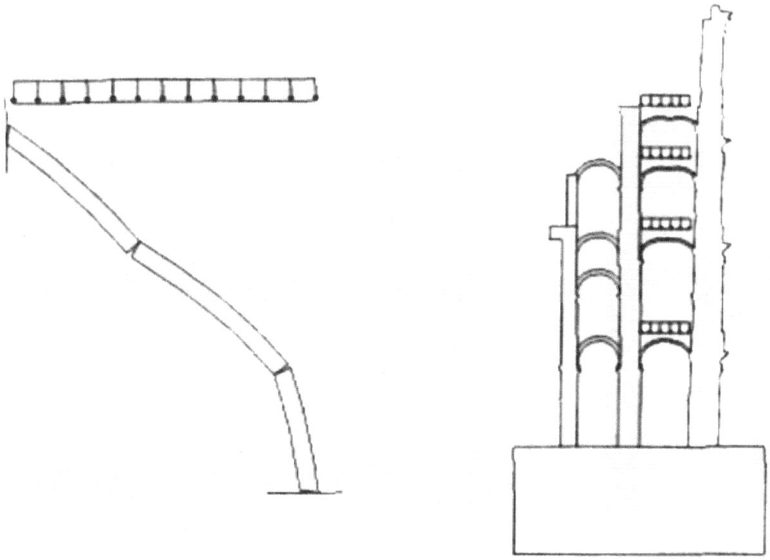

**Fig. 25.** An analogy between the collapse mechanism of the hemisperical dome and Colosseum

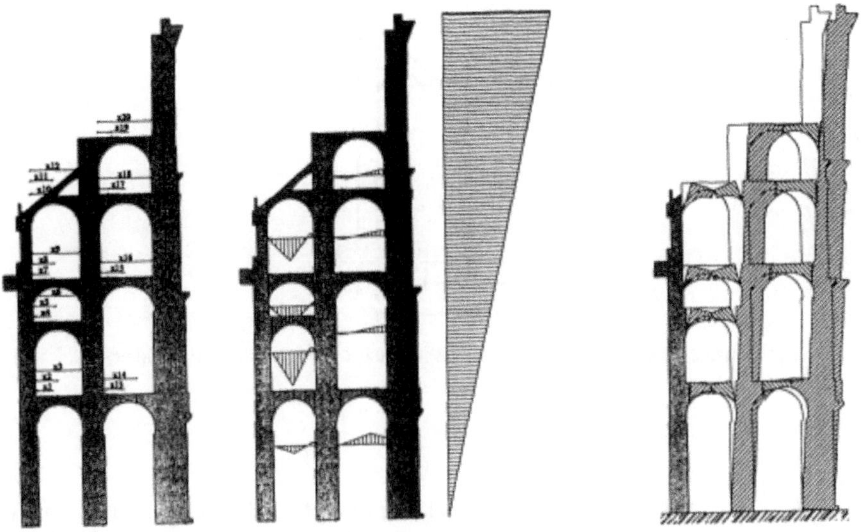

**Fig. 26.** The first global mechanism

**Fig. 27.** The second global mechanism

bridges showing the prevailing truss behaviour of the bridge. Bruno and Grimaldi [13] investigated the nonlinear static behaviour of cable-stayed bridges using both a continuous and a discrete model of the bridge, and showed the strong influence of nonlinearities for long spans. Moreover, the dynamic behaviour of cable-stayed bridges has been investigated by Bruno [14] who analyzed the effects of moving loads, and by Bruno and Leonardi [15] who analyzed aerodynamic instability problems.
In above studies the fan-shaped cable-stayed bridges was studied using both a continuous and a discrete model of the bridge, and the dominant truss behaviour of the bridge was found. In particular, the influence of the dynamic properties and geometric nonlinearities of the structure are included in the analysis.
The main aspects related to aerodynamic stability of cable stayed and suspension bridges are discussed by M. Como [16].
Recently, many projects with central span exceeding largely the longest existing suspension and cable stayed bridges have been proposed. Studies of new bridges propose central spans larger than 3500 mt for suspension solutions, and 1500 mt for cable stayed solutions. The main technical aspects limiting the effective feasibility of so long bridges are strictly related to their deformability and aerodynamic stability.
In the next simple qualitative evaluations of the feasibility of extreme span ca-

ble supported bridges are given by using continuous models of the bridges and comparisons with more refined numerical models. Simple analytical results focus the more relevant technical parameters that influence the behaviour of the bridge and the feasibility limitations.
It is shown that more than the static deformability, the aerodynamic behaviour is the relevant technical limit to the increasing of the span length.
Finally, some applications and examples of long span bridges are given. In particular, the geometric aspect ratio $L_c/H$ between the main span length and the tower height, the loads ratio p/g between live and dead loads, the relative flexural stiffness between girder and cables, are taken into account.
In conclusion the qualitative results of the models show that the geometrical parameters and the dead load value and distribution, are the main factors that determine the behaviour and the feasibility of superlong cable supported bridges.

## 3.1 Bridge models and deformability analysis.

At first the fan shaped cable stayed bridge scheme is examined. We refer to the scheme of Fig.28 in which the girder is simply supported at its ends and is hung to the tops of H-shaped towers by means of two stays curtains.
It is assumed that the stays spacing $\Delta$ is a small quantity compared to the central span length $L_c$.
The aspect ratios $r_1=L_c/H$; $r_2=L_s/H$ of span lengths to the tower height are usually obtained on the basis of economic considerations and of the anchor cable stability condition.
The longitudinal vertical plane y,z is assumed to be a symmetrical one; in addition, the bridge is also symmetrical with respect to the midspan cross plane.
According to the usual erection procedures, girder and towers are assumed to be free from bending under dead load g. Then, the cross sectional areas A and $A_0$ of the couple of diffused stays and of the anchor stays, respectively, are obtained by referring to the truss scheme of the bridge.
We assume that towers and girder's axial elongations are negligible, and we apply the beam model for bending and torsion of the girder.
As far as the stays behaviour is concerned, the Dischinger modulus $E_s{}^*=E/(1+\gamma^2 l_0{}^2E/12\sigma_o{}^3)$ is used, where E is the Young modulus, $\gamma$ is the specific weight, $l_o$ is the horizontal projection length of the stay and $\sigma_o$ is the initial tension. The tower is characterized by the flexural stiffness k and the torsional stiffness $k_t$, while the girder is characterized by the inertia I and the torsional stiffness factor $C_t$. It is convenient to introduce the following non dimensional quantities which represent the main parameters governing the bridge behaviour [12,13]:

$$a = \frac{\gamma^2 H^2 E}{12\sigma_g^3} \quad \epsilon = \left(\frac{4I\sigma_g}{H^3 g}\right)^{\frac{1}{4}}; \quad \tau = \left(\frac{C_t\sigma_g}{Eb^2 H_g}\right)^{\frac{1}{2}}; \frac{E}{\sigma_a}; \quad P = \frac{p}{g} \quad (2)$$

that is:

-the parameter a which accounts for the Dischinger effect of stays;
-the girder relative flexural stiffness parameter $\epsilon$;
-the girder relative torsional stiffness parameters $\tau$;
-the ratio $E/\sigma_a$ between the elastic modulus and the allowable stress of the material stays;
-the ratio $P$ between live load and dead load $g$,
where $\sigma_g = \sigma_a/(1+P)$, $\sigma_a$ being the allowable stress.

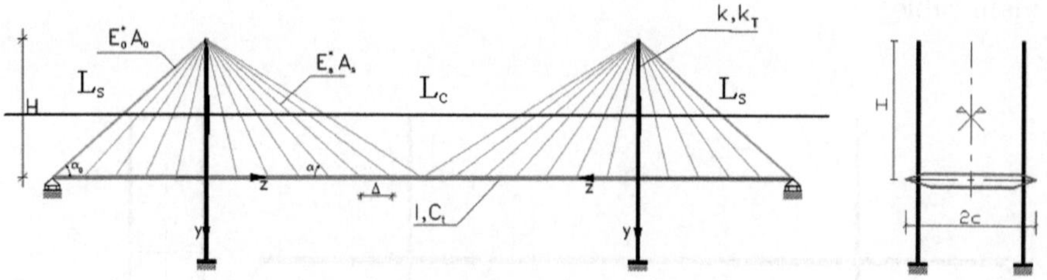


**Fig. 28.** Cable stayed bridge scheme

It can be observed that for given material stays and load ratio $P$, the parameter **a** can assume the meaning of bridge span parameter.
It is possible to show [12,13] that the statical behaviour of the bridge can be studied by using a continuous structural model which gives the main bridge deformation and stress parameters.
In Fig.30 some results relative to the deformation of the cable-stayed bridge scheme are given, where both an analytical continuous model and a FEM discrete one of the bridge are employed.
The results refers the following geometrical and material parameters of the bridge: $r_1=2.5$; $r_2=5/3$; $E/\sigma_a=2.1\text{x}10^6/7200$; $k/g = 50$.
In this figure the case of high live loads $(P=p/g=1)$, as in railway bridges, is considered. It can be observed that the transverse deflection of the bridge is practically unaffected by the tower shape, on the contrary, the torsional deformation is strongly influenced by the tower shape. The results are given in dimensionless form, where the quantity $\mu_o$ represents the nondimensional torsional couple m: $\mu_o = H\sigma_g m/Egb^2$.
We consider now the suspension bridge scheme depicted in Fig. 29. The meaning of symbols is the same of that used for the cable stayed bridge scheme previously described.

It can be shown that for this scheme the following main parameters can be assumed

$$K_v = \frac{gL_c^3}{EI}; \qquad K_t = \frac{2gL_cb^2}{C_t}; \qquad P = \frac{p}{g}, \qquad \frac{E}{\sigma_g}; \tag{3}$$

that is:

-the ratio between the shape flexural stiffness of the main cable and that of the girder;

-the ratio between the shape torsional stiffness of the main cable and that of the girder;

-the ratio $P$ between live load p and dead load g;

-the ratio $E/\sigma_g$ between the elastic modulus and the initial tension of the main cable.

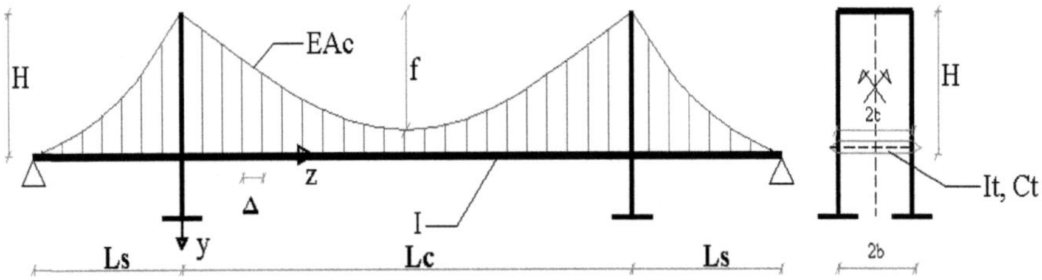


**Fig. 29.** Suspension bridge scheme

Also in this case it is assumed that the hangers spacing $\Delta$ is a small quantity compared to the central span length $L_c$. The aspect ratio $f/L_c$ of cable sag to the central span length is substantially related to static and economic considerations which lead to the mean value $f/L_c = 0.1$, while the aspect ratio $L_S/L_C$ of lateral span length to the central span length is usually related to aesthetic and environmental requirements. In our applications the values $f/L_c = 0.1$; $L_S/L_C = 0.25$ will be used. The bridge is assumed to be symmetric with respect to the longitudinal vertical plane $y,z$ and with respect to the midspan cross plane. Also in this case girder and towers are assumed to be free from bending under dead load g. Moreover, a diffused arrangement of hangers along the deck is assumed, which enables to develop a continuous model of the bridge leading to appropriate analytical solutions.

In Fig 31,32 some results relative to bridge deformation are given, where both an analytical continuous model and a FEM discrete one of the bridge are employed.The results refer to the following geometrical and material parameters of the bridge: $f/L_c = 0.1$; $L_5/L_c$=0. *25;* $E/\sigma$ =*2.1x10*$^6$*/5000.*

In these figures a parametric analysis of the bridge flexural and torsional deformation is developed as a function of the relative flexural and torsional stiffness parameters for several values of load ratio $\overline{\mathrm{p}} = \mathrm{p/g}$ ($\overline{m} = m/2gb$) . In particular, as far as the flexural deformation analysis is concerned the live load p acting on the right (or left) half central span is considered to give the maximum transverse deflection $\delta$ of the bridge, while a uniform torsional couple $m$ acting on the whole central span is considered to obtain the maximum midspan torsional rotation $\varphi$. From these figures emerges that, contrarily to that found for short spans, the influence of the relative stiffness parameters $\mathrm{K}_v$ and $\mathrm{K}_t$ on the nondimensional deformations of the bridge $\mathrm{V} = \delta/\mathrm{L}_c$ and $\Phi = \varphi\, \mathrm{b}/\mathrm{L}_c$ is practically negligible and the overall stiffness of the bridge can be attributed to that of the main cable only. In addition, it can be observed that the deformations are almost linear with respect to the load parameter $\overline{\mathrm{p}} = \mathrm{p/g}$ with a strong influence of this parameter on the deformability of the bridge.

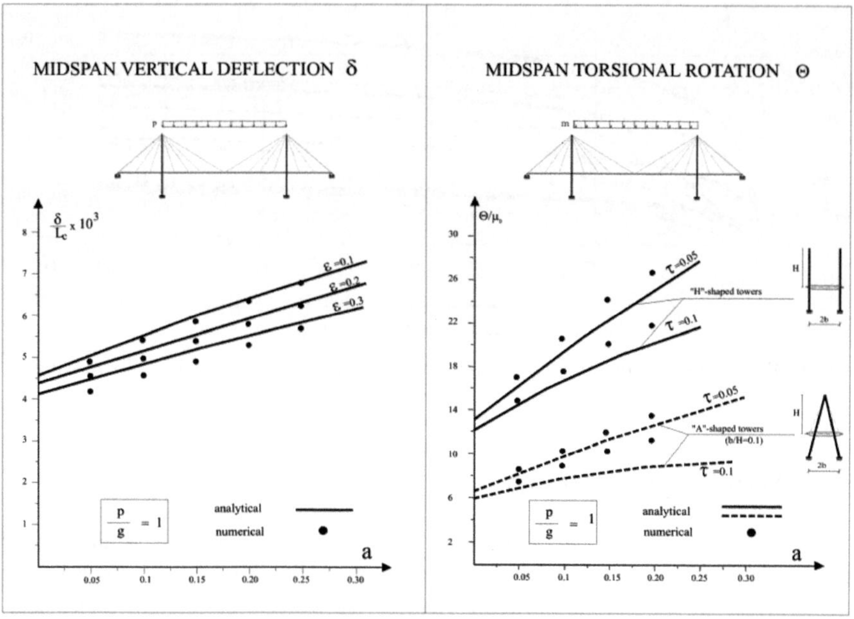


**Fig. 30.** Cable stayed bridge scheme Statical analysis: Flexural and Torsional deformations

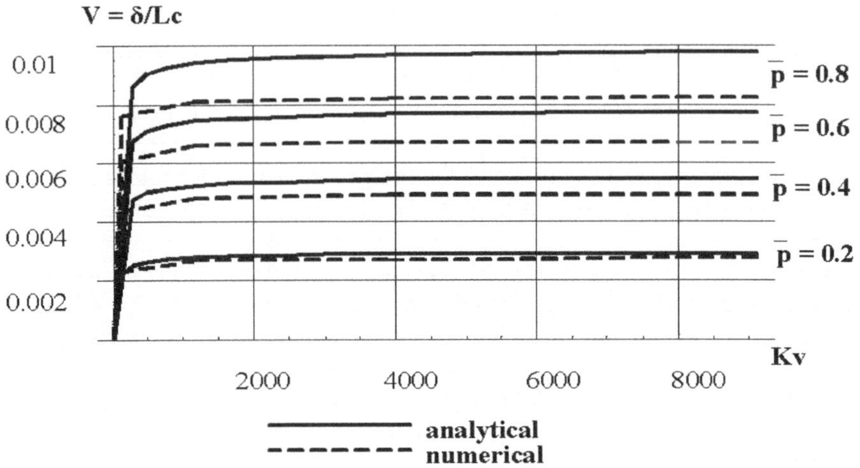


**Fig. 31.** Cable stayed bridge scheme Statical analysis: Flexural deformation

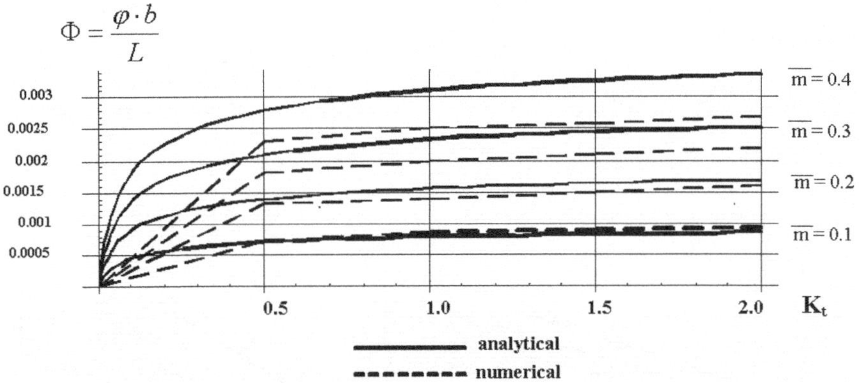


**Fig. 32.** Cable stayed bridge scheme Statical analysis: Torsional deformations

### 3.2 Aerodynamic stability

As well known among the numerous problems involving the behaviour of long span bridges, aerodynamic instability is the most complex and relevant one because of his strong influence on the feasibility aspects. In particular, the flutter instability can represent the more dangerous phenomenon in long spans. To study this phenomenon the dynamic properties of the bridge are needed to know, like that free oscillations in still air. In particular, the ratio $\phi = T_v/T_\theta$ between vertical and torsional natural periods is the key parameter with respect to flutter instability.

We now discuss about free vibrations of the fan shaped cable stayed bridge scheme of Fig. 28. In the case of long span bridges, the tower stiffness can be neglected with respect to that of the anchor stays; this implies that the first antisymmetric oscillation degenerates into a rigid transition of the girder, while the fundamental period of the symmetrical vertical oscillations is almost equal to the second period of antisymmetric oscillation *[16-17]*. In particular, the fundamental period of the symmetrical vertical oscillations can be put in the following form *[16-17]:*

$$T_{ov} = \left(\frac{\mu L_c}{g}\right)^{\frac{1}{2}} \left(1 + \frac{p}{g}\right)^{\frac{1}{2}} f\left(\frac{L_c}{2H'}, \frac{L_s}{H}, a, \frac{\sigma_a}{E}\right) \tag{4}$$

where $\mu$ is the unit length mass of the girder.
Moreover, the fundamental period of the torsional oscillations can also be put in the form:

$$T_{o\vartheta} = \left(\frac{I_o L_c}{b^2 g}\right)^{\frac{1}{2}} \left(1 + \frac{p}{g}\right)^{\frac{1}{2}} f\left(\frac{L_c}{2H'}, \frac{L_s}{H}, a, \frac{\sigma_a}{E}\right) \tag{5}$$

where $I_o$, is the polar moment of inertia of the cross section of the girder. From above results it is now possible to evaluate the ratio $\phi = T_v/T_\Theta$ between the fundamental periods of vertical and torsional oscillations, i.e.

$$\phi = \frac{T_{ov}}{T_{o\vartheta}} = \left(\frac{b^2 \mu}{I_o}\right)^{\frac{1}{2}} \tag{6}$$

The same result is obtained in the case of the suspension bridge scheme. In fact, with reference to Fig.33 where the fundamental the symmetric and antisymmetric vertical oscillations are sketched, it is possible to evaluate the fundamental period of the antisymmetric vertical oscillations involved in the flutter analysis *[16]:*

$$T_{ov} = \left(\frac{\mu L_c^2}{H_c}\right)^{\frac{1}{2}} \frac{1}{1 + 4\pi^2 \frac{EI}{H_c L^2}} \cong \left(\frac{\mu L_c}{g}\right)^{\frac{1}{2}} \left(\frac{8f}{L_c}\right)^{\frac{1}{2}} \tag{7}$$

While the fundamental period of the torsional oscillations is given by

$$T_{o\vartheta} = \left(\frac{I_o L_c^2}{C_t + H_c b^2}\right)^{\frac{1}{2}} \cong \left(\frac{I_o L_c}{b^2 g}\right)^{\frac{1}{2}} \left(\frac{8f}{L_c}\right)^{\frac{1}{2}} \tag{8}$$

where $H_c = \frac{gL_c^2}{8f}$ is the axial force in the cable index dead load g. Hence, for the suspension bridge scheme, the parameter $\phi$ is given by:

$$\phi = \frac{T_{ov}}{T_{o\vartheta}} = \left(\frac{b^2\mu}{I_o}\right)^{\frac{1}{2}} \tag{9}$$

which is the same of that previously found for the cable stayed bridge scheme.

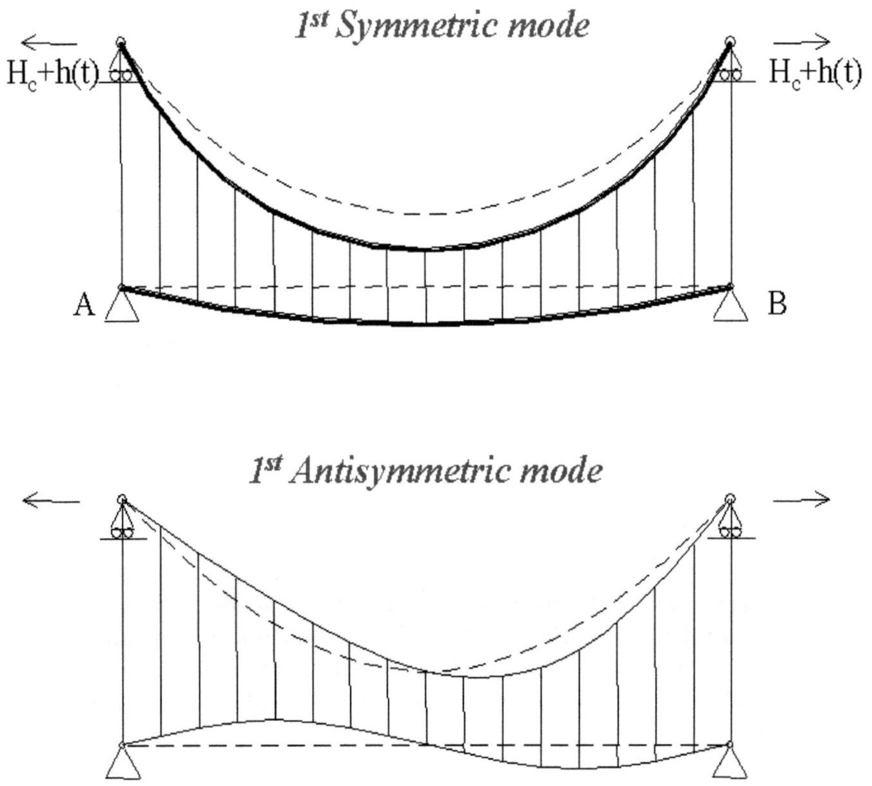


**Fig. 33.** Suspension bridge scheme: free oscillations

The coupled flutter velocity of the wind, both for suspension and cables stayed bridges is given by

$$V_c = \zeta_c \frac{c}{T_{o\Phi}} \frac{1}{\sqrt{1 + \beta\gamma A_3^* \zeta_c}} \tag{10}$$

where

$$\beta = \frac{\rho c^2}{\mu} \qquad \gamma = \frac{\mu c^2}{I_o} \tag{11}$$

are mass distribution coefficients of the bridge, $\rho$ is the ordinary air density, $c$ is the semi-width of the girder, $\mu$ is the unit length mass of the girder, comprehensive of the main cables and girder mass, $I_0 = I_{oGinder} + \mu_{cabl} b^2$ is the total polar moment of inertia, comprehensive of the inertia of the girder and of the cables, $A_3$* is the aerodynamical torsional stiffness of the bridge, appropriate function of the reduced frequency

$$k = \frac{c\varpi}{V} \tag{12}$$

$\zeta_c$ is the critical reduced frequency, i.e. the smallest solution of the critical equation

$$\beta A_3^* \left(\Phi^2 H_1^* + \gamma A_2^*\right) = A_2^* \left(\Phi^2 - 1\right) \tag{13}$$

with $H_1$* and $A_2$* aerodynamical vertical and torsional damping coefficients. For long span bridges the main flutter parameter is the ratio

$$\phi = \frac{\varpi_{o\theta}}{\varpi_{ov}} = \frac{T_{ov}}{T_{o\theta}} \tag{14}$$

between torsional and vertical frequencies of the considered oscillation mode. In fact the critical reduced frequency $\zeta_c$, is such that

$$lim_{\phi \to 1} \zeta_c = 0 \tag{15}$$

The flutter wind speed goes to zero when the ratio $\phi \to 1$ For superlong spans the ratio $\phi$ takes approximately the form

$$\phi = \frac{1 + \frac{\mu_{gird}}{\mu_{cabl}}}{1 + \frac{I_{ogird}}{\mu_{cabl}} b^2} \tag{16}$$

and, with the traditional planning of the bridge, fatally goes near to the unity. In this case an help comes out if we can realize higher values of the ratio $\mu_{gird}/\mu_{cabl}$. Lighter cables made with CFRP materials, can be in this case very convenient.

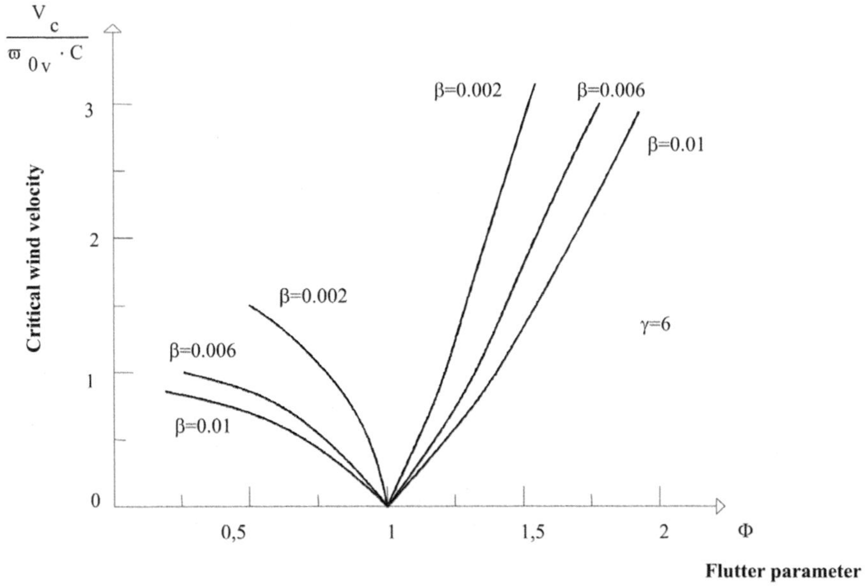


**Fig. 34.** Critical wind speed $V_C$ versus flutter parameter $\Phi$

## References


1. Heyman, J. 1966. The atone skeleton. *Journ. of Solids and Struct.*, 20
2. Heyman, J. 1977. *Equilibrium of shell structures.* Oxford: Clarendon Press.
3. Heyman, J. 1982. *The Stone skeleton* Cambridge: Cambridge Press.
4. Heyman, J. *1995. The Masonry Arch.* Cambridge: Cambridge Univ. Press.
5. Como M., Grimaldi, A., *Analisi limite di pareti murarie sotto spinta*, Dipartimento di Ingegneria Civile, Univ. di Roma Tor Vergata, 1984
6. Como, M., Grimaldi, A. *1985. An Uniateral Modei for the Limit Analysis of Masonry Walls.* lntem. Congr. on 'Unilatera Probl. in Struct. Anal, Ravello 1983, CISM, Springer Verlag.
7. Como, M. 1992. *On the equilibrium and collapse of masonry structures. Meccanica* 27, Netherland: Kluwer Academie Publishers.
8. M.Como, D. Abruzzese, G. Latini, *Some Results on the Strength Evaluation of Vaulted Masonry Structures* inArchitectural Studies, Materials and Analysis STREMA *95,* Computational Mechanics Publications, Boston, 1995
9. Como, *Multiparameter Loadings and Settlements in Masonry Structures*, Atti del Convegno naz.le La meccanica delle Murature tra teoria e Progetto, Messina, Sett. 1996, ed. Pitagora, Bologna, 1996
10. Como *Su di un antico Restauro Statico della cupola di S.Pietro in Roma* nel volume in corso di stampa Le cupole: Forma e Costruzione a cura di V.De Feo, Electa, 1997 M.Como, *Costruzione e forma statica del Pantheon,* Restauro, 1 Edizioni Scientifiche Italiane, Napoli, 51/2000

11. S.Coccia *Analisi dei dissesti delle strutture del Colosseo*, Tesi di Laurea, Rel.M.Como, U.Ianniruberto, Facoltà di Ing, univ.di Roma Tor Vergata, a.a. 1999-2000. M.Como, M.Imbimbo, U.Ianniruberto, Analisi Limite delle strutture del Colosseo Quaderni del Colosseo, 1,2000
12. Como, M., Grimaldi, A., Maceri, F. (1985). *Statical Behaviour of Long-Span Cable Stayed Bridges*mt. J. Solids andStruct., 21(8), 83 1-850.
13. Bruno, D., Grimaldi, A. (1985). *Nonlinear Behaviour of Long-Span Cable-Stayed Bridges. Meccanica,* Voi. 20, N. 4, 303-3 13.
14. Bruno D. *(1985). On the dynamical behaviour of long-span cable stayed bridges under moving loads, Costruzioni Metalliche,* N. 2.
15. Bruno D., Leonardi A. (1998). *Aerodynamic Instability of Long Span Cable-Stayed Bridges*. IABSE Symposium, Kobe, Japan.
16. Como M. (2000). *Stabilità Aerodinamica dei ponti di grande luce, Ingegneria delle Strutture*, Ed. da Elio Giangreco, Edizioni UTET 2000.
17. Bruno D., Leonardi A, *Natural Periods ofLong-Span Cable-Stayed Bridges,* Journal of Bridge Engineering-ASCE, Vol.2, N.3 ,1997.

# Convex Difference Algorithm and Applications to Some Mechanical Problems

Pierre Alart, Stéphane Pagano

Laboratoire de Mécanique et Génie Civil,
cc 048, Université Montpellier II,
F-34095 Montpellier Cedex 5, France

**Abstract.** The splitting of functional into two convex functions provides a specific algorithm to determine the local minima. This approach is applied to two mechanical problems with non convex potentials: the modeling of shape memory alloys and the buckling of thin beams.

## 1 Introduction

In non linear mechanics, the equilibrium states minimize the potential energy which is not necessary convex. Consequently solve such a mechanical problem consists in finding the local minima of the potential energy. The functions which may be written as the difference of two convex functions constitute a first set of non convex functions. Some properties associated with the convexity may be used in order to formulate some extremality characterizations [9] and to define appropriate solution algorithms. Such an approach is not very easy to handle in infinite spaces because the convexity is a power tool to get compactness theorems in such a way to prove existence of solutions of a minimization problems. Then the numerical results presented in the following have to be considered with caution; they may depend on the mesh and not necessary converge towards a continuous solution when the discretisation step tends to zero. But we have to try new algorithms to reach possible equilibrium states in non linear mechanics often characterized by instability; these possible solutions have to be discussed from a mechanical point of view a posteriori; these paper dedicated to two applications is a first attempt in this way.

## 2 CD formulation and CD algorithm

Let consider a boundary problem on an open domain $\Omega$ of the space. The potential energy $\Pi$ associated with this mechanical problem is written,

$$\Pi(\mathbf{v}) = \Phi_1(\mathbf{v}) - (\Phi_2 \circ \mathbf{D})(\mathbf{v}), \tag{1}$$

where $\Phi_1(\mathbf{v}) = \int_\Omega \phi_1(\mathbf{v}, \nabla \mathbf{v})\, d\mathbf{x} - l(\mathbf{v})$, $\Phi_2(\mathbf{D}\mathbf{v}) = \int_\Omega \phi_2(\mathbf{D}\mathbf{v})\, d\mathbf{x}$ and $\mathbf{D}$ is a linear partial differential operator.

A *critical point* $\mathbf{u}$ of the potential energy $\Pi$ is defined by,

$$\mathbf{0} \in \bar{\partial}\Pi(\mathbf{u}) = \partial\Phi_1(\mathbf{u}) - \partial(\Phi_2 \circ \mathbf{D})(\mathbf{u}). \tag{2}$$

where $\bar{\partial}$ denotes the generalized gradient introduced by Clarke [2] for the locally Lipschitz functions and $\partial$ denotes the subdifferential of a convex function. Since $\Phi_1$ is a strictly differentiable function [2] the set equality in Eq. 2 holds and the previous inclusion is equivalent to,

$$\partial(\Phi_2 \circ \mathbf{D})(\mathbf{u}) \cap \partial\Phi_1(\mathbf{u}) \neq \emptyset.$$

Among the critical points, a *local minimum* $\mathbf{u}$ satisfies,

$$\partial(\Phi_2 \circ \mathbf{D})(\mathbf{u}) \subset \partial\Phi_1(\mathbf{u}). \tag{3}$$

But this relation characterizes a local minimum only if $\Phi_2 \circ \mathbf{D}$ is a piecewise affine function [4,5]. This last property is not true in general for the continuum problem but can be verified in finite dimensional cases, i.e. after finite element approximation. This decomposition leads to define a type II Lagrangian [1,5,9] depending on two fields,

$$\mathrm{L}_{\mathrm{II}}(\mathbf{v}, \boldsymbol{\tau}_2) = \Phi_1(\mathbf{v}) + \Phi_2^*(\boldsymbol{\tau}_2) - \int_\Omega \mathbf{D}\mathbf{v} : \boldsymbol{\tau}_2 dx, \tag{4}$$

where the density $\phi_2^*(\mathbf{v}, \boldsymbol{\tau}_2) = \sup_{\mathbf{e}} \{\mathbf{e} : \boldsymbol{\tau}_2 - \phi_2(\mathbf{e})\}$ is the classical Fenchel conjugate function. This Lagrangian is separately convex in each of the dual variables, but the convexity of $\mathrm{L}_{\mathrm{II}}$ is not guaranteed in general; a very simple example is illustrated in the Fig. 1. Consequently a min-min problem is naturally related to this Lagrangian whose the arguments are called $\partial$-critical points (cf. Table 2).

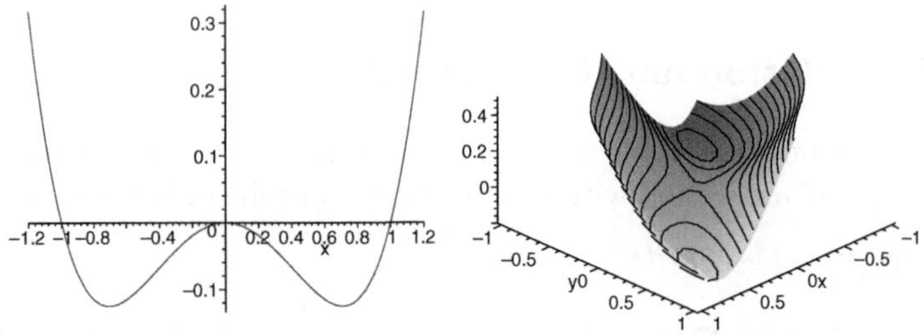


**Fig. 1.** $\Pi(\mathbf{v}) = \frac{\mathbf{v}^4}{2} - \frac{\mathbf{v}^2}{2}$ and $\mathrm{L}_{\mathrm{II}}(\mathbf{v}, \boldsymbol{\tau}) = \frac{\mathbf{v}^4}{2} + \frac{\boldsymbol{\tau}^2}{2} - \mathbf{v}\boldsymbol{\tau}$

*CD Algorithm*

- Initialization of the algorithm with $(\mathbf{u}^0, \boldsymbol{\sigma}_2^0)$,
- with $(\mathbf{u}^{n-1}, \boldsymbol{\sigma}_2^{n-1})$ known, we have to determine $\mathbf{u}^n, \boldsymbol{\sigma}_2^n$ as follows,

$$\text{step 1}: \mathbf{u}^n = \arg\min \mathrm{L}_{\mathrm{II}}(., \boldsymbol{\sigma}_2^{n-1})$$

$$\text{step 2}: \boldsymbol{\sigma}_2^n \in \arg\min \mathrm{L}_{\mathrm{II}}(\mathbf{u}^n, .)$$

Since $D$ is a partial differentiable operator, by duality the field $\boldsymbol{\tau}_2$ belongs to a less regular space than $\mathbf{u}$. Then the *second minimization*, involving eventually a non differentiable $\phi_2$ function, may be performed *locally*, i.e. in each finite element for instance. On the contrary, the *first minimization* always stays *global* on the structure and may take advantage to be quadratic in order to restrict this step to a linear solution. This situation (global but linear first step / non linear but local second step) compares the CD algorithm to augmented Lagrangian techniques based on this splitting.

## 3 Two applications

### 3.1 A set of models for Shape Memory Alloys (SMA)

In the first example we use the model of Shape Memory Alloys (SMA) given in [3,5,7] in an isothermal case. To do this, let us consider a decomposition into the difference between two convex functions $\Phi_1$ and $\Phi_{2\mathcal{G}}$ of the potential energy $\Pi$ associated with the mechanical problem,

$$\Pi(\mathbf{v}) = \Phi_1(\mathbf{v}) - (\Phi_{2\mathcal{G}} \circ \mathbf{e})(\mathbf{v}), \tag{5}$$

where $\Phi_1(\mathbf{v}) = \int_\Omega \phi_1(\mathbf{e}(\mathbf{v}))\, d\mathbf{x} - l(\mathbf{v})$ and $\Phi_{2\mathcal{G}}(\mathbf{e}) = \int_\Omega \phi_{2\mathcal{G}}(\mathbf{e})\, d\mathbf{x}$. According to the temperature range ($c = c(T)$) we have two behaviours; a pseudo-elastic behaviour with five regimes and a quasi-plastic one with three regimes, if $c + \frac{\mathcal{G}}{2} > 0$,

$$\phi_{2\mathcal{G}}(\mathbf{e}) = \begin{cases} -\alpha\boldsymbol{\tau} : \mathbf{e} - c & \text{if } \alpha\boldsymbol{\tau} : \mathbf{e} \le -c + \frac{\mathcal{G}}{2} \\ \frac{-1}{2\mathcal{G}}(\alpha\boldsymbol{\tau} : \mathbf{e} + c + \frac{\mathcal{G}}{2})^2 & \text{if } |\alpha\boldsymbol{\tau} : \mathbf{e} + c| \le -\frac{\mathcal{G}}{2} \\ 0 & \text{if } \alpha|\boldsymbol{\tau} : \mathbf{e}| \le c + \frac{\mathcal{G}}{2} \\ \frac{-1}{2\mathcal{G}}(\alpha\boldsymbol{\tau} : \mathbf{e} - c - \frac{\mathcal{G}}{2})^2 & \text{if } |\alpha\boldsymbol{\tau} : \mathbf{e} - c| \le -\frac{\mathcal{G}}{2} \\ \alpha\boldsymbol{\tau} : \mathbf{e} - c & \text{if } \alpha\boldsymbol{\tau} : \mathbf{e} \ge c - \frac{\mathcal{G}}{2} \end{cases} \tag{6}$$

if $c + \frac{\mathcal{G}}{2} \le 0$,

$$\phi_{2\mathcal{G}}(\mathbf{e}) = \begin{cases} -\alpha\boldsymbol{\tau} : \mathbf{e} - c & \text{if } \alpha\boldsymbol{\tau} : \mathbf{e} \le \mathcal{G} \\ \frac{-1}{2\mathcal{G}}(\alpha\boldsymbol{\tau} : \mathbf{e})^2 - c - \frac{\mathcal{G}}{2} & \text{if } \alpha|\boldsymbol{\tau} : \mathbf{e}| \le -\mathcal{G} \\ \alpha\boldsymbol{\tau} : \mathbf{e} - c & \text{if } \alpha\boldsymbol{\tau} : \mathbf{e} \ge -\mathcal{G} \end{cases} \tag{7}$$

where $l(\mathbf{v})$ is the work of the external loading, $\alpha$, $\boldsymbol{\tau}$, $c$ are thermo-mechanical constants, $\mathbf{e}$ is the small strain tensor and $\mathcal{G}$ is a parameter characterizing the scale the phase transition is modelized ($\mathcal{G} = 0$ for the microscopic scale and $\mathcal{G} = \mathcal{G}_c$ for a macroscopic scale [5,6]). This decomposition leads us quite naturally to introduce a type II Lagrangian which depends on the displacement field $\mathbf{u}$ and on the "stress" phase transformation field $\boldsymbol{\sigma}_2$; in that case $\Phi_{2\mathcal{G}}^*$ is easy to compute and is given in [5,7]. The CD algorithm may catch the local minima and so reproduce hysteretic processes and phase propagations [6] (an augmented Lagrangian technique finds a global minimum which is more homogeneous).

### 3.2 Buckling of rods by the Elastica theory

We are interested to modelize the buckling of flexible and inextensible rods. It is well known that a force applied to the ends of the rod and oriented according to the rod may lead to different equilibrium states if the magnitude of the force is large enough. The critical values and buckling modes around the trivial solution may be obtained by a stability theory using a linear modeling of the rod; but the deformed shape may be reached only by a non linear modeling. Moreover the possible buckling shapes, if unilateral conditions have to be considered (cf. Fig. 5), are an open problem (cf. [8]). The existence of local minima in buckling problems constitutes a motivation to try a CD approach. In a first approach we use the Elastica theory where a single scalar field $\theta$ must be found. We have to minimize the following potential where $P$ is an horizontal force, $\theta$ the angle between the deformed rods and horizontal axe, $E$ the young modulus and $I$ the inertial momentum. We remark that the non quadratic part is due to the work of the external force since the bulk energy is still a quadratic form,

$$\inf_{\theta} \int_0^{\ell} \frac{EI}{2}\,\theta'^2 \; - P\,\cos\theta\, d\mathbf{x}, \quad P < 0. \tag{8}$$

We postulate the following CD splitting by introducing a parameter $\beta$,

$$\phi_1(\theta) = \frac{EI}{2}\,\theta'^2 \; - \beta\, P\,\frac{\theta^2}{2}, \tag{9a}$$

$$\phi_2(\theta) = -P\left(\beta\,\frac{\theta^2}{2} \; - \; \cos\theta\right). \tag{9b}$$

The first energy density is convex if $\beta \geq 0$ because $P < 0$ and the second one is convex if $\beta \geq 1$ for all $\theta$. By restricting $\theta$ to the interval $\left]\frac{\Pi}{2}, \frac{\Pi}{2}\right[$ – which may be relevant in some cases – the convexity holds for $\beta$ non negative. The previous decomposition gives an optimal CD algorithm in the following meaning : the first minimization consists of a linear global problem since the second one is local and consists only in deriving $\phi_2$ to update the $\boldsymbol{\sigma}_2$ multiplier. The two numerical examples show the ability of the CD algorithm to reach equilibrium states if they are local minima of the potential. But only a first mode is obtained (even if the load permits others modes). In the first case (cf. Fig. 2 and Table 1) the first critical value is equal to $\frac{\Pi^2 EI}{4\ell^2} \simeq 25$ and the buckling may be recovered only for $\beta$ equal to 0. For $P$ less than 50, the solution satisfies the condition $\theta \in \left]\frac{\Pi}{2}, \frac{\Pi}{2}\right[$ and the CD algorithm converges for $\beta$ non negative. On the contrary, for $P$ bigger than 50 it is necessary to choose $\beta \geq 1$. The convexity assumption for $\phi_2$ is therefore essential. But the parameter $\beta$ has not to be too big otherwise the convergence rate slows down.

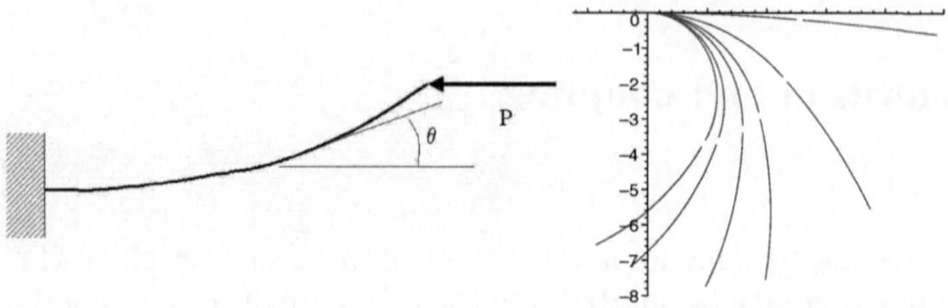


**Fig. 2.** The first buckling test and some deformed shapes (two elements)

The second example is quite different : a kinetic condition is imposed at the end of the beam (cf. Fig. 3) in such a way the first buckling mode may occur for $P$ equal to $\frac{\Pi^2 EI}{\ell^2} \simeq 100$. With the Elastica theory this condition is satisfied implicitly by imposing the slope at the end opposite to the slope at the origin. This fact is a severe limit to use Elastica theory in more complex boundary problems. However the conclusions of the parametric study on $\beta$ and $P$ are the same than the previous case as illustrated in the Table 1 and in Fig. 3.

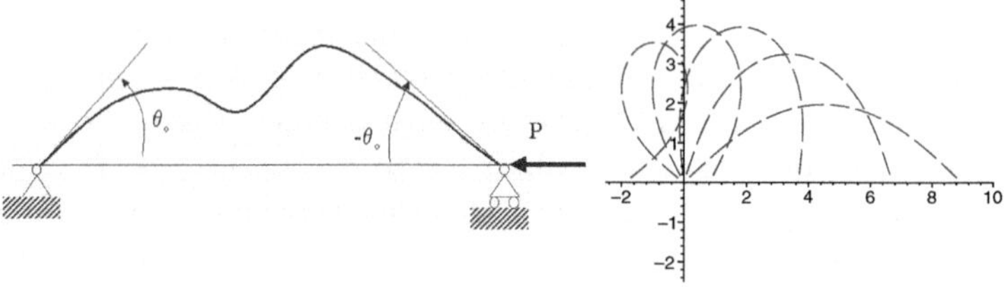


**Fig. 3.** The second buckling test and some deformed shapes (ten elements)

**Table 1.** Number of iteration for the CD algorithm. * : No buckling

| P | Example 1 | | | | | | Example 2 | | | | |
|---|---|---|---|---|---|---|---|---|---|---|---|
| $\beta$ | 26 | 30 | 40 | 50 | 75 | 100 | 105 | 120 | 150 | 200 | 300 |
| 0 | 1053 | 30 | 9 | 19 | $\infty$ | $\infty$ | 104 | 34 | 14 | *36* | $\infty$ |
| 0.25 | * | 41 | 14 | 9 | *25* | $\infty$ | 133 | 46 | 32 | 13 | *45* |
| 1 | * | 71 | 31 | 23 | 16 | 13 | 219 | 81 | 42 | 28 | 22 |
| 2 | * | 111 | 54 | 41 | 32 | 30 | 334 | 129 | 72 | 51 | 41 |

## 4 CD algorithm and coupling

It may be interesting to compare a non convex optimization problem whose the objective function may be splitted into the difference of convex functions and a convex optimization problem with constraints (cf. Table 2). Indeed the two formulations underline a part of the objective function as a convex perturbation which is either added or subtracted. Moreover this perturbation may often involve a linear operator by composition. Consequently a Lagrangian approach leads to similar problems, the first as a Inf Sup problem and the other as a Inf Inf problem. The arguments of such problems are either saddle points or $\partial$-critical points according to the definition of Auchmuty [1]. The algorithms associated with these formulations (Uzawa algorithm and CD algorithm) consist of two steps: the first step is a classical minimization problem according to the primal variable, the second one is either an updating of the Lagrange multipliers or a minimization problem according to the multipliers. A last common point concerns the formulation of the critical points of the augmented Lagrangian of a convex constrained problem and of critical points of the initial CD function; but this fact is out of our purpose here.

Such a analysis leads to consider a set of algorithms to solve the problems which couples a CD formulation with convex constraints, by combining the previous CD and Uzawa algorithms. The following example concerns the buckling of rods submitted to unilateral contact conditions. A previous study about coupling sleeves using SMA structure with contact was based on a similar approach. We consider a rod submitted to a concentrated force at the end and an obstacle. A distance $e$ separates the obstacle from the initial configuration of the rod. To account for the eventual contact of the rod with the obstacle, concentrated reactions are introduced at the nodes of the finite elements. The problem consists then in minimizing the new functional in (10) under the constraints (11) which are written as complementarity conditions on the gap $h_i(\theta)$ at each node and the contact reaction $R_i$.

$$\inf_{\theta} \int_0^{\ell} \left( \frac{EI}{2}\, \theta'^2 \;- P\, \cos\theta \right) d\mathbf{x} \;+ \sum_{i=1}^{n_p} \int_0^{x_i} \; R_i\, \sin\theta d\mathbf{x}, \tag{10}$$

$$h_i(\theta) \;=\; e \;-\; \int_0^{x_i} \; \sin\theta d\mathbf{x} \geq 0; \quad R_i \;\leq\; 0; \quad h_i(\theta)R_i \;=\; 0. \tag{11}$$

The CD splitting has to be completed as follows with the same conditions for $\beta_i,\ i = 1, .., n$ as for $\beta$,

$$\phi_1(\theta) = \frac{EI}{2}\, \theta'^2 \; - \beta\, P\, \frac{\theta^2}{2} \; - \; \sum_{i=1}^{n_p} \; R_i\; \beta_i\; \frac{\theta^2}{2} \chi_{[0,x_i]}, \tag{12}$$

$$\phi_2(\theta) = -P \left( \beta\, \frac{\theta^2}{2} \; - \; \cos\theta \right) \; - \sum_{i=1}^{n_p} R_i \left( \beta_i\, \frac{\theta^2}{2} - \sin\theta \right) \chi_{[0,x_i]}. \tag{13}$$

where $x_i$ is the abscisse of the i$^{th}$ node and $\chi_{[0,x_i]}$ is the characteristic function of the interval $[0, x_i]$. The coupled algorithm consists then of overlapped loops, the Uzawa loop containing the CD algorithm as summarized in Fig. 4.

Uzawa loop

CD loop

i) $\theta^n \;=\; \arg\min \mathrm{L}_{\mathrm{II}}(., \boldsymbol{\sigma}_2^{n-1})$

ii) $\boldsymbol{\sigma}_2^n \;\in\; \arg\min \mathrm{L}_{\mathrm{II}}(\mathbf{u}^n, .)$

$R_i \longleftarrow \min(R_i + \rho h_i(\theta); 0)$

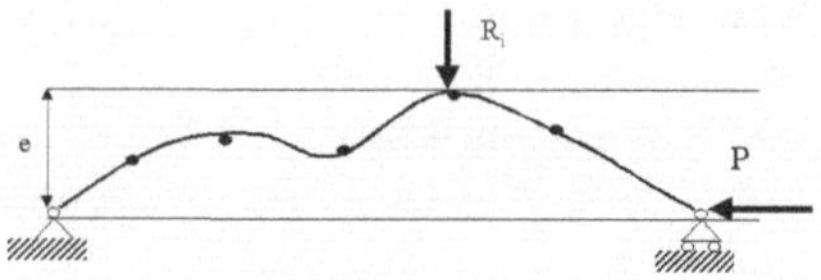


**Fig. 4.** Coupled Uzawa-CD algorithm and discrete nodal contact reactions $R_i$

The figure 5 shows three situations for different positions of the obstacle. For the lowest position no contact occurs and the low mode of buckling is obtained. For an higher position contact occurs only at the end of the rod and for the highest position the algorithm converges to the upper fundamental mode of buckling without contact because the solution with contact is unstable. The numerical experiments show that such an algorithm is unstable (because the process is unstable too). Moreover the contact treatment is not easy to handle in the context of the Elastica theory which is essentially dedicated to situations with concentrated forces. An other approach is performed today based on the modeling of inextensible rods by the bidimensional displacement field and a non convex constraint imposing the inextensibility condition.

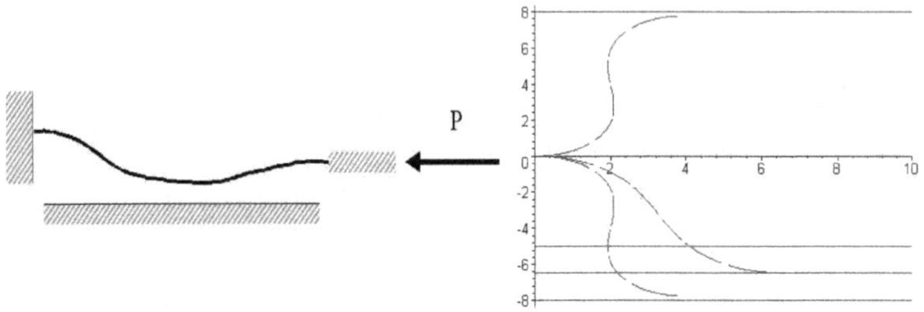


**Fig. 5.** A buckling test with contact and some deformed shapes (six elements)

**Table 2.** Comparison between a non convex problem and a convex problem with constraints. $N$ is the trace operator on the contact boundary $\Gamma_c$, $C^+$ is the admissible displacements convex set and $C^-$ is the polar cone set for the multiplier

| | |
|---|---|
| Non smooth convex problem (contact),<br>$\inf_{\mathbf{v}}\ \Phi_1(\mathbf{v}) + I_{C^+}(\mathbf{N}\mathbf{v})$. | Non convex problem (SMA),<br>$\inf_{\mathbf{v}}(loc)\ \Phi_1(\mathbf{v}) - \Phi_2(\mathbf{D}\mathbf{v})$ |
| Lagrangian,<br>$\mathrm{L}(\mathbf{v},\gamma) = \Phi_1(\mathbf{v}) - I_{C^-}(\gamma) + < \mathbf{N}\mathbf{v}, \gamma >$. | Type II Lagrangian,<br>$\mathrm{L}_{\mathrm{II}}(\mathbf{v},\boldsymbol{\tau}_2) = \Phi_1(\mathbf{v}) + \Phi_2^*(\boldsymbol{\tau}_2) - < \mathbf{D}\mathbf{v}, \boldsymbol{\tau}_2 >$ |
| Saddle point of L,<br>$\inf_v \sup_\gamma \mathrm{L}(\mathbf{v},\gamma)$ | Critical point of $\mathrm{L}_{\mathrm{II}}$,<br>$\inf_v \inf_{\boldsymbol{\tau}_2} \mathrm{L}_{\mathrm{II}}(\mathbf{v},\boldsymbol{\tau}_2)$ |
| Uzawa algorithm,<br>i) $\mathbf{u}^{k+1} = \arg\min \mathrm{L}(.,\boldsymbol{\lambda}^k)$,<br>ii) $\boldsymbol{\lambda}^{k+1} \in \arg\min -\mathrm{L}(\mathbf{u}^{k+1},.)$ | CD algorithm,<br>i) $\mathbf{u}^{k+1} = \arg\min \mathrm{L}_{\mathrm{II}}(.,\boldsymbol{\sigma}_2^k)$,<br>ii) $\boldsymbol{\sigma}_2^{k+1} \in \arg\min \mathrm{L}_{\mathrm{II}}(\mathbf{u}^{k+1},.)$ |

## References


1. Auchmut ,G. (1989) Dua ity algorithms for non-convex variational principles, Numerical Functional Analysis and Optimization, **10**, 211–264
2. Clarke, F.H. (1993) Optimization and Non-smooth Analysis, A Wiley-interscience publication
3 Fremond, M. (1987) Matériaux à mémoire de forme, C. R. Acad. Sci. **304**, série II, 239 244, Pa is
4. Hiriart-Urruty, J.B. (1989) From convex optimization to nonconvex optimization, Nonsmooth O timization and Related opics, 219–239
5 Pagano, S. (1997) (Quelques aspects de la) Modélisation numérique du com ortement des al iages à mémoire de forme par des potentiels non convexes, PhD Thesis, Univiversité Montpellier II
6. Pagano, S., Alart, P., Maisonneuve, O. (1998) Solid-Solid phase transition. Local and Global minimizations of non convex and relaxed potentials. Isothermal case for shape memory alloy , Int. Jour. of Ing. Sc., **36**, 1143–117
7. Pagano, S., Alart, P. (1999) Solid-Solid phase transition modelling : relaxation procedures, configurational energies and thermomechanical behaviours, Int. Jour. of Ing. Sc , **37**, 1821–184

8. Nguyen, Q.S. (2000) Stabilité et mécanique non linéaire, Hermès Science Publications, Paris
9. Stavroulakis, G.E., Panagiotopoulos, P.D. (1993) Convex multilevel decomposition algorithms for non-monotone problems, Int. J. Num. Meth. Engrg., **36**, 1945–66.

# Adhesively Bonded Joints in Civil Engineering: some Physico-chemical Aspects Related to the Mechanical Behaviour

Karim Benzarti, Yves Mouton, Thierry Chaussadent

Laboratoire Central des Ponts et Chaussées
75732 Paris Cedex, France

**Abstract.** The gluing technique is a promising assembling method in Civil Engineering, since it opens new horizons in the design of complex concrete or steel structures. However, at the present time, several breaks are slowing down its industrial development:

- it is a delicate technique which requires a qualified staff in order to respect basic rules of implementation,
- recurrent interrogations concerning the long-term durability of adhesively bonded joints are still not answered.

A physical-chemical approach of the problem may provide elements of solution. The first part of this article propose some rules of implementation derived from physical-chemical considerations, which are necessary to ensure safety of the adhesive bond. In the second part, experimental results are presented, giving an improved description of the microstructure of bonded joints. These results illustrate the interest of a physical-chemical study for the identification of parameters susceptible to govern mechanical strength and durability of the adhesive bond.

**Keywords:** Civil engineering; gluing; adhesive bond; joint; implementation; time of reticulation in place; durability; epoxy resins; concrete; steel; micro-thermal analysis; interfacial areas; mechanical modelling.

## 1 Introduction

Gluing is a well-known assembling technique which is used in many fields of human activity but is still not considered as reliable by the specialists of building technology. New arguments may be put forward from a physical-chemical approach to launch the debate and try to rehabilitate the gluing technique in construction, especially for applications such as assembling of steel or concrete structures.

The main reasons why civil engineers or architects are reserved for the use of adhesive bonding are identified:

- the lack of knowledge concerning the durability of glued assemblies (resistance to vibrations and fatigue, influence of a permanent stress level, environmental effects, etc..);
- problems of interfacial strength related to the poor mechanical properties of the concrete surface;
- the difficulty to perform an efficient control of the polymer joint by using non destructive tests;
- the complexity and the cost of a repair in case of damaged joint, etc.

Moreover, the design of adhesively bonded joints is generally achieved by means of a mechanical modelling. Such models are unable to take into account the chemical behaviour of polymer adhesives during implementation or the evolution of the physical-chemical properties with time. However, it is commonly admitted that such parameters play a determinant role towards the bond durability.
The purpose of this paper is not to solve all these problems, since no simple solutions are available. The aim is rather to argue in favour of a physical-chemical approach, complementary to the mechanical modelling, which may help engineers to design durable glued structures in the near future.
The first part of this article gives a general outline of the parameters governing quality of the adhesively bonded joint. Basic rules of gluing implementation required to obtain a good durability of the bond are reported, and a special attention is paid to the assembling of concrete structures.
In the second part, results of an experimental study are presented, providing a fine description of the microstructure at the adhesive/substratum interface. This section illustrates how a physical-chemical approach can contribute to the understanding of interfacial phenomena that may affect durability and mechanical properties of glued assemblies.

## 2 General points about adhesive bonding

### 2.1 Implementation of the gluing technique

From the macroscopic point of view, bonding can be considered as effective when the surfaces in contact are totally covered by the polymer adhesive and when this adhesive has reached its final physical-chemical equilibrium (i.e. after cooling, chemical reticulation, etc..). These conditions require a good wetting of the surfaces by the adhesive and a controlled evolution of the polymer towards its equilibrium state.

**Surface treatments and wetting dynamics** Gluing two materials together is an easy task if the surfaces are sound, well cleaned and accessible, and if the polymer adhesive is well formulated.
In a concrete substratum, the external surface is mainly constituted of cement

paste and includes only few aggregates. This surface layer shows poor cohesion with the inner material and exhibits low mechanical properties. For that reason and before any coat of glue, the concrete surface is generally cleaned by sand blasting in order to remove the weak layers and create microscopic roughness. The concrete surface can also be treated by a primer (in general an extremely fluid epoxy resin) in order to fill up the porosity and increase the cohesion. Another way of solving the problem is to use high or very high performance concrete with enhanced mechanical properties.
For metallic materials, specific treatments are necessary to remove weak layers of oxides, to prevent subsequent corrosion and enhance adhesion. This is generally performed by sanding surfaces, by use of chemical treatments (acidic solutions, galvanisation ... ) and sometimes by coating the metallic surface with adhesion promoters (organosilanes).

When the substratum is coated with the glue, a physical parameter characteristic of the gluing efficiency is the wetting ability of the surface (adherent) by the adhesive. This parameter takes into account:

- the level of physical interactions between adhesive and adherent, which depends on the relative surface energies;
- the dynamic viscosity of the adhesive;
- the spreading ability of the adhesive on the adherent;
- the roughness of the adherent.

A good wetting of the substratum is required in order to get an intimate contact between adherent and adhesive. Considering a concrete substratum and usual polymer adhesives, this condition is generally achieved if concrete surfaces are dry and well cleaned. However, this is not sufficient since the liquid adhesive must also be able to evolve towards its equilibrium solid state.

**Conditions of cure** Adhesives used in construction are mainly bi-component thermosetting systems, such as epoxy formulations. When the resin and the hardener have been mixed, chemical reactions (polymerization or cure) induce a progressive evolution of the polymer from a liquid glue towards a solid macromolecular network. Conditions of cure, especially temperature and time, govern this structural evolution and therefore control the properties of the resulting material.
First, temperature has an influence on the viscosity of the liquid glue. At low temperature (for instance in winter), the glue becomes too thick and the kinetics of polymerization is considerably slowed down, therefore implementation is not possible. The range of temperatures compatible with a good implementation is usually specified for each adhesive and should be respected. Then, at a given temperature (which is generally the ambient temperature), the viscosity increases with time as the macromolecular network builds-up. When the viscosity reaches a critical value, the glue can not be worked any

more and implementation becomes impossible. For this reason, a Usual Practice Time (UPT) is defined as the time while a polymer mix (resin + hardener) can be worked at a given temperature [1]. Operators should take care of the UPT during implementation.
After solidification of the adhesive, the mechanical properties continue to increase slowly with a temperature dependent kinetics. Usual epoxy resins (cured with aliphatic amine hardeners) require 12 days curing at ambient temperature (10-25°C) in order to reach an equilibrium state with quasi-stable mechanical properties. This observation leads to the notion of Time of Reticulation in Place (TRP) [2] which is the necessary period for a two-components mix to chemically react and give a solid material with specified mechanical performances. For example, an epoxy resin used for crack injections requires a cure of 12 hours at 20°C in order to be able to support mechanical testing [3]. To summarize, a glued structure is able to support a specified load when the corresponding TRP has been respected.
During the cure of the adhesive, polymerization mechanisms can also be affected by moisture from the surface of the adherent. In facts, water molecules may modify the reaction kinetics or induce secondary reactions susceptible to affect the resulting structure of the polymer joint. A complete cure is hardly achieved with a very moisture sensitive glue, and interfacial areas may present mechanical weakness in such situation [4]. For this reason, the choice of the glue formulation is usually different if the adherent is wet or dry.

### 2.2 Durability of adhesively bonded joints – control methods

Even if implementation of the gluing technique has been realized in the accurate conditions and if the bond is supposed effective, the long-term durability of the glued assembly remains a serious preoccupation. The main objective for civil engineers is to control the evolution of the mechanic behaviour in order to ensure the safety of the assembling.
First, it is well known that polymers behave as viscoelastic materials and that their mechanical properties evolve as a function of time, temperature and external loads. For this reason, a careful attention must be paid:

- to the creep behaviour of adhesive joint which are submitted to permanent loads;
- to the fatigue behaviour of glued assemblies submitted to periodic loads.

For bridge elements, which support both the permanent weight of the structure and the periodic load related to the traffic, creep and fatigue behaviours are both important parameters.
Then, the adhesive bond may be sensitive to the environmental conditions. In the long-term, moisture and chemical attacks (salts, pollution ... ) may induce breaking of the chemical bonds at the polymer/adherent interface, and

in extreme situations lead to an adhesive failure of the assembly. Moreover, in glued concrete structures, the polymer joint can be sensitive to the elevated pH and the alkalinity of the concrete.
Fortunately, most of those problems can be technically solved by using an adapted formulation of adhesive. Nowadays, chemists are able to formulate products susceptible to satisfy all technical requirements, if such requirements are compatible each other. But one must keep in mind that the universal product does not exist.
It remains finally a still unresolved problem: how is it possible to control the adhesively bonded joint along the whole interface by non destructive tests? At the present time, available methods are indirect and not quite reliable (mainly ultrasonic techniques). In the future, an improved knowledge of the interfacial areas may lead to the development of "in situ" control methods, opening new horizons for "smart" materials. The scientific community is clearly questioned about that subject.

## 3 Experimental study

As it was mentioned in the first part of the paper, organic materials and specially polymer resins are getting more and more popular in civil engineering [5]. Epoxy resins are usually applied in the repair of concrete structures (bridges, walls, etc ... ) : they can be injected in cracks in order to restore integrity of the damaged concrete, or used as adhesives to paste composite or steel plates to the structure in order to improve the stiffness [6]. Those adhesives also open up new horizons for the design of bridges, since part of the structure elements could be assembled by adhesive bonding in the future. The major drawback of the gluing technique is the insufficient control of the long term durability of the adhesively bonded joint. Therefore, it is necessary to acquire basic knowledge about the mechanisms of adhesion between concrete and polymers in order to identify parameters that govern the durability of glued structures. Micro-thermal analysis constitutes an interesting technique for such an investigation [7], and the following section presents some experimental results concerning the microstructure of interfacial areas between an epoxy adhesive and concrete, or between the same adhesive and steel.
Micro-thermal analysis is based on the recent developments of scanning probe microscopy. It combines the high spatial resolution of the conventional AFM technique with the characterization capabilities of thermal analysis. It is now possible to measure the thermal response of a polyphased material in a localized region rather than on a macroscopic scale. Therefore, $\mu$TA constitutes a "micro-equivalent" technique of the differential scanning calorimetry (DSC). Principles and possibilities of this technique are detailed in several papers by Price and al. [8-10].
This technique has been applied with success to the characterization of various multiphased materials based on organic components such as polymer

blends [9], pharmaceutical products [8, 11], multi-layered packaging films [10], composite laminates [12], adhesively bonded joints [13], etc . . . .In these applications, the technique imaged the internal structure of the samples and the contrast in thermal conductivity allowed the visualization of the spatial distribution for each individual constituent. In a second stage, the calorimetric properties of those constituents (glass transition or melting transition temperatures) were evaluated by local thermal analysis. This provides precious information about the microstructure in the interfacial areas. For example, Jansen and al. [13], who studied the microstructure of aluminium/epoxy interfaces, found that the glass temperature (Tg) of the epoxy can be 20°C higher in the vicinity of the aluminium than that in the bulk polymer. These authors also showed that surface treatments of aluminium influence Tg of the polymer in the interfacial areas. It means that interfacial properties strongly depend on the surface chemistry of the substratum. Moreover, complementary mechanical tests showed that the shear strength of the bonded joint may be correlated to the interfacial microstructure. This example taken from the literature gives a first illustration of the influence of physical-chemical properties on the mechanical behaviour of glued assemblies.

### 3.1 Materials

In our study, two different interfaces have been studied by micro-thermal analysis:

- first, an interface between an epoxy polymer and a cement paste was chosen. The epoxy system was constituted of a DGEBA based resin (diglycidyl ether of bisphenol A) and a diethylene triamine hardener, while the cement was an ordinary Portland cement. The polymer/cement paste assembly was prepared by setting a piece of cement paste on the bottom of a mould and by casting the epoxy resin into the mould. After polymerization, the surface of the sample was polished (the roughness was about 1 $\mu$m);
- then, an interface between epoxy polymer and steel was studied. The sample was elaborated with the previous epoxy system and with a cylindrical piece of steel. Stages for the preparation of the sample are the same as those previously described.

### 3.2 Technique

The instrument used was a 2990 Micro-Thermal Analyser from TA Instruments. The heart of the system is a small temperature sensor consisting of a platinum wire heater/thermometer constructed from a Wollaston wire and bent into a V-shaped point. This sensor is mounted on the end of a cantilever which is scanned across the surface of the sample.

In a first stage, micro-thermal analysis acts both as a thermal microscope and as an atomic force microscope (AFM), providing topographic pictures of the sample surface as well as contrast in thermal conductivity.
In a second stage, specific features on the images can be selected for characterization by local thermal analysis. In this mode, a ramp of temperature is applied to the tip in contact with the sample, and the deflection of the tip is recorded as a function of temperature. Any phase transition of the material leads to a downward deflection of the tip. It makes possible a local evaluation of the glass transition temperature (Tg) at a specific location of the sample.

### 3.3 Results and discussion

Figure 1 shows the topography and the contrast in thermal conductivity provided by micro-thermal analysis in the neighbourhood of an epoxy/cement interface. On the topographic picture, the line border between the two phases is clearly identified. On the thermal picture, the cement phase appears in pale grey (or in yellow for the coloured version of the paper) and the bulk polymer appears in black. An intermediate medium can be identified inside the polymer phase, in dark grey (or in red colour). This feature is typical of a modified polymer zone (called interphase) with a thickness of 10-15 $\mu$m. Thermal local analyses were then performed in the polymer at various distance from the interface line. Figure 2 shows the evolution of the glass transition temperature (Tg) in the polymer as a function of the distance to the cement border. Tg keeps a constant value in the bulk polymer, but shows a progressive rise in the modified polymer zone when one get closer to the cement phase. A maximum Tg deviation of 10°C was observed between the interphase and the bulk polymer. This phenomenon reveals a reduced motion ability of the polymer chains near the concrete substratum, and could be related to a gradient of crosslink density of the epoxy network within the interphase. This local modification of the crosslink density may be induced by chemical or physical interactions between the epoxy polymer and the cement surface:

- either a preferential adsorption of some components of the resin by the cement, which could modify locally the stoechiometry of the reactive system;
- or the influence of active species from the cement surface (water molecules, functional groups) which could induce secondary chemical reactions;
  - or in more simple way, the influence of the substratum roughness which could trap macromolecular chains and reduce their motion ability.

Further investigations by FTIR microscopy are required in order to confirm one of these explanations.
Similar analyses were performed for the epoxy/steel joint. Figure 3 shows the contrast in thermal conductivity in the neighbourhood of the epoxy/steel

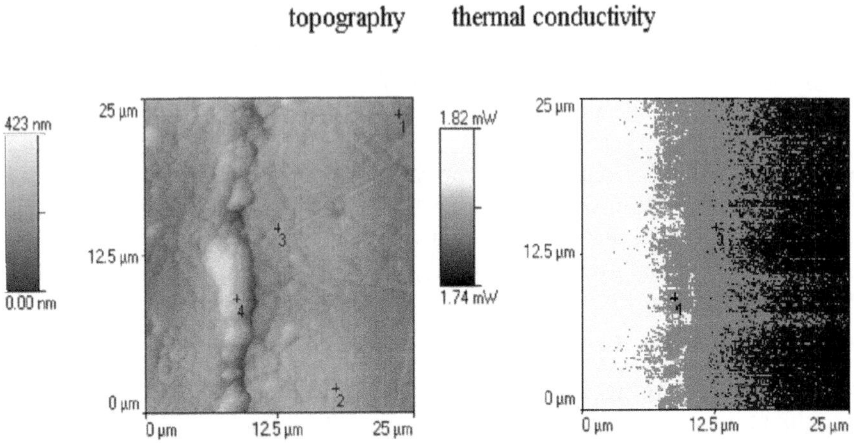


**Fig. 1.** Image of the topography and contrast in thermal conductivity at the epoxy/cement paste interface

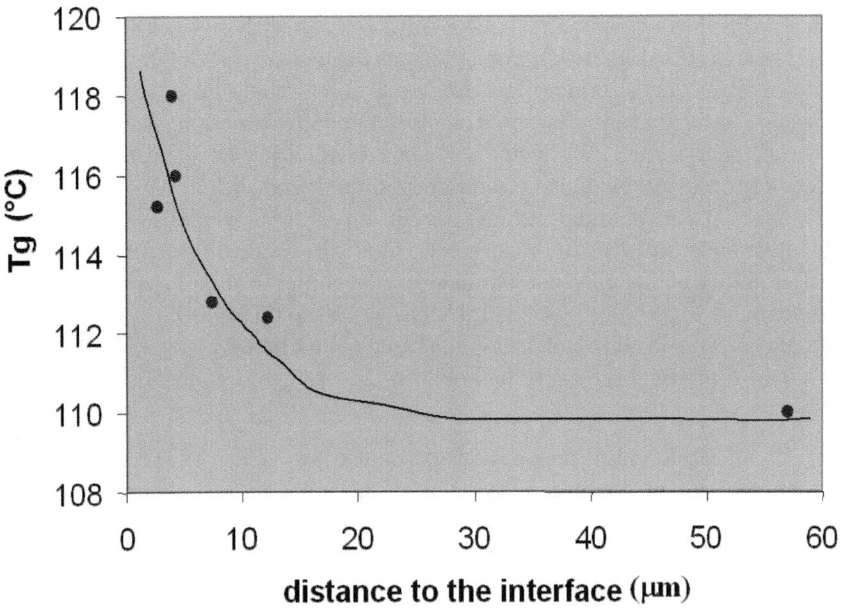


**Fig. 2.** Evolution of the glass transition temperature (Tg) in the polymer as a function of the distance to the epoxy/cement interface

interface. The steel phase appears in pale grey (or in yellow for the coloured version) and the bulk polymer in black. As previously, an intermediate modified polymer zone is identified in dark grey (red colour), but the thickness of the interphase is about 5 $\mu$m. Complementary local thermal analyses were also performed. Again, Tg of the polymer was found constant in the bulk polymer and showed a progressive increase in the interphase area close to the steel border. The maximum deviation of Tg between the interface zone and the bulk polymer is about 20°C. The same physical-chemical mechanisms as in the previous configuration may explain the gradient of crosslink density in the interphase area. Nevertheless, the thickness of the interphase and the maximum Tg deviation of the modified polymer zone are different in the epoxy/cement paste and in the epoxy/steel systems. Therefore, the polymer/substratum interactions and the properties of the interphase seem to be strongly dependent on the nature of the substratum (chemical species on the surface, roughness ... ). This is consistent with results found in the literature [13]. This interphase, which exhibits different properties from that of the bulk polymer, may have a significant influence on the mechanical behaviour of the bonded joint, and therefore on the durability of the overall glued assembly. The development of a mechanical modelling taking into account the existence of such an interphase is suitable and may lead to refinement in the design of adhesively bonded joints.

## 4 Conclusions

During implementation of the gluing technique, the formation of the adhesive bond is a physical-chemical process, which involves different stages:

- wetting and impregnation of the solid substratum by the adhesive,
- evolution of the liquid glue towards an equilibrium solid state, through chemical reactions (polymerisation).

In the first part of this study, the conditions of implementation and their influence on the final properties of the polymer joint are discussed. Basic rules are given in order to optimize both the wetting of the adherent and the mechanical strength of the bond. Those specifications are concerned with the surface preparation of the adherent (specially for concrete and steel) and the conditions of cure (temperature and time of reticulation).
In the second part of the article, the microstucture of epoxy joints has been investigated by a physical-chemical approach. Micro-thermal analyses, performed on epoxy/concrete and epoxy/steel assemblies, have revealed the existence of a modified polymer zone in the neighbourhood of the substratum. This interphase exhibits properties which are different from that of the bulk polymer, and its thickness seems to depend on the nature of the substratum. This modified polymer zone is suspected to play a significant role towards interfacial adhesion and durability of the bond. Such a result shows that a

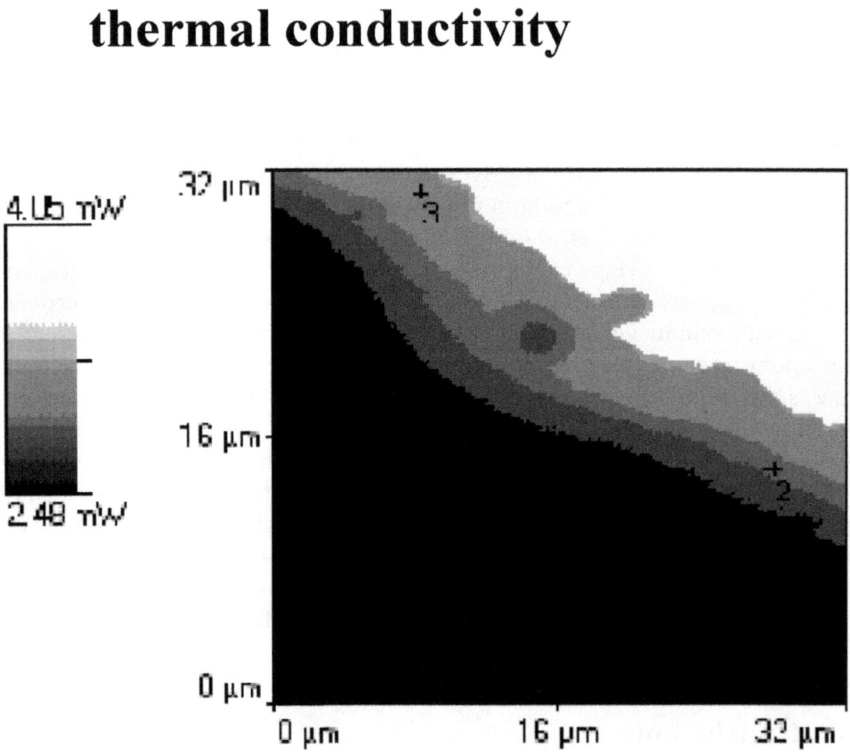


**Fig. 3.** Contrast in thermal conductivity at the epoxy/steel interface

physical-chemical approach may be complementary and even necessary to the mechanical point of view, in order to improve the basic knowledge on adhesive bonding.

## References

1. French Standard NF P 18 810. Special products for hydraulic concrete constructions - Synthetic based products - Determination of usual practice time (Détermination de la durée pratique d'utilisation) (D.P.U.) (1989)
2. MOUTON Y., Le temps de réticulation en place (T.R.P.) des résines réactives utilisées pour la réparation d'ouvrages soumis à la circulation, Coll. Eur. Construction et réhabilitation. Apport des polymères organiques, Lyon, Sept.1992, pp.169-175
3. MOUTON Y., SADI A., The injection into cracks of concrete structures : the behaviour of epoxy-amine resins at early stages, Sem. Int. RILEM, Capri, Mai 1989

4. BONOMI A., BORREILH J., MOUTON Y., Influence de l'eau sur les liants époxydes utilisés en réparation par injection, Sympos. Intern. RILEM ISAP'86, Aix en Provence, Sept.1986.
5. Choix et application des produits de réparation et de protection des ouvrages en béton, Guide Technique Editions du Laboratoire Central des Ponts et Chaussées (1996)
6. LUYCKX J. Techniques de l'Ingénieur, C5 440, pp 1-10 (1998)
7. BENZARTI. Le Vide, Proceeding of 5th European Conference on Adhesion, vol 54, n° 296, pp. 497-502 (2000)
8. PRICE D.M., READING M., HAMMICHE A., POLLOCK H.M. International Journal of Pharmaceutics, vol. 192, pp 85-96 (1999)
9. PRICE D.M., READING M., LEVER T.J. J. of Thermal Analysis and Calorimetry, vol. 56, pp 673-679 (1999)
10. PRICE D.M., READING M., HAMMICHE A., POLLOCK H.M., BRANCH M.G. Termochemica Acta, vol. 332, pp 143-149 (1999)
11. ROYALL P.G., CRAIG D.Q.M., PRICE D.M., READING M., LEVER T.J. International Journal of Pharmaceutics, vol. 192, pp 97-103 (1999)
12. HAMLER R., zur MUHLEN E. Thermochemical Acta, vol 361, pp 113-120 (2000)
13. JANSEN I., HAMLER R., KLEINERT H., . Le Vide, Proceeding of 5th European Conference on Adhesion, vol 54, n° 296, pp. 482-486 (2000)

# On the Role of Interface Energies in the Description of Material Behavior

Gianpietro Del Piero

Dipartimento di Ingegneria
Università di Ferrara
44100 Ferrara, Italia

## Abstract

At the level of constitutive equations relating macroscopic stress and strain measures, *real* materials exhibit a variety of behaviors. In the years, there was a tendency to focus on some prominent properties of material response and to study them individually; the result was the development of independent, sometimes unrelated, branches of Continuum Mechanics, such as Fracture Mechanics, Damage Mechanics, and many theories sharing the name of Plasticity.

Some hope for recovering a unified view is provided by a simple model, based on the assumption that the total energy of a body is the sum of two parts, a *bulk part* representing the elastic strain energy and a *surface part* associated with defects occurring both at the macroscopic and at the microscopic level. In the model, defects are represented by discontinuities in the displacement field. Such discontinuities may be large or small, concentrated at a single surface or distributed over surface-like regions diffused throughout the body. The first case is typical of fracture, and the second of plastic deformation.

At the present stage, the model is restricted to the one-dimensional case. Two or three dimensional generalizations are far from trivial, due to mathematical difficulties. Within this limit, fracture was studied in [1], plasticity in [4], and damage and other forms of progressive failure were considered in [3], [5]. A first comprehensive account was given in [2]. The present communication can be viewed as an updated review of the state of the research; mainly, it is based on the contents of a paper in progress [6], in which emphasis is given to the mathematical structure of the model. The complete text of the communication is reported in [7].

## References

1. Choksi R., Del Piero G., Fonseca I., Owen D.R. (1999) Structured deformations as energy minimizers in models of fracture and hysteresis, Math. and Mech. of Solids 4, 321-356.

2. Del Piero G. (1998) Towards a unified approach to fracture, yielding, and damage. In: Inan E., Markov K.Z. (Eds.), Continuum Models and Discrete Systems, World Scientific, Singapore, 67-692.
3. Del Piero G., Truskinovsky L. (1998) A one-dimensional model for localized and distributed failure, J. Phys. IV France 8, Pr8, 95-102.
4. Del Piero G. (1999) One-dimensional ductile-brittle transition, yielding, and structured deformations. In: Argoul P., Frémond M., Nguyen Q.S. (Eds.) Variations of domains and free boundary problems in solid mechanics, Kluwer, Dordrecht, 203-210.
5. Del Piero G., Truskinovsky L. (2001) Macro and micro cracking in one-dimensional elasticity, Int. J. of Solids and Structures 38, 1135-1148.
6. Del Piero G., Truskinovsky L., Elastic bars with cohesive energy, in preparation.
7. Del Piero G., On the role of interface energies in the description of material behavior. In: Antonić N., van Duijn C.J., Jäger W., Mikelić A. (Eds.) Proceedings of the Conference "Multiscale problems in science and technology. Challenges to mathematical analysis and perspectives", Dubrovnik, Croatia, Sept 2000.

# Equilibrium of Masonry Vaults

Maurizio Angelillo, Antonio Fortunato

Dipartimento di Ingegneria Civile
Università di Salerno
84084 Fisciano, Italia

**Abstract.** This paper describes a new method for the analysis of masonry vaults based on the masonry-like material model. In the present work, restricted to the equilibrium of vaults, we look for a compressive thrust surface, contained within the extrados and intrados surfaces of the structure, carrying the applied loads and transferring them to the abutments of the vault. The optimal shape of such surface is obtained through minimization of a convenient form of relaxed energy.

## 1 Introduction

As a first approximation to the real behaviour of masonry structures the *masonry like material* has been proposed. This crude model that describes the material as elastic in compression but incapable of sustaining tensile stresses (zero-tension material), was first rationally introduced by Heyman in [1].
The idea of a zero-tension material underlies more or less consciously the design of masonry structures since antiquity (see Benvenuto [2], Sinopoli et al. [3]), particularly for vaulted masonry structures and arches.
For example, according to the well known method of Mery, the safety of the arch depends on the existence of any compressive thrust line, in equilibrium with the applied loads, within the rig of the arch.
In the present work, concerned with the equilibrium of vaults, we look for a compressive thrust surface $S$, contained within the extrados and intrados surfaces of the structure, carrying the applied loads and transferring them to the abutments of the vault.
Since a membrane is a statically determined structure, if $S$ were given and composed of conventional elastic material, the equilibrium stress state, with proper boundary conditions, would be uniquely determined by the equilibrium equations alone. On adding the unilateral constraints on stress, at least in regions where the constraints are active, both the shape of $S$ and the loads cannot be given arbitrarily and become part of the solution. In such regions the structure reduces to an underdetermined system, as a cord or a net.
Keeping $S$ fixed the load must be changed to verify equilibrium on the given shape. If loads are given the surface $S$ must adapt in order to balance the prescribed loads.
In the following we shall assume that the loads are fixed and consider $S$ as an unknown smooth surface, constrained to lie in between the extrados and intrados surfaces of the vault.

The numerical strategy we have in mind to get a solution is to start with a tentative shape $S^\circ$ and evolve it by minimizing a convenient form of energy with side constraints. The tool we are going to use is the powerful minimization program named *Surface Evolver*, developed by Brakke [4]. Starting from an initial faceted surface the *Evolver* evolves the surface toward minimum energy by a conjugate gradient descent method. We have been lately involved in using such a tool to solve large scale minimization problems concerning nets [5] and axially compressed thin tubes [6].

## 2 Geometry of the problem

We assume that $S$ can be given in the *Monge* description:

$$\mathbf{y} = x_1\mathbf{e}_1 + x_2\mathbf{e}_2 + f\mathbf{e}_3 \tag{1}$$

where $\mathbf{y}$ is the position vector of points on $S$, $\{\mathbf{e}_1,\ \mathbf{e}_2,\ \mathbf{e}_3\}$ a given orthonormal triad, $(O,\ x_1\ ,\ x_2,\ x_3)$ a Cartesian frame coherent with such basis, $(x_1\ ,\ x_2) \in \Omega$ are curvilinear coordinates on $S$ and $f = f(x_1, x_2)$ is a smooth function of its arguments.

$\Omega$ is a simply connected plane region, called the planform of $S$, whose boundary is endowed a. e. with a unit normal vector $\mathbf{n}$.
Though we assume that the shape $S$ is not given, its planform will be considered as fixed.
The covariant bases associated to $(x_1\ ,\ x_2)$ are:

$$\mathbf{a}_1 = \mathbf{e}_1 + f_{,1}\mathbf{e}_3\ , \tag{2}$$

$$\mathbf{a}_2 = \mathbf{e}_2 + f_{,2}\mathbf{e}_3\ . \tag{3}$$

Here a comma followed by an index, say $i$, stands for partial differentiation with respect to $x_i$ . The unit normal to $S$ is then

$$\mathbf{a}_3 = \frac{\mathbf{a}_1 \times \mathbf{a}_2}{|\mathbf{a}_1 \times \mathbf{a}_2|} = \frac{-f_{,1}\mathbf{e}_1 - f_{,2}\mathbf{e}_2 - \mathbf{e}_3}{J}, \tag{4}$$

where $J = |\mathbf{a}_1 \times \mathbf{a}_2|$ is the Jacobian determinant. The reciprocal bases are:

$$\mathbf{a}^1 = \frac{1}{J^2}\Big((1 + f_{,2}^2)\,\mathbf{e}_1 - f_{,1}\,f_{,2}\,\mathbf{e}_2 + f_{,1}\,\mathbf{e}_3\Big)\ , \tag{5}$$

$$\mathbf{a}^2 = \frac{1}{J^2}\left(-f_{,1}\,f_{,2}\,\mathbf{e}_1 + (1 + f_{,1}^2)\,\mathbf{e}_2 + f_{,2}\,\mathbf{e}_3\right)\ . \tag{6}$$

## 3 Forces and equilibrium

We consider the problem of equilibrium of $S$ subject to a given surface load per unit area of :

$$\mathbf{b} = b_{(1)}\mathbf{e}_1 + b_{(2)}\mathbf{e}_{21} + b_{(3)}\mathbf{e}_3 \tag{7}$$

Calling

$$\mathbf{T} = T^{\alpha\beta} \mathbf{a}_\alpha \otimes \mathbf{a}_\beta \ , \quad , \ \alpha, \beta = 1, 2, \tag{8}$$

the membrane stress on $S$ , the equilibrium equations read

$$\frac{\partial}{\partial x_\gamma} \left( T^{\alpha\beta} a_\alpha \otimes a_\beta \right) a^\gamma + b = 0 \ , \tag{9}$$

Projecting (9) on the three non-coplanar directions $\mathbf{e}_1$, $\mathbf{e}_2$, $\mathbf{a}_3$ , after some algebra, we obtain:

$$\left( J T^{11} \right)_{,1} + \left( J T^{12} \right)_{,2} + J\, b_{(1)} = 0 \ , \tag{10a}$$
$$\left( J T^{21} \right)_{,1} + \left( J T^{22} \right)_{,2} + J\, b_{(2)} = 0 \ , \tag{10b}$$
$$J\, T^{\alpha\beta} f_{,\alpha\beta} - f_{,\alpha} J\, b_{(\alpha)} + J\, b_{(3)} = 0 \ . \tag{10c}$$

Introducing the *pseudo-stresses:*

$$S^{\alpha\beta} = J T^{\alpha\beta} \ , \tag{11}$$

and denoting:

$$\mathbf{b}^\circ = J\, \mathbf{b} \ , \tag{12}$$

the surface load per unit planform area, we can rewrite (10) as:

$$S^{\alpha\beta}_{,\beta} + b^\circ_{(\alpha)} = 0 \ , \tag{13a}$$
$$S^{\alpha\beta} f_{,\alpha\beta} - f_{,\alpha} b^\circ_{(\alpha)} + b^\circ_{(3)} = 0 \ . \tag{13b}$$

that is essentially *Pucher's form* of membrane equilibrium.

## 4 Stress function *F*

To simplify analysis we assume (vertical loads):

$$b^\circ_{(1)} = b^\circ_{(2)} = 0 \ , \quad b^\circ_{(3)} = -p \tag{14}$$

In this case the general solution of equations (13a), is:

$$S^{11} = F_{,22}, \quad S^{22} = F_{,11}, \quad S^{12} = -F_{,12} \ . \tag{15}$$

Introducing the curvature coefficients:

$$a_{11} = f_{,22} \ , \quad a_{22} = f_{,11} \ , \quad a_{12} = -f_{,12} \ , \tag{16}$$

we can rewrite (13b) in the form

$$a_{\alpha\beta} F_{,\alpha\beta} - p = 0 \ , \tag{17}$$

that is a second order partial differemtial equation to be solved with proper side conditions.

We restrict to the elliptic case ($S$ concave) and consider the boundary conditions:

$$\frac{dF}{dn} = l(s) \ , \quad \text{over } \partial\Omega \tag{18}$$

## 5 Material restrictions

The restriction on stress that characterize the material as masonry–like are:

$$\det \mathbf{T} \geq 0 \ , \quad tr\mathbf{T} \leq 0 \ , \tag{19}$$

which are equivalent to the zero-tension assumption:

$$\hat{\mathbf{m}} \cdot \mathbf{T}\,\hat{\mathbf{m}} \geq 0 \ , \quad \forall \hat{\mathbf{m}} \ s.t. \ \hat{\mathbf{m}} \cdot \mathbf{a}_3 = 0 \ . \tag{20}$$

In contravariant components these restrictions rewrite as:

$$J^2(T^{11}T^{22} - (T^{12})^2) \geq 0 \ , \ g_{\alpha\beta}T^{\alpha\beta} \leq 0 \ , \tag{21}$$

where $g_{\alpha\beta} \equiv \mathbf{g}_\alpha \cdot \mathbf{g}_\beta$ are the metric tensor components.

Conditions (21) translate into the following equivalent conditions in terms of pseudo-stresses:

$$\mu(\mathbf{S}) = S^{11}S^{22} - (S^{12})^2 \geq 0 \ , \tag{22a}$$

$$\nu(\mathbf{S}) = (1 + f_{,2})S^{11} + (1 + f_{,1})S^{22} - 2f_{,1}f_{,2}S^{12} \leq 0 \ . \tag{22b}$$

The inequality (22a) defines as admissible the stress points ($S^{11}$, $S^{22}$, $S^{12}$) contained inside the right circular cone of equation $S^{11}S^{22} - (S^{12})^2 = 0$ , with vertex in the origin, whose axis is the axis of isotropy, $S^{11} = S^{22}$ , and whose wedge angle is $\pi/2$. Inequality (22b) defines as admissible the points belonging to a semi–space defined by the plane passing through the origin and whose equation is:

$$(1 + m_2)\,S^{11} + (1 + m_1)\,S^{22} - 2\,m_1\,m_2\,S^{12} = 0 \ , \tag{23}$$

where we put $m_1 = f_{,1}$ , $m_2 = f_{,2}$ . It is easy to see that for any value of $m_1$, $m_2$ , the plane and the cone meet only in the origin. Therefore the linear inequality (22b) can be substituted with the easier inequality:

$$\nu^*(S) = S^{11} + S^{22} \leq 0 \ . \tag{24}$$

In terms of $F$ the unilateral constraints on stress reduce to the form:

$$F_{,11} + F_{,22} \leq 0 \ , \tag{25}$$

$$F_{,11}\, F_{,22} - (F_{,12})^2 \geq 0 \ . \tag{26}$$

which are equivalent to require that the stress surface whose graph is $F$ be concave.

# 6 Non fractured and potentially fractured regions

For any statically admissible stress state $\mathbf{S}$, $\Omega$ can be splitted into three parts:

$$\Omega \equiv \{\mathbf{x} \in \Omega,\ tr\mathbf{S} < 0 \text{ and } \det \mathbf{S} > 0\}\ ,\ (\mathbf{S}\ \textit{biaxial})\ , \tag{27a}$$

$$\Omega \equiv \{\mathbf{x} \in \Omega,\ tr\mathbf{S} < 0 \text{ and } \det \mathbf{S} = 0\}\ ,\ (\mathbf{S}\ \textit{uniaxial})\ , \tag{27b}$$

$$\Omega \equiv \{\mathbf{x} \in \Omega,\ tr\mathbf{S} = 0 \text{ and } \det \mathbf{S} = 0\}\ ,\ (S = 0)\ . \tag{27c}$$

In $\Omega_1$ the stress is biaxial and strictly compressive, fractures cannot arise.
In $\Omega_2$ the stress is uniaxial and compressive, fractures may arise in the direction orthogonal to compression lines.
In $\Omega_3$ the material is inert and any kind of fracture may occur.

It can be shown that in $\Omega_2$ the compression lines form a family of straight lines. The proof of this statement is similar to an analogous argument concerning the plane case (see [7]). The compression rays are lines potentially fractured. In the present context this means that, under the previous assumptions on the load, the projection of the fracture lines over the planform must be straight.

# 7 Variational formulation

To the previous boundary value problem we give the following variational formulation:

$$\mathcal{E}(f^\circ, F^\circ) = \min_{\substack{F \text{ concave} \\ f_i \le f \le f_e}} \mathcal{E}(f, \mathrm{F})\ , \tag{28}$$

where the energy $\mathcal{E}$ is defined as follows:

$$\mathcal{E}(f,\, F) = \frac{1}{2} \int_{\Omega} a_{\alpha\beta} F_{,\alpha} F_{,\beta} da + \int_{\Omega} pF da - k \int_{\partial\Omega} F ds\ , \tag{29}$$

$$k \equiv \frac{1}{ar(\Omega)} \int_{\Omega} p da\ .$$

In order to approximate both the surface $S$ and the stress surface with simplicial surfaces we adopt a relaxation strategy based on the introduction of a family of relaxed functionals $E_h$ whose minima converge to the minimum of the functional $E$ as $h \to 0$ . This is done in a way similar to that adopted in paper [8] where the LSM-method for elastic structures was introduced. To this end we consider a triangulation of the planform $\Omega$ (primal mesh):

$$\Pi_h = \{\Omega_m,\, m \in 1,\, 2,\, ...,\, M\}\ , \tag{30}$$

and a dual mesh:

$$\boldsymbol{\Pi}_h = \{ \boldsymbol{\Omega}_n,\, n \in 1, 2, ..., N\}, \tag{31}$$

formed by polygons enclosing the nodes of the primal mesh, as shown in Fig.1. Here $h$ represent a characteristic length of the primal mesh.

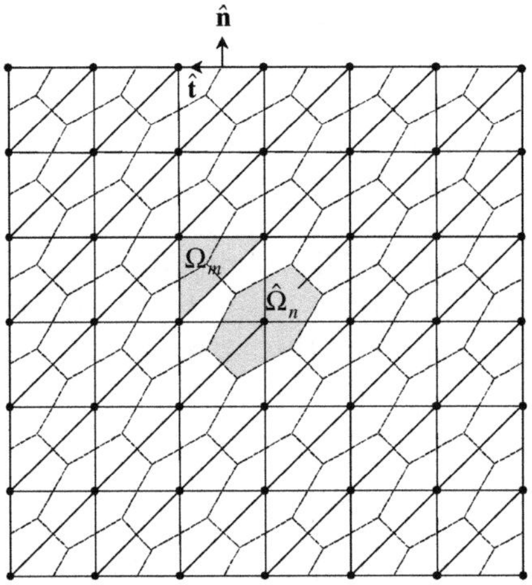


**Fig. 1.** Primal and dual mesh over a rectangular planform.

The form of relaxed energy we consider is then:

$$\mathcal{E}_h(f,F) = \frac{1}{2}\sum_{n=1}^{N} \frac{1}{ar(\boldsymbol{\Omega}_n)} \int_{\boldsymbol{\Omega}_n} a_{\alpha\beta} da \int_{\boldsymbol{\Omega}_n} F_{,\alpha} F_{,\beta} da + \int_{\Omega} pF da - k \int_{\partial\Omega} F ds \ . \tag{32}$$

An example of the way in which both the surface of the vault and the stress function are approximated is shown in Fig 2.

## References


1. Heyman, J. (1966) The stone skeleton. Int. J. Solids Structures **2**, 249-279.
2. Benvenuto, E. (1991) An introduction to the history of structural mechanics. Vol. **2**, Springer Verlag.
3. Sinopoli, A., Corradi, M., Foce, F. (1997) Modern formulation for preelastic theories on masonry arches. J. Eng. Mech., ASCE, **123**, 3, 204-213.

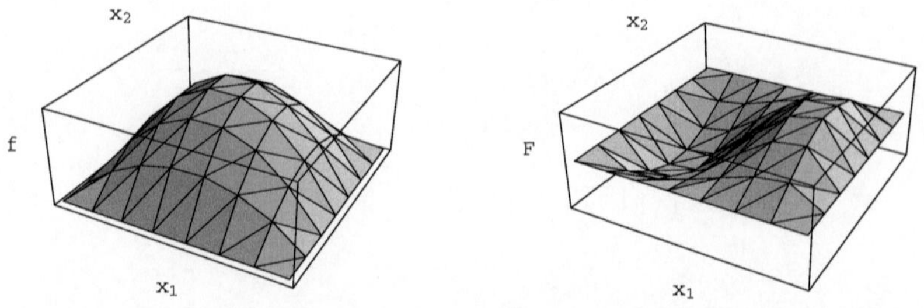


**Fig. 2.** Discretization of f and F over a rectangular planform.

4. Brakke, K. (1999) The surface evolver – A program for the modelling surfaces shaped by various forces and constraints, Mathematics Department, Susquehanna University, Selinsgrove.
5. Angelillo, M., Fortunato, A. (2000) Equilibrium problems for hexagonal nets. In preparation.
6. Angelillo, M., Fortunato, A. (2000) Folding deformations of thin tubes. In preparation.
7. Angelillo, M., Olivito, R. S. (1995) Experimental analysis of masonry walls loaded horizontally in plane. Masonry International. **8**,3, 91-100.
8. Fraternali, F., Angelillo, M., Fortunato, A. (2002) A Lumped Stress Method for plane elastic problems and the Discrete-Continuum approximation. Int. J. Solids Struct. **39**,25, 6211-6240.

# Micromechanical Analysis of Reinforced Masonry

Sonia Marfia, Elio Sacco

DiMSAT
Università di Cassino
03043 Cassino, Italia

**Abstract.** In this paper a homogenization procedure is developed for a one-dimensional reinforced masonry problem, considering the progressive damage and plasticity of the mortar and the block. The brittle failure of the FRP reinforcement is accounted for. The delamination effect of the composite sheets from the masonry element is also modeled. A numerical procedure is developed to study the behavior of the reinforced masonry. Numerical applications regarding the axial and the bending response of the material are presented.

## 1 Introduction

Fiber reinforced plastic (FRP) composite materials appears to be good candidates to repair and to reinforce masonry constructions, since they are light, very simple to install and are also removable. Moreover, composite materials are characterized by high strength, good resistance to corrosion, durability and reduced installation and maintenance costs. As matter of fact, composite materials have been successfully used in several fields of structural Civil Engineering (Neale and Labossière, 1992; El-Badry, 1996), mainly for strengthening concrete and wood structures. Triantafillou and Fardis (1995) and Triantafillou (1996) studied the applications of advanced composites for strengthening historical masonry structures. The behavior of walls, reinforced by carbon fiber sheets or conventional woven fabric bonded on the masonry surfaces, have been investigated with experimental tests by Schwegler (1994). Luciano and Sacco (1996, 1998) proposed a model for studying the behavior of masonry panels reinforced by FRP sheets.

Although applications of advanced composites are successfully adopted to restore masonry structures, a lack of appropriate models and computational procedure able to predict the response of reinforced masonry can be pointed out. Thus, the aim of this paper is to investigate on the overall behavior of the masonry reinforced by means of FRP composite materials.

In the present paper, the micromechanics and the homogenization technique for deriving a reinforced masonry overall model. The proposed approach has been succesfully adopted for modeling the masonry, regarded as a heterogeneous material, with periodic microstructure, realized by bricks in a matrix of mortar (Kraly et al., 1991; Pietruszczac and Niu, 1992; Luciano and Sacco, 1995).

Herein a one-dimensional beam model is studied (Marfia and Sacco, 2001). The model accounts for damage in tension and damage and plasticity in compression for the block and the mortar. The reinforcement brittle failure and the delamination process of the FRP from the masonry, is also considered. A numerical procedure, based on the arc-length technique with an appropriate choice of the control parameters, is developed. Numerical applications regarding the study of the axial and the bending response of the reinforced masonry under different loading histories are presented. The beneficial effect of FRP reinforcement on the masonry behavior is emphasized.

## 2 Masonry and reinforcement modelling

The attention is focussed on the study of a beam obtained as a repetitive sequence of blocks and mortar. Moreover, FRP sheets are partially glued on the top and on the bottom of the beam. In Fig. 1 the unit cell representing one-half of the repetitive microstructure is reported; the total length of the unit cell is $L = L_b + L_m$, where $L_b$ and $L_m$ denote the block and mortar half lengths.

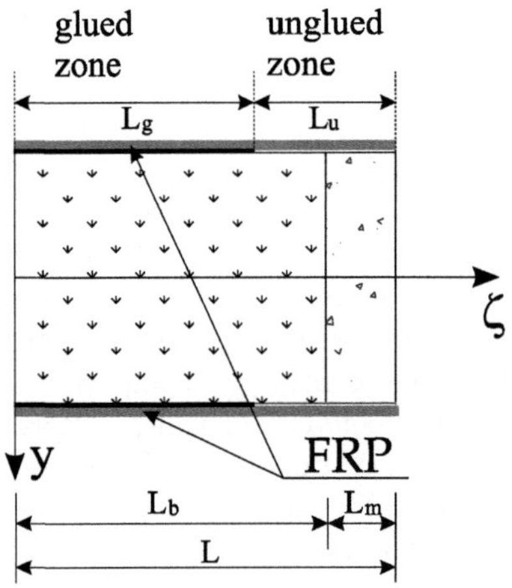


**Fig. 1.** Unit cell obtained from the periodicity and symmetry of the masonry.

The masonry has constant rectangular cross-section $A_M = b \times h$. The reinforcements on the top and the bottom of the masonry are characterized by the areas $A_R^+$ and $A_R^-$, respectively. With reference to Fig. 1, it is assumed that the reinforcement is in perfect adhesion on the block for a length equal to $L_g$ and it is completely unglued in the zone corresponding to the mortar, characterized by the length $L_u = L - L_g$.

With the aim of deriving the overall response for the defined unit cell in terms of resultant axial force and bending moment as function of the average axial strain and curvature, one-dimensional continuous stress-strain relations for the mortar and the block are introduced and a FRP damage mechanism is considered.

A simple one-dimensional constitutive relation is chosen both for the mortar and the block. It is characterized by a damage response in traction and damage-plasticity law in compression. In particular, the constitutive equation is:

$$\sigma = (1 - D)\, E\, (\varepsilon - \varepsilon_p) \tag{1}$$

where $D$ is the damage parameter and $\varepsilon_p$ is the plastic strain, so that $\varepsilon - \varepsilon_p = \varepsilon_e$ represents the elastic strain. The Young modulus $E$ of the block and mortar are denoted as $E_b$ and $E_m$, respectively.

A damage evolution law, inducing a linear softening in the stress-strain relationship, is assumed:

$$0 \le \dot{D} = \begin{cases} 0 & \text{for } \varepsilon_d < \varepsilon_c^{\pm} \\ \dfrac{\varepsilon_c^{\pm}\varepsilon_u^{\pm}}{\left(\varepsilon_u^{\pm} - \varepsilon_c^{\pm}\right)\varepsilon_d^2}\dot{\varepsilon} & \text{for } \varepsilon_c^{\pm} < \varepsilon_d < \varepsilon_u^{\pm} \\ 0 & \text{for } \varepsilon_u^{\pm} < \varepsilon_d \end{cases} \tag{2}$$

where $\varepsilon_c^{\pm}$ and $\varepsilon_u^{\pm}$ are the elastic strains corresponding to the undamaged and completely damaged material, respectively. The superscript $^{\pm}$ is $^{+}$ in traction and $^{-}$ in compression. The governing evolution parameter $\varepsilon_d$ is set in a different way in traction and in compression. In fact, it is assumed that the damage in traction depends on the elastic stress, i.e. $\varepsilon_d = \varepsilon_e$, while in compression it depends on the total strain, i.e. $\varepsilon_d = \varepsilon$.

The mortar and block response in traction, i.e. for $\sigma \ge 0$, can be considered as purely cohesive, without any plastic effect. Thus, the plastic strain rate is set to be zero:

$$\dot{\varepsilon}_p = 0 \qquad \text{for} \qquad \sigma \ge 0 \tag{3}$$

The mortar and block response in compression, i.e. for $\sigma < 0$, presents a damage-plastic behavior. Introducing the effective stress $\tilde{\sigma}$ as:

$$\tilde{\sigma} = \frac{\sigma}{1 - D} \tag{4}$$

the following yield function with hardening is considered:

$$f(\tilde{\sigma}, \alpha) = -\tilde{\sigma} - (\sigma_y + K\,\alpha) \tag{5}$$

where $\alpha$ is the internal hardening variable and $K$ is the plastic hardening parameter. The evolutive equations are:

$$\dot{\varepsilon}_p = \dot{\lambda}\,\frac{\partial f}{\partial \tilde{\sigma}} = -\,\dot{\lambda} \tag{6}$$

$$\dot{\alpha} = |\dot{\varepsilon}_p| = -\dot{\varepsilon}_p \tag{7}$$

and the Kuhn-Tucker conditions result:

$$\dot{\varepsilon}_p \leq 0 \qquad f \leq 0 \qquad \dot{\varepsilon}_p \, f = 0 \tag{8}$$

The composite material behavior is characterized by a linear elastic response with quite brittle failure Barbero (1999).

In the present work, the composite is considered as a linear elastic material with Young modulus $E_c$ until a brittle failure, i.e. the composite is assumed to collapse suddenly when the tensile or the compressive stress reaches a threshold stress $f_R^+$ in tension or $f_R^+$ in compression.

Moreover, the degradation is allowed also at the masonry-composite interface. The delamination of the composite sheet from the masonry element can be considered as brittle and mainly due to the shear fracture (i.e. mode II fracture) of the masonry, rather than to the damage of the glue.

It is assumed that at the virgin state the adhesion between the laminate and the mortar is negligible, and thus the FRP is perfectly glued only to the block. Hence, it is considered the presence of an initial defect of adhesion in correspondence of the mortar. The fracture propagation is ruled by the classical Griffith criterion, setting the critical release rate energy $G_c$ as the one associate to the fracture mode II of the block material (Bazant and Planas, 1998). Note that the fracture evolution is coupled with the damage and plasticity.

Let $(x, y, \zeta)$ be a Cartesian coordinate system such that $\zeta$ lies on the center-line axis of the undeformed beam, as reported in Fig. 1. The kinematics of the cross-section is defined by the elongation $e$ and the curvature $\chi$, such that the strain at a typical point of the beam is $\varepsilon = e + y\,\chi$.

In order to evaluate the overall behavior of the reinforced masonry beam, the unit cell is considered subjected to an axial force $N_{tot}$ and to a bending moment $M_{tot}$. The elongation and the curvature for the masonry and the reinforcement are piecewise constant. In particular, it can be set:

| | $0 < \zeta < L_g$ | $L_g < \zeta < L_b$ | $L_b < \zeta < L$ |
|---|---|---|---|
| Masonry | $e_g\ \chi_g$ | $e_b\ \chi_b$ | $e_m\ \chi_m$ |
| Reinforcement | $e_g\ \chi_g$ | $e_u\ \chi_u$ | |

Note that the unglued reinforcement elongation and curvature $e_u$ and $\chi_u$ can be evaluated as:

$$\begin{aligned} e_u &= \eta_b\, e_b + \eta_m\, e_m \\ \chi_u &= \eta_b\, \chi_b + \eta_m\, \chi_m \end{aligned} \tag{9}$$

where $\eta_b = (L_b - L_g)/L_u$ and $\eta_m = L_m/L_u$.

The reinforced masonry elongation $e_{tot}$ and curvature $\chi_{tot}$ are determined as:

$$\begin{aligned} e_{tot} &= \frac{[e_g\, L_g + e_b\, (L_b - L_g)] + e_m\, L_m}{L} = \frac{e_g\, L_g + e_u\, L_u}{L} \\ \chi_{tot} &= \frac{[\chi_g\, L_g + \chi_b\, (L_b - L_g)] + \chi_m\, L_m}{L} = \frac{\chi_g\, L_g + \chi_u\, L_u}{L} \end{aligned} \tag{10}$$

The resultant normal forces and bending moments in the masonry beam, computed by integrating the normal stress in the cross sections, are $N_g^M$ and $M_g^M$ in the glued block, $N_b$ and $M_b$ in the unglued block and $N_m$ and $M_m$ in the mortar. Analogously, the resultant normal forces and bending moments are $N_g^R$ and $M_g^R$ in the glued part of the composite reinforcement and $N_u$ and $M_u$ in the unglued part.

The total resultant axial forces and the bending moments in the three parts of the reinforced masonry are obtained as:

$$\begin{array}{ll} N_1 = N_g^M + N_g^R & , M_1 = M_g^M + M_g^R \\ N_2 = N_b + N_u & , M_2 = M_b + M_u \\ N_3 = N_m + N_u & , M_3 = M_m + M_u \end{array} \tag{11}$$

Finally, as the axial force and the bending moment are constant along the whole beam, the six equilibrium equations are:

$$\begin{array}{lll} N_1 = N_{tot} & , & M_1 = M_{tot} \\ N_2 = N_{tot} & , & M_2 = M_{tot} \\ N_3 = N_{tot} & , & M_3 = M_{tot} \end{array} \tag{12}$$

The delamination of the reinforcement from the masonry occurs when the critical energy $G_c$ is equal to the energy release rate:

$$G_c\, b\, \delta L_u = \delta W - \delta \mathcal{E} - \delta \Phi - \delta \Psi \tag{13}$$

where $\delta W$ is the external work performed during the process, $\delta \mathcal{E}$ is the elastic energy variation, $\delta \Phi$ and $\delta \Psi$ are energies dissipated for the plasticity and damage effects, respectively.

## 3 Computational procedure and numerical applications

Because of the damage, the plasticity, the brittle failure of the reinforcement and the possible delamination of the composite from the masonry, the solution of equations (12) is not straightforward. Thus, a computational procedure is developed. The numerical algorithm is summarized in the following.

- The equilibirum equations (12) are written in the equivalent residual form and solved using a Newton algorithm;
- the time discrete model is solved using a return map algorithm, adopting a predictor-corrector method (Simo);
- the integration over the cross-section to determine the residuals and its derivatives is performed by discretizing the cross-section in stripes and applying the Gauss integration formula within each stripe;
- the delamination effect is studied through a discretization of the evolutive equation according to a finite difference scheme, in the framework of the virtual crack method;

- the cylindrical as well as the linearized arc-length methods Crisfield, 1991) with local control are developed for the particular problem under consideration.

Applications are developed for a reinforced masonry element characterized by mortar and block with the following material properties:
mortar

$$E_m = 5000MPa \; \sigma_{y,m} = 3MPa \; K_m = 500MPa$$
$$\varepsilon^+_{c,m} = 1E-4 \quad \varepsilon^+_{u,m} = 4E-4 \; \varepsilon^-_{c,m} = 10E-4 \; \varepsilon^-_{u,m} = 40E-4$$

block

$$E_b = 15000MPa \; \sigma_{y,b} = 10MPa \; K_b = 1500MPa$$
$$\varepsilon^+_{c,b} = 1E-4 \quad \varepsilon^+_{u,b} = 6E-4 \; \varepsilon^-_{c,b} = 15E-4 \; \varepsilon^-_{u,b} = 60E-4$$

The masonry geometrical parameters are:

$$L_b = 25mm \; L_m = 5mm$$
$$b = 130mm \; h = 250mm$$

which correspond to a typical masonry with blocks $50 \times 130 \times 250mm^3$ and mortar layers of $10mm$ width.

A carbon fiber-reinforced plastic composite material with Young's modulus $E_c = 200000MPa$ is considered. Note that, although the strength of the FRP sheets is generally different in traction and in compression, in the application developed in the following, it is set $f^+_R = f^-_R = 2500MPa$. It is assumed partially glued to the masonry so that $L_g = 22mm$.

Initially, the axial response of the reinforced masonry is investigated, neglecting the possible delamination. In Fig. 2 the axial force $N_{tot}$ versus the strain $e_{tot}$ is plotted in tension and in compression, respectively, for different values of the reinforcement area, with $A_R = A^+_R = A^-_R$.

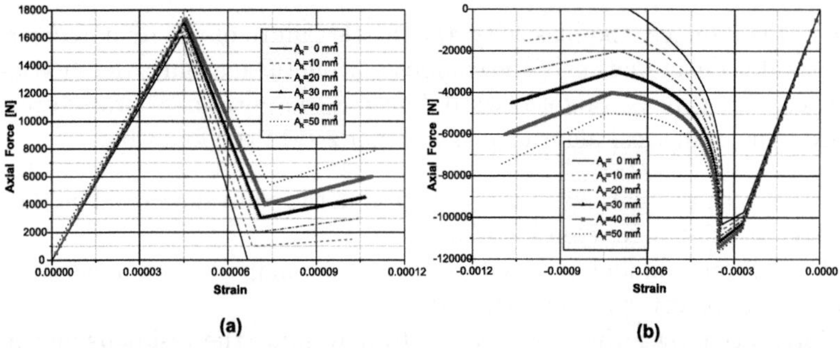


**Fig. 2.** Axial force - strain (a) in tension and (b) in compression for different amounts of FRP reinforcement.

It can be noted that the presence of the reinforcement improves the mechanical response of the masonry. In fact, the reinforcement allows to transfer the external axial force to the block, also when the mortar is completely damaged.

In Fig. 3 the total bending moment $M_{tot}$ is plotted versus the curvature $\chi_{tot}$ for different amounts of FRP reinforcement, during the initial damage of the mortar. As for the axial problem, also for the bending case it is apparent the beneficial effects of the reinforcement in improving the mechanical response, and in particular the ductility, of the masonry.

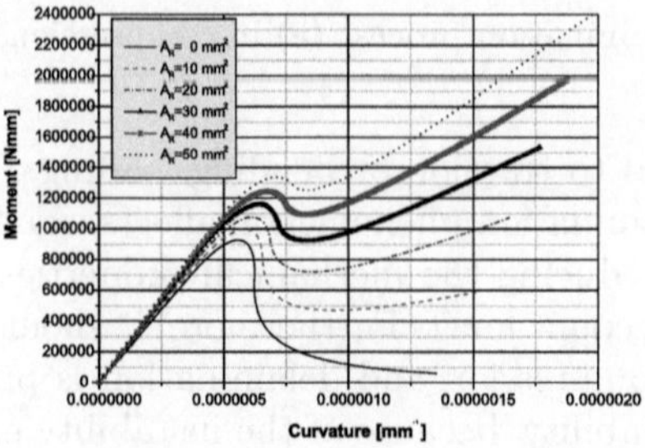


**Fig. 3.** Bending moment - curvature relation for different amounts of FRP reinforcement.

Two different values of the critical energy release rate governing the delamination phenomenon are assumed $G_c = 0.002 \div 0.04$ N/mm; these values correspond to possible critical energies for the block in mode II. In Fig. 4(a), the axial force $N_{tot}$ is plotted versus the strain $e_{tot}$ in compression using lines for different fixed values of the glued length $L_g$ of the reinforcement. In the same figure, the axial force-strain relation, which takes into account the delamination process, is represented with a thick line. The delamination occurs during the mortar damage, when the block still behaves elastically.

In Fig. 4(b), the bending moment $M_{tot}$ is plotted versus the curvature $\chi_{tot}$ for different fixed values of $L_g$ with thin lines, while the two curves with thick lines represent the material response taking into account the delamination process for two different values of the fracture energy $G_c$. It can be pointed out the significative influence of the delamination effects on the mechanical response of the reinforced masonry. Of course, it results that for higher values of the fracture energy the delamination occurs later.

## 4 Conclusions

The obtained results show the beneficial effects of the presence of the reinforcement on the overall response especially when the masonry is subjected to tension and to bending. Hence, it can be deduced that the reinforcement of the masonry should be designed in order to work in tension, i.e. when the

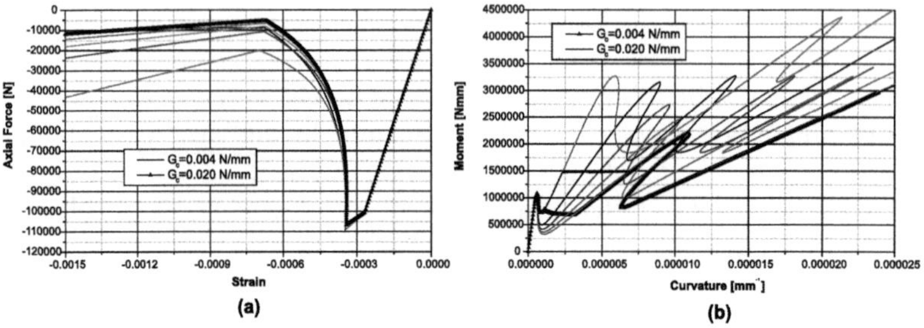


**Fig. 4.** Delamination process (a) in compression, (b) bending.

masonry is subjected to traction or bending loading. Computations demonstrates that the delamination phenomenon affects significantly the reinforced material behavior, reducing the mechanical properties. For this reason it is very important to account for it. Furthermore, it should be noted that, when the masonry is in compression and delamination is present, the FRP could lose any loading capability, because of the instability effects of the reinforcement.

## References

1. El-Badry, M. (1996) Advanced Composite Materials in Bridges and Structures, 2nd International Conference, Montréal, Québec, Canada.
2. Triantafillou, T.C. (1996) Innovative strengthening of masonry monuments with composites. Proceedings of 2nd International Conference Advanced Composite Materials in Bridges and Structures. Montréal (Canada), August 11-14.
3. Schwegler, G. (1994) Masonry construction strengthened with fiber composites in seismically endangered zones. In: Proceedings of 10th European Conference on Earthquake Engineering, Vienna, Austria.
4. Luciano, R., Sacco, E. (1996) Behavior of masonry panels reinforced with FRP composites. In: Proceedings of 2nd International Conference Advanced Composite Materials in Bridges and Structures. El-Badry M. (Editor). Montréal (Canada), August 11-14.
5. Luciano, R., Sacco, E. (1998) Damage of masonry panels reinforced by FRP sheets. International Journal of Solids and Structures **35**, 1723–1741.
6. Luciano, R., Sacco, E. (1997) Homogenization technique and damage model for old masonry material. Int. J. Solids and Structures **34**, 3191-3208.
7. Marfia, S., Sacco, E. (2001). Homogenization technique and damage model for old masonry material. Int. J. Solids and Structures **34**, 3191–3208.
8. Bazant, Z.P., Planas, J. (1998) Fracture and Size Effects in Concrete and Other Quasibrittle Materials. CRC Press LLC.
9. Simo, J.C., Hughes, T.J.R. (1998) Computational Inelasticity, Springer-Verlag, New York.
10. Crisfield, M.A. (1991) Non-linear Finite Element Analysis of Solids and Structures, Vol. 1. John Wiley & Sons, England.

# Localization Analysis in Dilatant Elasto-Plastic Solids by a Strong-Discontinuity Method

Carlo Callari[1], Alessio Lupoi[2]

[1] Dipartimento di Ingegneria Civile
Università di Roma "Tor Vergata"
00133 Roma, Italia

[2] Dipartimento di Ingegneria Strutturale e Geotecnica
Università di Roma "La Sapienza"
00100 Roma, Italia

**Abstract.** In this paper, we present an analysis of plane-strain associative dilatant plasticity in presence of displacement discontinuities. The obtained results are amenable to straightforward particularization to any dilatant model and a Drucker-Prager yield criterion is considered as a model example. We formulate an enhanced finite element method incorporating the (unregularized) discontinuous fields. Numerical results of a plane-strain localization analysis are discussed.

## 1 Introduction

Civil engineering problems are often characterized by the concentration of strains in narrow bands, typically indicated as "shear bands". Such a strain localization is frequently observed in geomaterials (e.g. soils, rock, concrete) and plays a crucial role in failure mechanisms. As a consequence, study of localization has many important applications in the analysis of structural [7] and geotechnical engineering problems (e.g. foundation collapse, excavation and slope stability, tunnel stability [4,5]).

The so-called "strong-discontinuity" approach to localization analysis has been developed for general rate-independent classic plasticity in [3,12,15]. In these works, solutions involving a discontinuous displacement field are considered, together with the corresponding singular distributions of strain. Enhanced finite element methods incorporating regularized [12,15] or unregularized [3] approximations of such discontinuous fields are formulated for the $J_2$ flow theory model problem.

An extension of the unregularized strong-discontinuity approach to the coupled case of a saturated poro-plastic solid is presented in [2,6].

In the present work, the general results of the analysis of strong discontinuities in uncoupled elasto-plastic models [15] are re-obtained in a form amenable to straightforward particularization to any associative dilatant plasticity model. In particular, under plane strain assumptions, we obtain the localized dilatancy expression, the localization condition and the relation between the stress and the displacement jumps at the discontinuity surface

(Section 2). These results are easily particularized for the case of a Drucker-Prager model, leading to a Mohr-Coulomb type localized softening law (Section 3). Following the approach proposed in [1,3], an enhanced finite element method is formulated for dilatant solids in plane strain conditions (Section 4). In this numerical treatment, *unregularized* discontinuous fields are considered, in contrast with the approach to dilatant models reported in [10,11]. In Section 5, representative numerical tests are considered to evaluate the performance of the implemented formulation.

Throughout the present paper, comparisons are made with the unregularized strong-discontinuity approach to Drucker-Prager model reported in [13].

## 2 Strong discontinuities in dilatant elasto-plastic media

In the following, we develop a plane-strain analysis of strong discontinuities in dilatant associative models.

### 2.1 Kinematics

In the solid $\Omega$, we denote by $\mathbf{u}$ the "large-scale" displacement field, satisfying standard regularity conditions. In a local neighborhood $\Omega_x$ of the point $\mathbf{x} \in \Omega$, we consider the following "small-scale" displacement field [1], exhibiting a discontinuity across a surface $\Gamma_x$ with unit normal $\mathbf{n}$ in $\mathbf{x}$ (Fig. 1):

$$\mathbf{u}_\mu = \mathbf{u} + \boldsymbol{\zeta}\, \Psi_{\Gamma_x} \quad \text{with} \quad \Psi_{\Gamma_x} = H_{\Gamma_x} - N_{\Gamma_x} \tag{1}$$

where $H_{\Gamma_x}$ is the Heaviside step function across $\Gamma_x$ and $N_{\Gamma_x}$ is a general smooth function. Note that the local field $\boldsymbol{\zeta}$ corresponds to the displacement jump: $[\![\mathbf{u}_\mu]\!] = \boldsymbol{\zeta}$.

The strain field corresponding to (1) is given in the distributional sense by:

$$\boldsymbol{\varepsilon}_\mu := \nabla^s \mathbf{u}_\mu = \underbrace{\boldsymbol{\varepsilon}(\mathbf{u}) + \Psi_{\Gamma_x}\, \nabla^s \boldsymbol{\zeta} - (\boldsymbol{\zeta} \otimes \nabla N_{\Gamma_x})^s}_{=:\bar{\boldsymbol{\varepsilon}}_\mu \quad \text{regular distribution}} + \underbrace{(\boldsymbol{\zeta} \otimes \mathbf{n})^s\, \delta_{\Gamma_x}}_{\text{singular distribution}} \tag{2}$$

with $\delta_{\Gamma_x}$ the Dirac delta distribution across $\Gamma_x$.

### 2.2 Strong discontinuities in dilatant solids

We consider the rate-independent associated elasto-plastic model defined by the following relations:

$$\begin{aligned} \dot{\boldsymbol{\sigma}} &= \mathbb{C}(\dot{\boldsymbol{\varepsilon}}_\mu - \dot{\boldsymbol{\varepsilon}}^p_\mu) \\ \dot{\boldsymbol{\varepsilon}}^p_\mu &= \lambda\, \partial_{\boldsymbol{\sigma}} f \qquad \mathcal{H}^{-1}\dot{q} := -\dot{\alpha}_\mu = -\lambda\, \partial_q f \end{aligned} \tag{3}$$

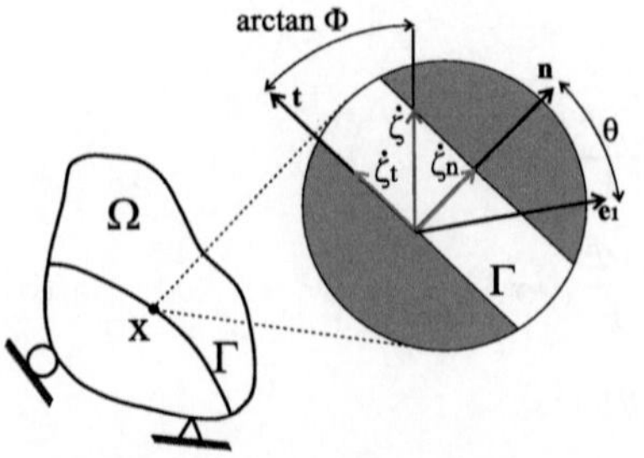


**Fig. 1.** Kinematics of strong discontinuities in plane strain conditions

with $\mathbb{C}$ the $4^{th}$ order elasticity tensor, $\boldsymbol{\varepsilon}^p_\mu$ the plastic strain tensor, $f(\boldsymbol{\sigma}, q)$ the yield function and $\mathcal{H}$ the isotropic hardening modulus. In (3), the strain-like and the stress-like conjugated hardening variables are denoted by $\alpha_\mu$ and $q$, respectively. The usual loading/unloading and consistency conditions are added to these equations.

In the following, a necessary condition for the appearance of the localized plastic flow $\lambda = \tilde{\lambda}\, \delta_\Gamma$ is obtained for the plane-strain case. Under this assumption, denoting by $\mathbf{t}$ the unit tangent to $\Gamma_x$, the *localized* dilatancy can be defined as the ratio $\Phi := \dot{\zeta}_n / |\dot{\zeta}_t|$, where $\zeta_n := \boldsymbol{\zeta} \cdot \mathbf{n}$ and $\zeta_t := \boldsymbol{\zeta} \cdot \mathbf{t}$ are the normal and the tangential components of the displacement jump rate, respectively (Fig. 1). From equations (2) and (3), the expression of the stress increment is obtained as:

$$\dot{\boldsymbol{\sigma}} = \underbrace{\mathbb{C}(\dot{\bar{\boldsymbol{\varepsilon}}}_\mu - \dot{\bar{\boldsymbol{\varepsilon}}}^p_\mu)}_{\text{regular distribution}} + \underbrace{\left[\mathbb{C}\Big((\dot{\boldsymbol{\zeta}} \otimes \mathbf{n})^s - \tilde{\lambda}\partial_{\boldsymbol{\sigma}} f\Big)\right] \delta_{\Gamma_x}}_{\text{singular distribution}} \tag{4}$$

Therefore, for the stress to be non singular, it must be:

$$\left[\text{sign}(\dot{\zeta}_t)\, \Phi\, (\mathbf{n} \otimes \mathbf{n}) + (\mathbf{t} \otimes \mathbf{n})^s\right] \dot{\zeta}_t = \tilde{\lambda}\; \partial_{\boldsymbol{\sigma}} f \tag{5}$$

The application of spheric and deviatoric operators on both members of (5) leads to, respectively:

$$\tilde{\lambda}\; \partial_{\boldsymbol{\sigma}} f : \mathbf{I} = \Phi\, |\dot{\zeta}_t| \qquad \tilde{\lambda}\; \|\text{dev}(\partial_{\boldsymbol{\sigma}} f)\| = \sqrt{2\Phi^2/3 + 1/2}\;\; |\dot{\zeta}_t| \tag{6}$$

The combination of these equations gives the localized dilatancy expression:

$$\Phi = \sqrt{\frac{1}{2} \frac{(\partial_{\boldsymbol{\sigma}} f : \mathbf{I})^2}{\partial_{\boldsymbol{\sigma}} f : \partial_{\boldsymbol{\sigma}} f - (\partial_{\boldsymbol{\sigma}} f : \mathbf{I})^2}} \tag{7}$$

Setting $\mathbf{n} = [\cos\theta \;\; \sin\theta \;\; 0]^T$ and $\mathbf{t} = [-\sin\theta \;\; \cos\theta \;\; 0]^T$, with $\theta$ the angle between $\mathbf{n}$ and the major (in plane) principal stress direction $\mathbf{e}_1$ (Fig. 1), the following three scalar equations are obtained from (5) for the case of isotropic yield function $f$:

$$\begin{aligned}
&\text{sign}(\dot{\zeta}_t)\ \Phi \cos^2\theta - \sin\theta\cos\theta = \frac{\text{sign}(\dot{\zeta}_t)\,\Phi}{\partial_{\boldsymbol{\sigma}} f : \mathbf{I}}(\partial_{\boldsymbol{\sigma}} f)_1 \\
&\text{sign}(\dot{\zeta}_t)\ \Phi \sin^2\theta + \sin\theta\cos\theta = \frac{\text{sign}(\dot{\zeta}_t)\,\Phi}{\partial_{\boldsymbol{\sigma}} f : \mathbf{I}}(\partial_{\boldsymbol{\sigma}} f)_2 \\
&\text{sign}(\dot{\zeta}_t)\ \Phi \sin\theta\cos\theta - \frac{1}{2}(\sin^2\theta - \cos^2\theta) = 0
\end{aligned} \tag{8}$$

where $(\partial_{\boldsymbol{\sigma}} f)_1$ and $(\partial_{\boldsymbol{\sigma}} f)_2$ are the in-plane eigenvalues of $\partial_{\boldsymbol{\sigma}} f$. Equation $(8)_3$ and the combination of $(8)_{1,2}$ lead to the expression of the angle $\theta$ and to the localization condition, respectively:

$$\theta = -\frac{1}{2}\arctan\left[\frac{\text{sign}(\dot{\zeta}_t)}{\Phi}\right] \qquad \frac{(\partial_{\boldsymbol{\sigma}} f)_1 - (\partial_{\boldsymbol{\sigma}} f)_2}{\partial_{\boldsymbol{\sigma}} f : \mathbf{I}} = \frac{\sqrt{\Phi^2+1}}{\Phi} \tag{9}$$

The consistency condition implies that both the regular and singular parts of $\dot{f}$ must be nil. The latter condition is satisfied as a consequence of (5). To verify the former condition, the softening law $(3)_2$ must be understood in the distributional sense:

$$\mathcal{H}^{-1}\dot{q} = -\dot{\alpha}_\mu = -\tilde{\lambda}\partial_q f\ \delta_{\Gamma_x} \qquad \Longrightarrow \qquad \mathcal{H}^{-1} = \tilde{\mathcal{H}}^{-1}\ \delta_{\Gamma_x} \tag{10}$$

where $\tilde{\mathcal{H}}$ is the *localized* softening modulus. The following expression, obtained for the localized plastic flow:

$$\tilde{\lambda} = \frac{\partial_{\boldsymbol{\sigma}} f : \dot{\boldsymbol{\sigma}}}{(\partial_q f)^2\ \tilde{\mathcal{H}}} \tag{11}$$

is introduced in (5) and $(6)_1$, leading to the localized softening relation:

$$\dot{\tau}_{\Gamma_x}\ \text{sign}(\dot{\zeta}_t) + \Phi\ \dot{\sigma}_{\Gamma_x} = \left(\frac{\partial_q f\ \Phi}{\partial_{\boldsymbol{\sigma}} f : \mathbf{I}}\right)^2 \tilde{\mathcal{H}}\ |\dot{\zeta}_t| \tag{12}$$

where $\tau_{\Gamma_x} := \mathbf{t}_{\Gamma_x} \cdot \mathbf{t}$ and $\sigma_{\Gamma_x} := \mathbf{t}_{\Gamma_x} \cdot \mathbf{n}$ are the tangential and the normal components of the traction vector $\mathbf{t}_{\Gamma_x} = \boldsymbol{\sigma}\mathbf{n}|_{\Gamma_x}$ on the discontinuity $\Gamma_x$.

**Remark 2.1.** In plane strain conditions, $(9)_2$ and (12) are respectively equivalent to loss of ellipticity of the elastic perfectly plastic tensor and to the localized softening law firstly reported in [15].

## 3 Model problem: associated Drucker-Prager

Results obtained in the previous section can be easily particularized for a dilatant model defined by the following Drucker-Prager yield function:

$$f(\boldsymbol{\sigma}, \alpha_\mu) = \|\mathbf{s}\| + \beta\frac{1}{3}\boldsymbol{\sigma} : \mathbf{1} + \sqrt{\frac{2}{3}}\ q(\alpha_\mu) \tag{13}$$

where $\mathbf{s} = \text{dev}(\boldsymbol{\sigma})$ is the deviatoric part of the stress tensor and the pressure coefficient $\beta$ is a material parameter.

For the considered model, the expressions for the localized dilatancy (7) and for the angle $\theta$ $(9)_1$ read, respectively:

$$\Phi = \frac{\sqrt{2}\,\beta}{2\sqrt{1-2\beta^2/3}} \qquad \theta = -\frac{1}{2}\arctan\left[\text{sign}(\dot{\zeta}_t)\frac{\sqrt{2(1-2\beta^2/3)}}{\beta}\right] \tag{14}$$

Denoting by $\sigma_1, \sigma_2$ the in-plane principal stresses ($\sigma_1 \geq \sigma_2$) and by $s_3$ the out-of-plane principal deviatoric stress, the localization condition $(9)_2$ is satisfied for:

$$\frac{\|\mathbf{s}\|}{r} = \sqrt{\frac{2}{1-\beta^2/6}} \qquad \text{with} \qquad r = \frac{\sigma_1 - \sigma_2}{2} \tag{15}$$

or, equivalently, for $s_3 + \beta/3 = 0$. The post-localization relation (12) between the stress and the displacement jump rate reads in this case:

$$\dot{\tau}_{\Gamma_x}\,\text{sign}(\dot{\zeta}_t) + \Phi\,\dot{\sigma}_{\Gamma_x} = \frac{\tilde{\mathcal{H}}}{3-2\beta^2}|\dot{\zeta}_t| \tag{16}$$

**Remark 3.1**. The *continuum* dilatancy is commonly defined as the ratio: $(\dot{\boldsymbol{\varepsilon}}^p : \mathbf{1}) \,/\, \|\text{dev}(\dot{\boldsymbol{\varepsilon}}^p)\|$. It is well known that in plane strain conditions, the following expression of the Drucker-Prager pressure coefficient

$$\beta = \sqrt{\frac{6\sin^2\varphi}{3+\sin^2\varphi}} \qquad \text{with } \varphi \text{ the friction angle} \tag{17}$$

leads to the same continuum dilatancy prediction of the Mohr-Coulomb model [7]. The substitution of this expression in equations (14) gives: $\Phi = \tan\varphi$ and $\theta = \pm(\pi/4 - \varphi/2)$, as expected. In particular, as a consequence of $\Phi = \tan\varphi$, relation (16) assumes the same form of the localized softening law obtained for the Mohr-Coulomb model in [2].

**Remark 3.2**. As shown in [1], it is consistent to employ two different constitutive laws to characterize the continuum and the localized response, respectively. This approach is followed in [13], where a Drucker-Prager model is adopted for the continuum solid, whereas an independent Mohr-Coulomb law is introduced for the post-localization state. On the contrary, in the present work, it is shown that a Mohr-Coulomb type localized softening law (i.e. eqn. 16) arises naturally from the continuum Drucker-Prager model.

## 4 Enhanced finite element formulation

On the basis of the approach presented in [1,3] for the $J_2$ flow theory, we develop in this Section an enhanced finite element method to solve the localized equations for dilatant elasto-plastic solids reported in Section 2.

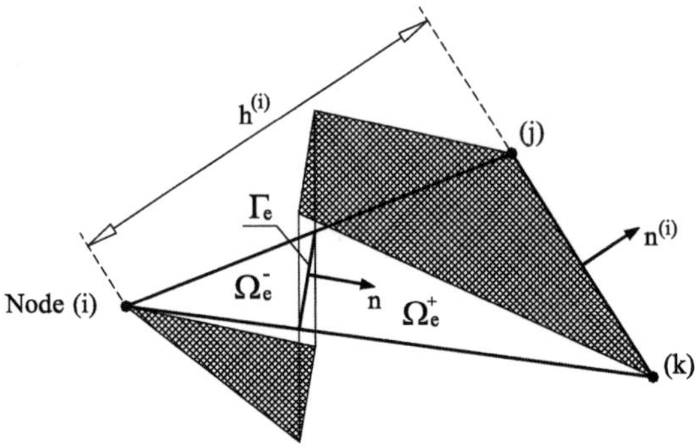


**Fig. 2.** Displacement discontinuous interpolation function $\Psi_{\Gamma e}$

The multi-scale framework considered in previous sections is implemented in the finite element method by identifying the local neighborhood $\Omega_x$ with the element $\Omega_{e,loc} \subset \Omega$ where localization has been detected. In particular, $\Psi_{\Gamma_x}$ in (1) is approximated by the *unregularized* discontinuous function: $\Psi_{\Gamma e}(\mathbf{x}) = H_\Gamma(\mathbf{x}) - N^{(i)}(\mathbf{x})$ where $N^{(i)}(\mathbf{x}) = 1 - (\mathbf{x}^{(i)} - \mathbf{x}) \cdot \mathbf{n}^{(i)}/h^{(i)}$ is the linear shape function associated to node $(i)$ sustaining the discontinuity line $\Gamma_e$ (Fig. 2). The corresponding assumed enhanced strain field is expressed in terms of the element nodal displacements $\mathbf{d}_e$ and of the displacement jump components $\mathbf{z}_e = [\zeta_{n_e} \; ; \; \zeta_{t_e}]^T$:

$$\boldsymbol{\varepsilon}_e = \underbrace{\mathbf{B}_e \mathbf{d}_e}_{\text{conforming}} \quad \underbrace{- \mathbf{G}_e \, \mathbf{z}_e + \mathbf{P}_e \, \mathbf{z}_e \, \delta_\Gamma}_{\text{enhancement}} \tag{18}$$

for the standard strain matrix $\mathbf{B}_e$ and for the enhanced operators: $\mathbf{G}_e = [(\mathbf{n} \otimes \mathbf{n}^{(i)})^s/h^{(i)} \; ; \; (\mathbf{t} \otimes \mathbf{n}^{(i)})^s/h^{(i)}]$, $\mathbf{P}_e = [\mathbf{n} \otimes \mathbf{n} \; ; \; (\mathbf{t} \otimes \mathbf{n})^s]$.

The following finite-element residual equation is obtained from the weak formulation of the continuum equilibrium:

$$\mathbf{r}_d = \mathbf{f}^{ext} - \mathop{\mathsf{A}}_{e=1}^{n_{elem}} \int_{\Omega_e} \mathbf{B}_e{}^T \boldsymbol{\sigma} \, d\Omega = \mathbf{0} \tag{19}$$

for the nodal external force vector $\mathbf{f}^{ext}$. The symbol $\mathsf{\mathbf{A}}$ denotes the assembly of all the $n_{elem}$ element contributions. The stresses $\boldsymbol{\sigma}$ are obtained through the relation $(3)_1$ in terms of the regular part $\bar{\boldsymbol{\varepsilon}}_e := \mathbf{B}_e \mathbf{d}_e - \mathbf{G}_e \, \mathbf{z}_e$ of the enhanced strain field (18).

We denote by $A_e$ the area of the localized element and by $l_{\Gamma_e}$ the discontinuity length within $\Omega_{e,loc}$. The weak formulation of the equilibrium on the discontinuity can be expressed in the form:

$$\int_{\Omega_{e,loc}} \delta\boldsymbol{\varepsilon}_{e,enh} : \boldsymbol{\sigma} \, d\Omega = 0 \qquad \text{with} \quad \delta\boldsymbol{\varepsilon}_{e,enh} = \left( \delta_\Gamma - \frac{l_{\Gamma_e}}{A_e} \right) \mathbf{P}_e \, \delta\mathbf{z}_e \tag{20}$$

The assumption of a piece-wise constant approximation of the enhanced and traction fields leads to the following finite-element residual equation:

$$\hat{\mathbf{r}}_{enh} = -\frac{1}{A_e}\int_{\Omega_{e,loc}} \mathbf{P}_e^T \boldsymbol{\sigma}\, d\Omega \;+\; \mathbf{t}_\Gamma = 0 \tag{21}$$

The relation existing between the two displacement jump components, i.e. the localized dilatancy expression (7), allows the reduction of the enhanced modes to the single $\zeta_{t_e}$ by application of the projection operator $\mathbf{D} = [\,\mathrm{sign}(\dot{\zeta}_{t_e})\Phi\,;\,1\,]$ to weak equation (21):

$$r_{enh} = -\frac{1}{A_e}\mathbf{D}\int_{\Omega_{e,loc}} \mathbf{P}_e^T \boldsymbol{\sigma}\, d\Omega \;+\; s_\Gamma = 0 \tag{22}$$

with $s_\Gamma := \mathbf{D}\mathbf{t}_\Gamma = \tau_\Gamma + \mathrm{sign}(\dot{\zeta}_{t_e})\,\Phi\,\sigma_\Gamma$.

In the generic time step $[t_n, t_{n+1}]$, nonlinear equations (19) and (22) are solved through a Newton-Raphson iterative procedure. Return-mapping algorithms are employed in the integration of the post-localization stress-displacement relation (29) and of the bulk elasto-plastic model.

Introducing the notation $\Delta(\cdot) = (\cdot)_{n+1}^{(k+1)} - (\cdot)_{n+1}^{(k)}$, with $(k)$ the iteration number, and assuming a constant localized dilatancy, the linearization of equation (19) can be written:

$$\mathbf{r}_d^{(k)} - \mathop{\mathsf{A}}_{e=1}^{n_{elem}} \left(\mathbf{K}_e\,\Delta\mathbf{d}_e - \mathcal{L}_{G_e}\,\Delta\zeta_{t_e}\right) = \mathbf{0} \tag{23}$$

for the standard element stiffness matrix:

$$\mathbf{K}_e = \int_{\Omega_e} \mathbf{B}_e{}^T \mathbb{C}\,\mathbf{B}_e\, d\Omega \tag{24}$$

and the enhanced operator:

$$\mathcal{L}_{G_e} = \int_{\Omega_{e,loc}} \mathbf{B}_e{}^T \mathbb{C}\,\mathcal{G}_e\, d\Omega \qquad \text{with} \quad \mathcal{G}_e = \mathbf{G}_e\mathbf{D}^T \tag{25}$$

Linearization of the local equation (22) leads to:

$$r_{enh}^{(k)} = \mathcal{L}_{P_e}^T\,\Delta\mathbf{d}_e - \mathcal{H}_e\,\Delta\zeta_{t_e} \tag{26}$$

for the enhanced operators:

$$\mathcal{L}_{P_e} = \frac{1}{A_e}\int_{\Omega_{e,loc}} \mathbf{B}_e{}^T \mathbb{C}\,\mathcal{P}_e\, d\Omega \qquad \mathcal{H}_e = \frac{1}{A_e}\int_{\Omega_{e,loc}} \mathcal{P}_e^T \mathbb{C}\,\mathcal{G}_e\, d\Omega + \frac{ds_\Gamma}{d\zeta_{t_e}} \tag{27}$$

with $\mathcal{P}_e = \mathbf{P}_e\mathbf{D}^T$.

In summary, the proposed finite element formulation implies the numerical solution of the system of equations (23) and (26). Solution of (26) gives local enhanced mode $\zeta_{t,e}$ which is eliminated through static condensation in equation (23). However, similarly to what proposed in [14], we adopt the *staggered* static condensation procedure described in the following.

- At the element level, the enhanced parameter $\zeta_{t,e}$ is obtained by solution of non linear equation (22) (i.e., by driving $r_{enh}$ to zero) for given displacements $\mathbf{d}_e^{(k)}$. Therefore, this local Newton iteration involves the scalar equation (26) with $\Delta\mathbf{d}_e = 0$.
- At the global level, the following equation is solved:

$$\mathbf{r}_d^{(k)} - \mathop{\mathsf{A}}_{e=1}^{n_{elem}} \left[ \left( \mathbf{K}_e - \mathcal{L}_{G_e} \mathcal{H}_e^{-1} \mathcal{L}_{P_e}^T \right) \Delta\mathbf{d}_e \right] = 0 \tag{28}$$

obtained substituting in (23) the solution of the local problem (22)($r_{enh}^{(k)} = 0 \Rightarrow \Delta\zeta_{t_e} = \mathcal{H}_e^{-1} \mathcal{L}_{P_e}^T \, \Delta\mathbf{d}_e$) .

In (22), $s_\Gamma = \hat{s}_\Gamma(\zeta_{t_e})$ is obtained by integrating the given localized softening law (12) on the discontinuity $\Gamma_e$, that is:

$$\dot{\hat{s}}_\Gamma(\zeta_{t_e}) = \left( \frac{\partial_q f \ \Phi}{\partial_{\sigma} f : \mathbf{I}} \right)^2 \tilde{\mathcal{H}} \ \dot{\zeta}_{t_e} \tag{29}$$

for the general case with constant localized dilatancy, and:

$$\dot{\hat{s}}_\Gamma(\zeta_{t_e}) = \frac{\tilde{\mathcal{H}}}{3 - 2\beta^2} \dot{\zeta}_{t_e} \tag{30}$$

for the particular case of the Drucker-Prager model (13). From this post-localization relation, also the term $ds_\Gamma / d\zeta_{t,e}$ of the enhanced tangent operator $(27)_2$ is obtained.

**Remark 4.1.** The proposed enhanced finite element method is implemented in a mixed "P2/P1" 6-noded triangle element (quadratic interpolation of the displacements and linear interpolation of the volumetric strain and stress [9]).

**Remark 4.2.** The strong discontinuity propagates through the mesh, without the spatial discretization knowing a priori its location. Following [8], the implemented procedure employs the localization condition (15) and the discontinuity direction $(14)_2$.

**Remark 4.3.** The enhanced test functions considered in [13] depend on the particular localized constitutive relation. On the contrary, in the present work, we adopt test functions of the general form $(20)_2$.

## 5 Numerical examples

To evaluate the performance of the proposed enhanced finite-element formulation, the numerical simulation of a plane-strain compression test is considered in this Section. Sample sizes are in the range typically adopted in these laboratory tests ($4 \times 14$ cm). Smooth basis are assumed and no confining pressure is applied on the unrestrained lateral boundaries. A vertical displacement is

**Table 1.** Material parameters considered in the numerical simulations

| | | | |
|---|---|---|---|
| Young modulus | $E$ | 20000 | $kPa$ |
| Poisson coefficient | $\nu$ | 0.25 | |
| pressure coefficient | $\beta$ | 0.56 | |
| initial yield stress | $\sigma_y$ | 63.41 | $kPa$ |
| localized softening modulus | $\tilde{\mathcal{H}}$ | −50000 | $kPa/m$ |
| continuum softening modulus | $\mathcal{H}$ | 0 | $kPa$ |

imposed to top base and bottom base is vertically restrained. We adopt the Drucker-Prager model described in Section 3 with a linear localized softening law. The overconsolidated clay parameters reported in Table 1 are considered. Note that, according to (17), the adopted $\beta$ is obtained for $\varphi = 24°$.

The test is performed with two different discretizations of the sample: $2 \times 4 \times 14$ (Fig. 3a,c) and $2 \times 8 \times 28$ (Fig. 3b,d) elements. To trigger localization, a 1 % initial yield stress imperfection is introduced in a lateral element, distant 3.0 cm from top base. To evaluate the effects of mesh orientation, computations are repeated for the "imperfect" element located at the right side (Fig. 3a,b) and at the left side (Fig. 3c,d) of the sample.

The solutions obtained with the two meshes and for the two different positions of the "imperfect" element are practically coincident (Fig. 4a). Hence, the strong-discontinuity formulation leads to a solution independent from both the size and the alignment of elements in the mesh. The horizontal displacement distributions shown in Fig. 3 point out the localized elements and demonstrate the ability of the enhanced method in capturing localization. After localization, a super-convergence is attained (e.g.: $\|\mathbf{r}_d^{(k)}\|/\|\mathbf{r}_d^{(0)}\| = 1.0 \cdot 10^0;\ 3.6 \cdot 10^{-1};\ 6.5 \cdot 10^{-13}$).

The test is also performed by means of standard finite-element method with continuum strain-softening ($\mathcal{H} = -200$ kPa). The obtained deformed configurations show that the standard formulation has difficulties in capturing localization (Fig. 5). In Figure 4b, a significant mesh-dependency of the standard F.E. solution is observed, in terms of the effects of both the size and the alignment of elements. Due to convergence problems, it is very difficult to perform the test after the attainment of the peak value of vertical reaction.

## 6 Concluding remarks

We have presented an analysis of strong discontinuities in dilatant elastoplastic solids and an enhanced finite-element method incorporating unregularized approximations of such discontinuous fields. The results of representative numerical simulations show the objectivity of the proposed enhanced method and its ability in capturing localization, in contrast with the response obtained with the standard finite-element formulation.

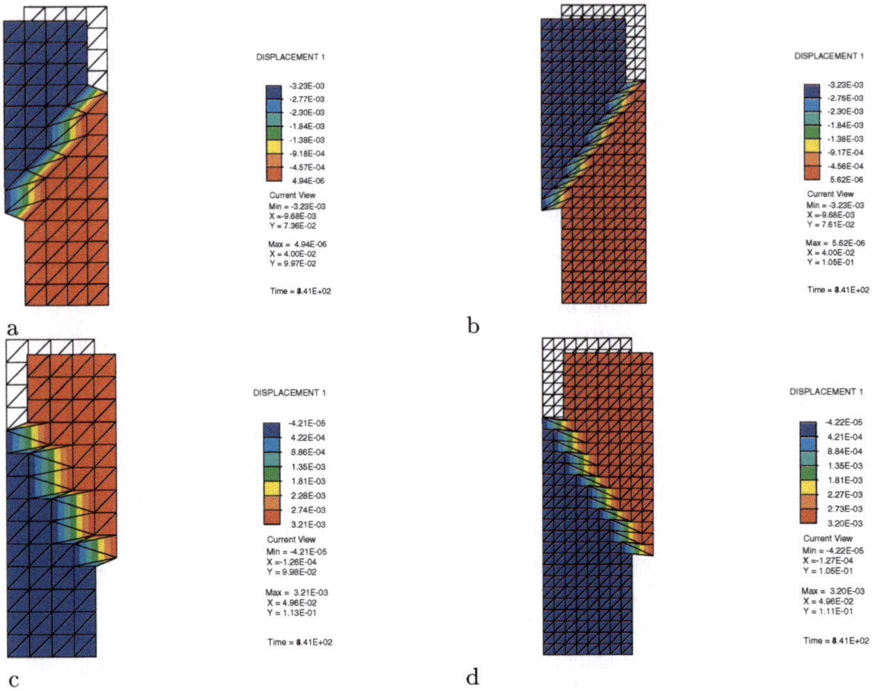


**Fig. 3.** Strong-discontinuity formulation. Deformed meshes and horizontal displacement distributions for the 2×4×14 (a, c) and 2×8×28 (b, d) discretizations. Imperfect element at the right (a, b) and at the left side (c, d) of the sample (top base vertical displacement: $-2.21 \cdot 10^{-3}$m; amplification factor: 3)

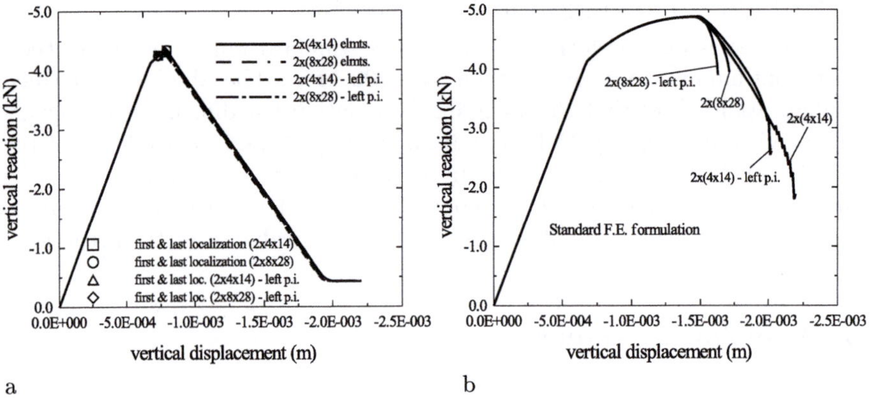


**Fig. 4.** Vertical reaction vs. vertical imposed displacement. Results corresponding to 2×4×14 and 2×8×28 meshes for the cases of imperfect element at the right and at the left side of the sample. a) Strong-discontinuity FE formulation. b) Standard FE formulation

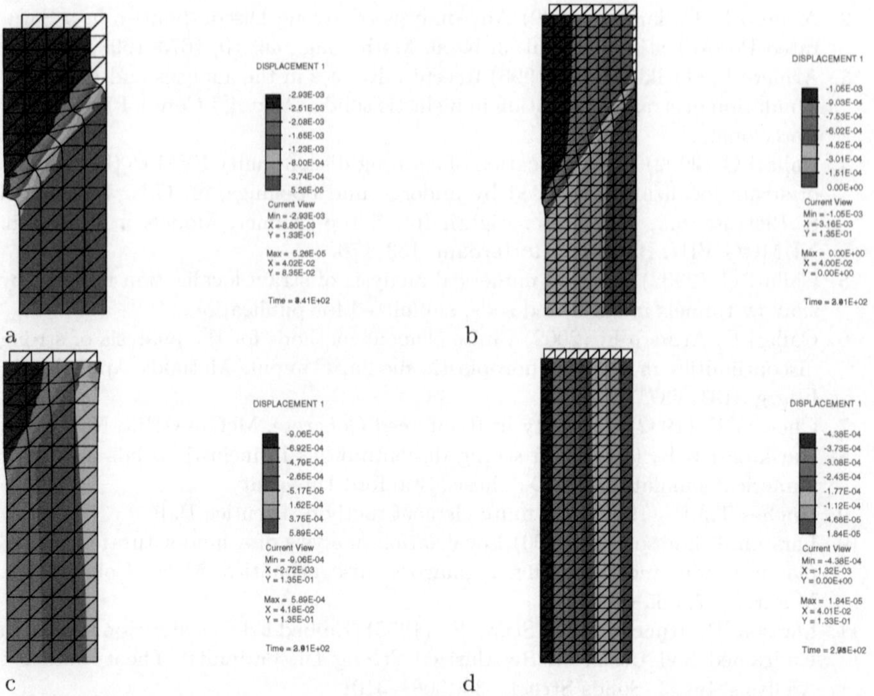


**Fig. 5.** Standard FE formulation. Deformed meshes and horizontal displacement distributions for the 2×4×14 (a, c) and 2×8×28 (b, d) discretizations. Imperfect element at the right (a, b) and at the left side (c, d) of the sample. Top base vertical displacement: a) $-2.21 \cdot 10^{-3}$m; b,c) $-1.70 \cdot 10^{-3}$m; d) $-1.46 \cdot 10^{-3}$m. Amplification factor: 3

**Acknowledgments.** The finite element method proposed in this work has been implemented in the general code FEAP, courtesy of Professor R.L. Taylor at UC Berkeley. Prof. F. Armero of UC Berkeley and F. Raggi of University "Tor Vergata" are gratefully acknowledged for their helpful suggestions. The research was partially supported by the "Progetto giovani ricercatori 2000" of the University of Rome "Tor Vergata" and by a CNR (Italian National Research Council) scholarship (Short Term Mobility Program, 1999).

*The present paper was submitted on February 2001 and posted on Internet in the same month. The notation and the bibliography have been revised on October 2002.*

## References

1. Armero F. (1999) Large-scale modeling of localized dissipative mechanisms in a local continuum: applications to the numerical simulation of strain localization in rate-dependent inelastic solids, Mech. Cohes.-Frict. Mater., 4, 101–131.

2. Armero F, Callari C. (1999) An Analysis of Strong Discontinuities in a Saturated Poro-Plastic Solid, Int. J. Num. Meth. Eng., 46, 10, 1673–1698.
3. Armero F, Garikipati K. (1995) Recent advances in the analysis and numerical simulation of strain localization in inelastic solids, Proc. $4^{th}$ Comp. Plast. Conf., Barcelona.
4. Callari C. (2002) The application of a strong-discontinuity FEM to the analysis of strain localization induced by underground openings, in: G.N. Pande and S. Pietruszczak, eds., Proc. Eighth Int. Symp. Numer. Models in Geomech. NUMOG VIII. Balkema, Rotterdam, 163–170.
5. Callari C. (2002) Coupled numerical analysis of strain localization induced by shallow tunnels in saturated soils, submitted for publication.
6. Callari C, Armero F. (2002) Finite element methods for the analysis of strong discontinuities in coupled poroplastic media, Comput. Methods Appl. Mech. Engrg., 191, 4371–4400.
7. Chen W.F. (1982) Plasticity in Reinforced Concrete, McGraw-Hill, New York.
8. Garikipati K.R. (1996) On strong discontinuities in inelastic solids and their numerical simulation, Ph.D. Thesis, Stanford University.
9. Hughes T.J.R. (1987) The finite element method, Prentice Hall.
10. Larsson J, Larsson R. (2000) Localization analysis of a fluid-saturated elastoplastic porous medium using regularized discontinuities, Mech. Cohes.-Frict. Mater., 5, 7, 565–582.
11. Larsson R, Runesson K, Sture S. (1996) Embedded Localization Band in Undrained Soil Based on Regularized Strong Discontinuity: Theory and FE Analysis, Int. J. Solids Struct., 33, 3081–3101.
12. Oliver J. (1996) Modelling Strong Discontinuities in Solid Mechanics via Strain Softening Constitutive Equations. Part 1: Fundamentals. Part 2: Numerical Simulation, Int. J. Num. Meth. Eng., 39, 3575–3623.
13. Regueiro R.A., Borja R.I.(2001) Plane strain finite element analysis of pressure-sensitive plasticity with strong discontinuity, Int. J. of Solids Struct., 38, 3647–3672.
14. Simo J.C., Armero F, Taylor R.L.(1993) Improved versions of assumed enhanced strain tri-linear elements for 3D finite deformation problems, Comp. Meth. Appl. Mech. Eng., 110, 359–386.
15. Simo J.C., Oliver J, Armero F.(1993) An analysis of strong discontinuities induced by strain-softening in rate-indipendent inelastic solids, Comp. Mech., 12, 277–296.

# Elasto-plastic Modelling of the Behaviour of Granular Soil-structure Interfaces

Vincenzo De Gennaro, Roger Frank

CERMES
Ecole Nationale des Ponts et Chaussées
77455 Marne–la-Vallée, Cedex 2, France

**Abstract.** It is well known that interfaces usually play a major role in the definition of the mechanical behaviour of engineering structures having interactions with the soil. In this paper the general framework of an elasto-plastic constitutive model developed on purpose for describing the interface behaviour is presented. The model is based on a Mohr-Coulomb failure criterion, including deviatoric hardening/softening, phase transformation state (compaction and dilatancy) and critical state. The choice of the constitutive parameters and their identification is first discussed. The predictions of the model are then presented and compared with available experimental data from various interface tests between sand and metal plates. The results of the numerical analyses emphasise the key role played by the volumetric behaviour of the interface (compaction and dilatancy), linked in some cases with the change in the normal stress acting on the structure surface and, consequently, controlling the shear resistance at the interface.

## 1 Introduction

When materials of different stiffness are in contact they undergo inevitably intense strain localization phenomena, stress concentration points are due to the inherently high stiffness heterogeneity between the adjacent media. This is the case of the classical granular soil-structure interaction problems. The serviceability of a wide range of engineering structures involving interfacing between structural elements and soil is highly dependent on the behaviour of the shear band forming close to the structure surface. This shear band, commonly referred to as the "interface", acts as a transition layer between the stiffer structural element and the softer soil medium. When strained in the tangential direction, it often exhibits a volumetric behaviour leading to compaction as well as dilatancy in the normal direction.
The major effects of the interface layer on the general response of engineering structures involving interactions with soil are well known. In order to perform a pertinent analysis of such problems, the experimental study of the essential features of the behaviour of interface and the formulation of reliable constitutive laws is mandatory.

Experimental data from modified direct shear boxes (Potyondy 1961) have been often employed to investigate the behaviour of granular soil-structure interface and to enlighten the factors controlling the interaction between the two media. This laboratory equipment has been the object of several modifications, especially related to the possibility of reducing its principal deficiencies (Wernick 1978, Desai et al. 1985, Boulon 1988, Hoteit 1990, Tabucanon and Airey 1992, Tabucanon et al. 1995). In this respect it is worth noting that a wide range of increasingly sophisticated testing devices are available to date. Among others, let us quote: the pull out apparatus (Brummund and Leonards 1973), the ring torsion apparatus (Yoshimi and Kishida 1981, Boulon 1988, Boutrif 1993), the simple shear apparatus (Uesugi and Kishida, 1986) and more recently the cyclic three-dimensional simple shear interface apparatus (Fakharian and Evgin 1996), and the ring simple shear apparatus (Lerat et al. 1997). One basic phenomenological aspect is however common to all the apparatuses for interface testing: the observable kinematic state variables required for the description of the interface behaviour are the relative displacements, rather than deformations as for a standard continuum medium.

A number of models have been presented in recent past for the description of the behaviour of interfaces. Most of them mainly concentrate on the behaviour of interfaces between granular soils and metal.
In a first class of models, on the basis of results concerning the modelling of rock joints, the material of the interface is supposed to be linear elastic, or non-linear with a stress-displacement relationship of hyperbolic type in the normal and tangential directions (Desai et al. 1984). Hardening is not taken into account and the condition of failure, in terms of shear strength, is brought back to the traditional relation dictated by the Mohr-Coulomb failure criterion. At failure the normal stiffness remains constant and the dilatancy of interface in the direction normal to the direction of shearing is neglected. More complex models, incorporating the concepts of dilatancy, compaction and damage, have been also proposed within this theoretical framework (Patton 1966, Goodman et al. 1968, Ladanyi and Archambault 1970 and 1972, Goodman and Dubois 1972).
A second class refers to the theory of elasto-plasticity. Formulations relate as well to the assumptions of a perfect plasticity as those of an hardening material (inter alia: Ghaboussi et al. 1973, Desai et al. 1984, Boulon and Nova 1990, Gens et al. 1990, Desai and Fishman 1991, Sharma and Desai 1992, Day and Potts 1994, Shahrour and Rezaie 1997). More sophisticated models have been proposed for the analysis of particular conditions such as cyclic loading (Aubry et al. 1990, Desai and Ma 1992) and softening (Desai and Ma 1992, Leong and Randolph 1994). Along with the new definition of the kinematic state variables, stresses are related to displacements, and the theoretical structure of these models remains identical to that of the majority of the elasto-plastic models suggested for the description of soil behaviour. Finally, also the use of polar continua (Tejchman and Tejchman 1990) and of directionally dependent rate type laws (Boulon 1991, Rouainia et al. 1992, Boulon et al. 1995) has been attempted.

Given the results of previous studies already published in the literature, the research presented herein addresses the possibility to model the behaviour of an interface subjected to monotonic loading using a purposely developed elasto-plastic constitutive model. The model has direct links with the Mohr-Coulomb failure criterion, includes deviatoric hardening/softening behaviour, and integrates the phase transformation state (concept of compaction-dilatancy) and the critical state at large tangential displacements (parallel to the direction of shear). The choice of the constitutive parameters and their identification is discussed. The potential of such an approach to describe the response of a real interface is assessed through the comparisons of model predictions with various laboratory interface test results performed under various conditions (constant normal stress, constant normal stiffness, constant volume).

## 2 Basic concepts of interface behaviour

Geometrically, the interface is defined as a thin zone of soil having constant thickness, which is formed close to the surface of the structure. Specific studies seem to converge towards a quantification of the thickness "t" of the interface related to the average diameter ($D_{50}$) of the grains; its determination can be only done by direct visualisation (Yoshimi and Kishida 1981, Uesugi and Kishida 1986, Uesugi et al. 1988, Hoteit 1990, Boutrif 1993, Hassan 1995, Tejchman and Wu 1995, Zong-Ze et al. 1995, Lerat 1996). The thickness of this zone is strictly related to the roughness of the surface. The latter is usually referred to the average grain dimension by means of the normalised roughness coefficient $R_n = R_{max}/D_{50}$ , $R_{max}$ being the maximum gauge depth of the surface (Yoshimi and Kishida 1981). For a rough surface, available experimental results suggest values ranging between 5 and 10 $D_{50}$.
If we focus our attention on the case of a two-dimensional problem (plane strain or axysimmetry) the kinematic state variables are the normal relative displacement $u_n$ and the tangential relative displacement $u_t$ , and the corresponding stress variables are the normal stress $\sigma_n$ and shear stress $\tau$. As already mentioned, the concept of deformation for an interface test is rather random. If one supposes a mechanism of deformation in simple shear the distortion $\gamma = u_t /t$ and the normal deformation $\epsilon_n = u_n /t$ depend on the thickness of the interface. The definition of this thickness is however variable, being related to the average size of the grains; it involves consequently an unquestionable scale effect. As a matter of fact the only measurable kinematic quantities are the displacements (normal or parallel to the interface layer).
Interface mechanical behaviour shows strong similarities with the mechanical behaviour of granular soils tested in the triaxial apparatus. This particular aspect has been skilfully discussed by Boulon and Nova (1990); it is beyond the scope of this paper to re-analyse such a typical results of interface tests. It can be stated that interfaces are both pressure-sensitive and density-

dependent, in the sense that: increasing levels of normal stress will induce higher values of maximum shear resistance; depending upon the initial density of the soil the interface either contracts or dilates. Compaction during loading is typical of loose interfaces and holds until failure is reached, for such a density state material hardens until failure. On the other hand, dense interfaces experience during loading low initial compaction and significant dilatancy at yield. As it is well known, such a density state is often associated to softening behaviour after the point of peak shear resistance. At large tangential displacements, the volumetric behaviour is characterised either by a normal displacement rate equal to zero, according to the general findings of the critical state theory, or by a second phase of compaction associated to wearing phenomena (Williams 1980, quoted in Leong and Randolph 1991, Plytas 1985, Hoteit 1990, Lerat 1996). All the elements which have been just introduced will have a considerable role in the formulation of the interface constitutive model.

## 3 Formulation of the elasto-plastic stress-displacement law

As already mentioned, the model has been formulated to describe the behaviour of the granular soil-structure interface subjected to monotonic loading within the framework of the theory of elasto-plasticity. Thereafter, the following assumptions will be adopted:

1. The interface represents the remoulded zone of soil adjacent to the surface of the structural element. In granular materials, in absence of a direct measurement, the thickness "t" of this layer can be estimated as being a multiple of the average diameter of the grains ($D_{50}$).
2. The formulation will be limited to the two-dimensional case (case of problems in plane strain or having a symmetry of revolution). Following the general approach of incremental elasto-plasticity the kinematic state variables are the normal relative displacement of the interface, $u_n$ , and the tangential relative displacement of the interface, $u_t$ . The associated stress variables are the normal stress, $\sigma_n$, and the shear stress, $\tau$, parallel to the direction of the interface, $\underline{\Sigma} = (\sigma_n, \tau)^T$ is the stress vector and $\underline{\mathrm{U}} = (\mathrm{u}_n , \mathrm{u}_t )^T$ is the relative displacement vector.
3. Stresses and displacements are taken as positive in compression, and considered homogeneous within the interface layer. The soil is assumed to be dry, so that analysis can be performed in terms of total as well as effective stresses.

In the following the general formulation briefly outlined, based on experimental evidences, will be applied to the specific problem of the soil-structure interface.

### 3.1 Elastic response

The elastic behaviour of the interface is given by the following linear relation:

$$\underline{d\Sigma} = K^e \underline{dU^e} \quad ; \quad K^e = \begin{bmatrix} K_n & 0 \\ 0 & K_T \end{bmatrix} \tag{1}$$

The $\mathbf{K}^e$ matrix contains the stiffness of the interface in the normal ($K_n$) and tangential ($K_t$) directions, which might be considered dependent on the initial normal stress and the initial density. For instance, possible expressions are:

$$K_n = k_n \sigma_{ni}^N \quad ; \quad K_t = k_t \sigma_{ni}^N \tag{2}$$

where $k_n$ , $k_t$ and N are three constitutive parameters of the model, $\sigma_{ni}$ is the initial normal stress acting on the interface. Note that the elastic response does not consider a coupling between normal and shear behaviour of the interface.

### 3.2 Yield criterion and hardening/softening law

Experimental evidences show that during interface tests under various conditions (i.e. constant normal stress, constant volume, constant normal stiffness) the relationship between normal and shear stresses at failure is well approximated by the Mohr-Coulomb model. Therefore, neglecting cohesion, the failure condition is given by:

$$\tau_f = \tan \delta_f \sigma_n = \mu_f \sigma_n \tag{3}$$

where $\delta_f$ is the friction angle of the interface at failure and $\mu_f = \tan \delta_f$ is the coefficient of friction. In perfect analogy with granular soils, interfaces experience during loading (shear) progressive hardening behaviour and significant reduction in shear stiffness until failure is reached. Such a hardening phase could either tends to a plateau (loose interfaces) or evolves into strain-softening and then lean towards a final plateau corresponding to the critical state (dense interfaces).

A schematic sketch of such a response, in terms of evolution of the stress ratio $\mu = \tau/\sigma_n$ versus tangential displacements $u_t$, is presented in Fig. 1(a). Continuous hardening, typical of loose interfaces, leads to a progressive mobilisation of the coefficient of friction $\mu$, which increases until reaching the limit value $\mu_f$ at failure. The latter coincides also with the residual value $\mu_r$ at large tangential displacements. The hardening/softening response of the

interface, on the other hand, corresponds to increasing values of the coefficient of friction $\mu$, which grows towards $\mu_f$, and then decreases to the asymptotic residual value $\mu_r$ corresponding to the coefficient of friction at constant volume (critical state). In the plane $\sigma_n$-$\tau$ such an evolution of the stress state during hardening, in agreement with the frictional failure criterion (3), corresponds to a counter-clockwise rotation of the locus $\tau = \mu\sigma_n$ , starting from the initial position coinciding with the axis $\tau = 0$, until the failure line ($\mu = \mu_f$) defined by equation (3), as shown in Fig. 1(b). The softening phase is well represented by a clockwise rotation of the same locus until the residual state ($\mu = \mu_r$). Consequently the yield mechanisms, whatever hardening or softening, are obtained by a generalisation of the Mohr-Coulomb failure criterion, as suggested by the deviatoric hardening concept (Poorooshasb and Pietruszczak 1985, Sadrnejad and Pande 1989, Pietruszczak and Niu 1993, Bencheikh 1991).
The rotation of the yield surface in the $\sigma_n$-$\tau$ stress plane due to deviatoric hardening/softening is assumed to be a function of the plastic relative displacements generated within the interface layer during shearing. The yield surface F is governed by the standard Mohr-Coulomb failure criterion; the equation adopted is:

$$F = \tau - \mu(\xi^p)\,\sigma_n = 0 \tag{4}$$

In this equation $\mu(\xi^p)$ is the hardening/softening function, giving the evolution of the mobilised friction coefficient during loading.
Based on the schematic diagram shown in Fig. 1(a), in the hardening regime, the function $\mu(\xi^p)$ is assumed of hyperbolic type; it can be defined explicitly as:

$$\mu(\xi^p) = \mu_o + (\mu_f - \mu_o)\frac{\xi^p}{A\left(\frac{\sigma_{ni}}{p_o}\right)t + \xi^p} \quad \textit{and} \quad u_t < u_t^f \tag{5}$$

$u_t^f$ being the tangential displacement of the dense interface at failure (i.e. peak in the diagram shown in Fig. 1(a)). In equation (5) $\mu_f$ is the coefficient of friction at failure, $\mu_o = \tan\delta_o$ is the friction coefficient delimiting the initial elastic region ($\delta_o$ is the initial friction angle). Inside this wedge-shaped region (Fig. 1(b)) only reversible relative displacements are permitted, given by inverting relation (1). The parameter t is the thickness of the interface layer, A is a parameter of the model governing the shape of the hardening fonction, $\sigma_{ni}$ is the initial normal stress and $p_o$ is a reference pressure. The introduction of t into equation (5) allows to consider an internal length parameter for the interface. The ratio $\frac{\sigma_{ni}}{p_o}$ , as it will be discussed later, is introduced in order to take into account the effect of the normal stress $\sigma_n$ on the shape of the curves of mobilised shear stress. The variable $\xi^p$ is the hardening parameter; it is given by the following relation:

$$\xi^p = \frac{1}{2}\sqrt{(u_n^p)^2 + (u_t^p)^2} \tag{6}$$

where $u_n^p$ *and* $u_t^p$ are the plastic normal and tangential relative displacements cumulated during loading. In the softening regime ($u_t > u_t^f$) the evolution of the yield surface is governed by the following equation:

$$\mu(\xi^p) = \mu_r + (\mu_f - \mu_r)\operatorname{sec}h\left[\frac{A_o}{t}\left(\xi^p - \xi_f^p\right)\right] \quad and \quad u_t > u_t^f \tag{7}$$

where two more parameters are added, namely: $A_o$ and $\mu_r$. The former controls the shape of the softening function (7), the latter defines the residual friction coefficient of the interface at large tangential displacements, i.e. at critical state. The variable $\xi_f^p$ corresponds to the value of the hardening parameter at failure (i.e. when $\mu = \mu_f$).
It has to be mentioned that softening behaviour has been considered in this work only with a view of modelling simplicity. There is no doubt that many factors are at the origins of this phenomenon, for instance shear banding instability (De Gennaro and Pande 1998) or other various aspects that have been fully described elsewhere and will be not be investigated here (Boulon and Jarzebowski 1991, Desai and Ma 1992, Leong and Randolph 1994).

### 3.3 Plastic potential and flow rule

Plastic relative displacements appear if the condition F = 0 and the condition of consistency dF = 0 are simultaneously fulfilled. Their magnitude and direction are given by the definition of the plastic potential function Q and of the flow rule. The pertinent choice of the plastic potential function, Q, is essential to reproduce the typical volumetric behaviour observed during interface tests, which is assumed here to be only plastic. Consequently, its expression should be suitable for the description of the following phenomena:
1) the presence of an initial compaction ($du_n>0$) in a test with constant normal stress, or correspondingly a reduction of the stress $\sigma_n$, normal to the interface layer, in a test with imposed constant volume or constant normal stiffness;
2) the existence of a threshold corresponding to the phase transformation from compaction ($du_n>0$) to dilatancy ($du_n<0$), which corresponds, in a test with imposed constant volume or constant stiffness, to an increase of the normal stress $\sigma_n$;
3) the stabilisation of the normal relative displacement $u_n$, or the normal

stress $\sigma_n$, on an asymptotic value for large relative tangential displacements of the interface (i.e. $\mathrm{d}u_n = 0$ or $\mathrm{d}\sigma_n = 0$). This corresponds to the salient features of the critical state theory, where constant volume conditions are assumed at failure.

Compaction (i.e. $\mathrm{d}u_n > 0$ or $\mathrm{d}\sigma_n < 0$) due to grain crushing at very large tangential displacements and high normal stresses is not considered in the present version of the model.

In order to describe the above-mentioned phenomena (points (1) to (3)), non-associated elasto-plasticity has been assumed. The plastic potential function is:

$$Q = \tau + \mu_c \sigma_n \ln \frac{\sigma_n}{\sigma_o} \tag{8}$$

where the parameter $\mu_c$ is the slope of the phase transformation line $\tau = \mu_c \sigma_n$ and $\sigma_o$ is defined by the current state of stress acting on the interface. Taking into account equation (8), $\sigma_o$ can be expressed as:

$$\sigma_o = \sigma_n \exp\left(\frac{\mu}{\mu_c}\right) \tag{9}$$

The plastic potential function introduced is analogous to those proposed for soils by Poorooshasb and Pietruszczak (1985), Sadrnejad and Pande (1989) and Pietruszczak and Niu (1992 and 1993). Typical plots of the function Q in the $\sigma_n$-$\tau$ plane are presented in Fig. 2. This function describes a series of continuous surfaces which expand progressively during deformations, changing in size but not in shape (i.e. with the same derivatives on points having the same coefficient of mobilised friction). Differentiating equation (8) and rearranging using equation (9), the plastic relative displacement increments are:

$$\begin{bmatrix} du_n^p \\ du_t^p \end{bmatrix} = d\lambda \begin{bmatrix} \mu_c - \mu \\ 1 \end{bmatrix} \tag{10}$$

Thus compaction holds if $\mu < \mu_c$ ( $du_n^p > 0$) and dilatancy takes place if $\mu > \mu_c$ ($du_n^p < 0$ ). The phase transformation state corresponds to the condition $\mu = \mu_c$, as a result $du_n^p = 0$ (points A, B, C in Fig. 2). The dilatancy of the interface, D, is given by the following relation:

$$D = \frac{du_n^p}{du_t^p} = \mu_c - \mu \tag{11}$$

It is worth noting that dilatancy at large relative tangential displacements is constant and tends towards the asymptotic value D = $\mu_c$ - $\mu_r$ ($\mu_r = \mu_f$ in a loose interface). Therefore, residual zero rate of volumetric deformation of the interface at large tangential displacements, in agreement with experimental observations (i.e. critical state), cannot be reproduced if equation (8) is considered.
In order to introduce this further important feature of the volumetric behaviour of the interface, a modified form of the potential function is now proposed.

### 3.4 Behaviour at critical state

One of the assumptions of the critical state theory is that at large tangential displacements associated with failure the rate of volumetric strain is zero. Considering this in terms of evolution of the void ratio, e, of the sand, this means that at critical state the void ratio is constant, equal to the critical void ratio $e_{cr}$ introduced by Casagrande in 1936. At large strains, the void ratios of loose and dense granular structures have the same asymptotic value, namely the critical void ratio $e_{cr}$. Since the early works of Taylor (1948) and Roscoe et al. (1958) quoted in Atkinson and Bransby (1978), there are a number of experimental findings obtained with various equipments that corroborate this assumption.
We examine now the conditions that allow to introduce the required zero dilatancy rate at the interface associated to the critical state. From equation (11) such a requirement is fulfilled if

$$u_t \to \infty \Rightarrow D = (\mu_c - \mu) \to 0 \tag{12}$$

In addition, at critical state, we also know that:

$$u_t \to \infty \Rightarrow \mu \to \mu_r = tan\delta_r = constant \tag{13}$$

and $\mu_r = \mu_f$ in the case of continuous hardening until critical state condition. From equations (12) and (13) it can be deduced that:

$$u_t \to \infty \Rightarrow D = (\mu_c - \mu_r) \to 0 \tag{14}$$

and condition (14) is checked if

$$\mu_c \to \mu_r \tag{15}$$

Consequently, the condition of zero dilatancy at the interface at critical state can be obtained if the coefficient $\mu_c$ , the stress ratio at phase transformation state, increases after phase transformation towards the final value $\mu_r$ (i.e. towards the stress ratio at critical state). Condition (15) corresponds to an evolution of the size of both compaction and dilation regions.
The mechanism is described in Fig. 3, on a typical stress path involving initial compaction, phase transformation and dilatancy (e.g. interface tests at constant volume), without taking into account for the time being softening behaviour ($\mu_f = \mu_r$). Following the indicated path, at point C, for a friction coefficient $\mu = \mu_{co}$ there is a transition from compaction ($\mu < \mu_{co}$) to dilatancy ($\mu > \mu_{co}$). At a generic point M, in the dilation region, two different potential surfaces are plotted. The first (Q'$_o$) is given by the family of functions in equation (8), admitting $\mu_{co}$ constant. The second ($Q_1$) has been plotted admitting an increase of the coefficient $\mu_c$ from the initial value $\mu_{co}$ to $\mu_{c1}$ according to condition (15). By comparing the direction of the plastic displacement vector at point M, it can be argued that dilatancy at the interface can be substantially reduced assuming $Q_1$ as the current potential function. The minimum value of dilatancy is obtained when the phase transformation surface $\tau = \mu_{c1}\sigma_n$ is close to the failure locus $\tau = \mu_f\sigma_n$, so that $D = (\mu_{c1} - \mu_f) \to 0$ (condition (14)). This is associated to a progressive shrinkage of the dilation region. If softening is expected, after reaching the maximum value $\mu_f$, the coefficient of friction reduces to $\mu_r$ while the dilation region slightly expands (Fig. 1b). In this case the minimum value of dilatancy is obtained when the phase transformation surface is close to the residual locus $\tau = \mu_r\sigma_n$ , so that D = $(\mu_{c1} - \mu_r) \to 0$ (condition (14) with $\mu_f = \mu_r$). Such a mechanism has a direct physical interpretation. Yield of dense interface layers due to shearing causes plastic dilation, resulting in an increase of the voids in the sample (shrinkage of the dilation region). On the other hand, shearing on loose interface layers causes an opposite effect, leading to an overall compaction of the interface and a reduction of the void ratio (shrinkage of the compaction region).
It is noticed that, within the theoretical framework of the model, this is equivalent to a change in shape of the surfaces given by the plastic potential function Q (equation (8)), which is implicitly taken into account by allowing for an evolution of parameter $\mu_c$. As suggested by the experimental results, in order to introduce such a mechanism in the model formulation, it has been chosen to consider the void ratio as the internal state variable of the potential function Q, enabling to represent the volumetric plastic behaviour of the sandy interface. It is then admitted that the parameter $\mu_c$ in equation (8) is not a constant during deformation but is related to the evolution of the void ratio e of the sample, so that:

$$\mu_c = \mu_c(e) \tag{16}$$

It is assumed that the void ratio e is a function of the plastic relative displacements normal ($du_n^p$ ) and tangential ($du_t^p$ ) to the interface via the hardening parameter $\xi^p$ (equation (6)). The evolution rule of the void ratio reads:

$$e(\xi^p) = e_{cr} + (e_{co} - e_{cr}) \exp[-\frac{B}{t} (\xi^p - \xi^o)] \tag{17}$$

when $\mu \geq \mu_{co}$ et $e_{co} > e_{cr}$ (i.e. loose interface), or

$$e(\xi^p) = e_{co} + (e_{cr} - e_{co}) \tan h \left[ \frac{B}{t} (\xi^p - \xi^o) \right] \tag{18}$$

when $\mu \geq \mu_{co}$ et $e_{co} < e_{cr}$ (i.e. dense interface).

In equations (17) and (18) t is the thickness of the interface, $\mu_{co}$ defines the extension of the initial contracting region of the interface (i.e. the slope of the phase transformation line), B is a constitutive parameter controlling the shape of the evolution rules (17) or (18), $\xi^o$ is the cumulated plastic displacement at phase transformation (when $\mu = \mu_{co}$), $e_{co}$ and $e_{cr}$ are the void ratio of the interface at phase transformation and the critical void ratio (asymptotic value of e when $u_t \to \infty$), respectively. Typical plots of equations (17) and (18) are presented in Fig. 4. Starting from the stress state corresponding to the phase transformation ($\mu = \mu_{co}$), dilatancy is given by equation (11), where now $\mu_c = \mu_c(e)$. In order to reproduce the condition of shearing at constant volume at large tangential displacements (i.e. critical state), the following expression is proposed for the parameter $\mu_c(e)$:

$$\mu_c(e) = \mu_{co} + (\mu - \mu_{co})\ D(e) \tag{19}$$

The rate of dilatancy at the interface is controlled by the function D(e), defined as

$$D(e) = 1 - \sec h \left[ C \left( \frac{\sigma_{ni}}{p_o} \right) Arc \tan h \left( \frac{e}{e_{cr}} \right) \right] \tag{20}$$

In equations (19) and (20) C is a constitutive parameter of the model, $\mu$ is the coefficient of friction mobilised during shearing, $\sigma_{ni}$ is the initial normal stress and $p_o$ is a reference pressure. The ratio $\frac{ni}{p_o}$ is introduced in order to take into account the observed reduction in dilatancy at higher normal stresses. Obviously, at phase transformation $\mu = \mu_{co}$, and equation (19) is

thus written $\mu_c(e) = \mu_{co}$. Figure 5 shows the evolution of $\mu$, $\mu_c$ and dilatancy $D = \mu_c - \mu$ versus the hardening parameter $\xi^p$ (cumulated plastic displacements).
In the case of a dense interface ($e_{co} < e_{cr}$) dilation is predominant (superscript D, Fig. 5), and at large cumulated plastic displacements the value of e increases asymptotically towards $e_{cr}$ (equation (18)). The coefficient of friction $\mu$ first increases towards $\mu_f$ and then decreases towards $\mu_r$ (equations (5) and (7), hardening/softening response), while D(e) increases towards unity (equation (20)). Therefore equation (19) at critical state can be written:

$$\mu_c(e) = \mu \tag{21}$$

and on the basis of equation (11) D = 0.
In the case of a loose interface ($e_{co} > e_{cr}$), compaction appears from the beginning of the loading process (superscript L, Fig. 5) and the value of e decreases towards the asymptotic value $e_{cr}$ (equation (17)). The coefficient of friction $\mu$ increases progressively until the limit value $\mu_r$ (equation (5)) and, as in the case of a dense interface, D(e) increases towards the unity (equation (20)). Again, based on a similar discussion, condition (21) is fulfilled at large tangential displacements and D = 0. It must be noted that for a loose interface the value of
$\mu_{co}^L$ is close to the limit value $\mu_r$. This leads to a reduced rate of mobilised dilatancy during shear and an overall compressive volumetric behaviour of the interface.

## 4 Identification of constitutive parameters of the model

As formulated the model allows for progressive refinements in order to account for various features of the interface behaviour. In the complete version it requires thirteen parameters; these are: $k_n$ , $k_t$ , $N$, $\mu_o$ , $\mu_f$ , $u_t^f$ , $\mu_r$ , $\mu_{co}$ , $A$, $A_o$ , $B$, $C$, $e_{cr}$. For their determination one can use results of interface tests at constant normal stress or constant volume and, at the occasion, the results of oedometric or isotropic compression tests. In the following, the methodology of parameters determination is briefly examined.

### 4.1 Elastic parameters: $k_n$ , $k_t$ and N

On the basis of relation (1), the normal and tangential stiffness of the interface depend on the values of parameters $k_n$ , $k_t$ and N. These relations are similar to existing empirical relationships proposed for Young's modulus. Both $k_n$ and $k_t$ are introduced in order to take into account the dependence of the stiffness on density, whereas the coefficient N allows to consider a possible dependence on the normal stress. If N = 0, the dependence on the normal stress is neglected. The parameter $k_t$ is found knowing the initial slope of the curve of mobilised shear stress ($\tau$) versus tangential displacement ($u_t$) of the interface. Such a slope, corresponding to the ratio $\frac{\tau}{u_t}$ , is the value of the tangential stiffness $K_t$ . The determination of the initial slope is rather delicate and often imprecise. In practice, one can either consider a secant stiffness or, if available, the slope of the unloading branch of the shear stress versus tangential displacement curve. For instance, in the first case, it can be assumed that:

$$K_t = \frac{\tau_f}{u_{to}} \tag{22}$$

where $u_{to}$ is the tangential displacement corresponding to the mobilisation of half of the maximum shear resistance $\tau_f$ and $\sigma_{ni}$ is the value of the initial normal stress acting on the interface. From equation (1) it can be deduced that:

$$k_t = \frac{K_t}{\sigma_{ni}^N} \tag{23}$$

where the value of parameter N is obtained through curve fitting of the values of the initial tangential stiffness plotted against the applied initial normal stresses $\sigma_{ni}$.The parameter $k_n$ depends on the compressibility of the interface. Its determination is thus possible from oedometric compression tests on the interface material. Obviously, this type of tests is not easy to realise, because of the difficulty to reconstitute samples having small thickness at a given initial density. Other factors, such as the density state, the normal stress level, the type of test (constant normal stress, constant volume) and the type of apparatus also affect the determination of interface normal stiffness. When appropriate, it is proposed to deduce $k_n$ from standard eodometric compression tests. Volume changes during oedometric compression are:

$$\frac{dv}{v_i} = \frac{de}{1+e_i} = \frac{dh}{h_i} \tag{24}$$

where $v_i$ is initial specific volume, $h_i$ is the initial height of the sample and $e_i$ is the initial void ratio. In oedometric compression:

$$dv = -C_s^{'} \frac{d\sigma_v}{\sigma_v} \tag{25}$$

where C'$_s$ is the swelling index giving the slope of the unloading branch of the vertical (normal) stress ($\sigma_v$) versus the specific volume changes curve. Then, from equations (24) and (25):

$$d\sigma_v = -\frac{\sigma_v\,(1+e_i)}{C_s' h_i}dh \tag{26}$$

Equation (26) allows to identify the normal stiffness K$_n$ ; we obtain:

$$K_n = \frac{\sigma_v\,(1+e_i)}{C_s' h_i} \tag{27}$$

where $\sigma_v$ is calculated at the beginning of unloading. It is then possible to evaluate k$_n$

$$k_n = \frac{K_n}{\sigma_{ni}^N} \tag{28}$$

### 4.2 Plastic parameters: $\mu_o$ , $\mu_f$ ,$u_t^f$ , $\mu_r$ and $\mu_{co}$

The value of the parameter $\mu_o$ = tan $\delta_o$ defines the extension of the initial elastic region of the interface, where $\delta_o$ is the minimum friction angle mobilised at the interface (Fig. 1).
The parameters $\mu_f$ and $\mu_r$ are the coefficients of friction of the interface at failure and at residual state, respectively. Usually $\mu_f = \mu_r$ in loose interfaces. They can be determined easily from interface tests at different constant normal stresses by linear interpolation of points ($\sigma_n$ , $\tau$) at failure and at residual state. Their values correspond to the slope of the Mohr-Coulomb failure line ($\mu_f$) and the ultimate linear envelope at residual state ($\mu_r$), as shown in Fig. 1b. The value of$u_t^f$ defines the position of peak shear resistance in the diagram u$_t$-$\tau$ (Fig. 1(a)). It should be mentioned that, due to the amplification of dilation characteristics of granular materials at low imposed normal stress levels, such a determination could lead to a small underestimation of $\mu_f$.The parameter $\mu_{co}$ is the coefficient of friction at the points of phase transformation (Fig. 2). At these points, in an interface test at constant normal stress, the rate of the normal relative displacement is zero (Fig. 5). The value of $\mu_{co}$ can be identified much easier from an interface test at constant volume. In fact, it can be determined at the point where the stress path in $\sigma_n$-$\tau$ plane changes of direction (point C in Fig. 3).

### 4.3 Parameters: A, $A_o$, B, C and $e_{cr}$

The parameter A appears in the expression of the hyperbolic hardening function (5). It allows for the control of the shape of the curve of mobilised friction at the interface. Differentiating equation (4), it is possible to write:

$$\frac{d\tau}{d\xi^p} = \frac{\sigma_{ni}\,(\mu_f - \mu_o)\,\beta}{(\beta + \xi^p)^2} \quad ; \quad \beta = A\frac{\sigma_{ni}}{p_o}t \tag{29}$$

Where $t$ is the thickness of the interface layer, assumed to be a multiple of the average grain diameter $D_{50}$. The value of A can be obtained by imposing the continuity of the value of the initial slope of the experimental curve ($u_t$, $\tau$) with the value of the analytical tangent $\frac{d\tau}{d\xi^p}$ given in equation (29), calculated when $\xi^p = 0$. The initial slope of the experimental curve ($u_t$, $\tau$) is the tangential stiffness $K_t$, so that

$$\frac{\sigma_{ni}\,(\mu_f - \mu_o)}{\beta} = K_t \tag{30}$$

As a result, the parameter A i

$$A = \frac{(\mu_f - \mu_o)\,p_o}{K_t t} \tag{31}$$

The parameters $A_o$ , B and C can be deduced following a procedure of optimisation by successive adjustments. $A_o$ controls the shape of the softening function in equation (7), B controls the shape of the evolution rule of the void ratio (equation (17) or (18)) and C controls directly the shape of the evolution rule of the parameter $\mu_c$ (equation (19)) and indirectly the rate of mobilisation of dilatancy at the interface (equation (11)).
The remaining parameter $e_{cr}$ can be easily obtained if tests characterising the critical state of sand are available, or can be roughly estimated from interface tests at constant normal stress. In this latter case, from the definition of the void ratio, we can write:

$$e_{cr} = \frac{V_o\left(1 + \frac{u_{nf}}{h_o}\right)\gamma_s}{W_s} - 1 \tag{32}$$

where $V_o$ and $h_o$ are respectively the initial volume and the initial height of the sample, $\gamma_s$ is the unit weight of grains and $W_s$ is the weight of the dry sample. The value of $u_{nf}$ in equation (32) is the asymptotic value of the normal relative displacement of the interface at large tangential displacement (Fig. 5).

# 5 Validation of the model

In order to check the capability of the model in predicting the interface behaviour, the simulations of interface tests at constant normal stress and constant volume are now presented.

## 5.1 Test at constant normal stress

Let us first consider the results of the interface tests at constant normal stress between a Fontainebleau sand and a rough metal plate ($R_n = 1$) performed in a modified direct shear box. The physical characteristics of this sand are shown in Table 1. The sand samples tested have an initial density index $I_D$ equal to 0.46 ($e_o = 0.753$) and an imposed normal stress ranging between 25 kPa and 400 kPa. Following the methodology described in section 4 the numerical computations have been performed assuming one set of parameters, namely the parameters determined from the results of tests BS3 at $\sigma_{ni} = 100$ kPa (Table 1). The other tests have been simulated just changing the initial conditions in terms of imposed normal stress $\sigma_n$. The results of simulations, are presented in Fig. 6. The comparison between the experimental results and simulations is satisfactory with regard to the evolution of the shear stress ($\tau$) versus the tangential displacement of the interface ($u_t$). It can be noted, as previously announced, that the value of the maximum shear stress predicted for the tests carried out at low levels of normal stress is slightly underestimated. This divergence between simulations and the experiments translates the effect of $\sigma_n$ on the value of $\mu_f$ (i.e. non-linearity of Mohr-Coulomb failure line in the $\sigma_n$-$\tau$ plane for the low values of $\sigma_n$). With regard to the evolution of the normal relative displacement ($u_n$) versus the tangential relative displacement ($u_t$), it can be observed that the transition from compaction to dilatancy reproduced by the model is more gradual than the observed experimental response. It should be mentioned, however, that the shapes of the experimental curves are rather singular. Indeed, where dilatancy appears (i.e. at low normal stress levels), the experimental results also show an extended initial phase where the change of normal relative displacement is zero (test BS1 with $\sigma_n = 25$ kPa), or a very prolonged phase where the rate of normal relative displacement is zero (test BS2 with $\sigma_n = 50$ kPa). Although it is possible to take into account the effect of $\sigma_n$ on the mobilisation of dilatancy at the interface (effect of the ratio $\frac{ni}{p_o}$ in equation (20)), the dilatancy predicted by the model is sligthly underestimated. In order to validate the model in the case of strong dilatancy at the interface we have examined the results of two interface tests presented by Tabucanon and Airey (1992). The tests were performed in a modified direct shear box on samples of siliceous Sydney sand. This sand has physical characteristics similar to Fontainebleau sand (Table 3). The tests were carried out at constant normal stress $\sigma_n =$ 150 kPa; the authors considered two different densities, namely $I_D = 0.15$

($e_o$ = 0.790, test A) and $I_D$ = 0,96 ($e_o$ = 0.580, test B). The complete set of parameters used for the numerical simulations has been obtained from the experimental results of test A; they are presented in Table 4. For the value of $e_{cr}$, due to the lack of available experimental results, we have taken the value obtained for Fontainebleau sand. This choice seems to be licit based on the similarity between the granulometric characteristics of the two sands. Since the results of isotropic (or oedometric) compression tests on Sydney sand were not available, we have assumed $k_n = 2\ k_t$. The comparison between the experimental results and simulations are presented in Fig. 7. The predictions are again satisfactory, particularly for the volumetric deformations. As a concluding remark, the capability of the model to capture the softening phase present in test B is emphasised.

### 5.2 Test at constant volume

In order to check the ability of the model on different stress paths, the results of interface tests at constant volume between Fontainebleau sand and rough metal plate are now analysed. These tests have been carried out on a ring simple shear apparatus (Lerat et al. 1997). Three test results corresponding to initial external radial pressure of 100 kPa (test CS1), 200 kPa (test CS2) and 400 kPa (test CS3) are simulated. The sand samples tested have an initial density index $I_D$ of about 0.49 ($e_o$ = 0.743). The numerical computations have been performed assuming the set of parameters given in Table 5, determined from the results of test CS1 at $\sigma_{ni}$ = 100 kPa. It can be seen that the computed responses match well the observed experimental results (Fig. 8). With regard to test CS3 a divergence between the predicted value of the shear stress at failure and the one obtained in the test can be observed. It is believed that the restrained contractancy due to grain crushing (local reduction of $\sigma_n$) for the higher level of normal (radial) stress acting on the interface could be at the origin of this difference. This aspect is not captured by the proposed model and could form the object of a further enhanced version.

## 6 Conclusions

This work addresses the description of the mechanical behaviour of granular soil-structure interfaces. To this purpose, the formulation of an elasto-plastic model able to describe the main features of the behaviour of the interface is proposed. The model allows for the description of the basic aspects identified in the interface tests, such as: hardening/softening mechanical response, phase transformation and critical state. It also includes the effect of the normal stress level on the mobilised dilatancy at the interface.

As formulated the model allows for progressive refinements in order to account for various features of the interface behaviour. The resulting small set

of parameters, which are thirteen in the complete version, is easy to found and parameters have a direct meaning. A methodology for their determination has been briefly outlined.
The validation of the proposed approach has been carried out on the experimental results of interface tests achieved by means of a modified direct shear box and of a ring simple shear apparatus. The model provides satisfactory predictions of the behaviour of the interface, for tests at imposed constant normal stress as well as for tests at imposed constant volume. However, the proposed model could be enhanced to account for damage (grain crushing) and for the effect of the normal stress level on some mechanical characteristics of granular materials (for example, friction angle at failure).

# 7 Tables and figures

| $\gamma_{d\,\min}(kN/m^3)$ | $\gamma_{d\,\max}(kN/m^3)$ | $\gamma_s(kN/m^3)$ | $e_{\max}$ | $e_{\min}$ | $D_{50}(mm)$ | Cu |
|---|---|---|---|---|---|---|
| 13.8 | 17.2 | 26.7 | 0.94 | 0.54 | 0.23 | 1.78 |

**Table 1.** Physical characteristics of Fontainebleau sand

| $k_n(m^{-1})$ | $k_t(m^{-1})$ | $\mu_0$ | $\mu_f$ | $\mu_r$ | $\mu_{c0}$ | A | $A_0$ | B(m) | C | $e_{cr}$ | $u_t^f(mm)$ | t(mm) |
|---|---|---|---|---|---|---|---|---|---|---|---|---|
| $0.86\times10^5$ | $0.37\times10^5$ | 0.27 | 0.7 | 0.7 | 0.46 | 0.0004 | - | 1.5 | 1 | 0.84 | - | 2 |

**Table 2.** Constitutive parameters used for the numerical study of interface tests between Fontainebleau sand and rough metal plate ($R_n = 1$) carried out on a modified direct shear box

| $\gamma_{d\,\min}(kN/m^3)$ | $\gamma_{d\,\max}(kN/m^3)$ | $\gamma_s(kN/m^3)$ | $e_{\max}$ | $e_{\min}$ | $D_{50}(mm)$ | Cu |
|---|---|---|---|---|---|---|
| - | - | 26.5 | 0.84 | 0.57 | 0.3 | 2 |

**Table 3.** Physical characteristicc of Sidney sand (Tabucanon and Airey 1992)

| $k_n(m^{-1})$ | $k_t(m^{-1})$ | $\mu_0$ | $\mu_f$ | $\mu_r$ | $\mu_{c0}$ | A | $A_0$ | B(m) | C | $e_{cr}$ | $u_t^f(mm)$ | t(mm) |
|---|---|---|---|---|---|---|---|---|---|---|---|---|
| $0.6\times10^6$ | $0.3\times10^6$ | 0 | 0.7 | 0.7 | 0.57 | 0.00008 | 20 | 0.021 | 15 | 0.84 | 0.26 | 3 |

**Table 4.** Constitutive parameters used for the numerical study of interface test between Sydney sand and rough metal plate ($R_n = 0.25$) carried out on a modified direct shear box

| $k_n(m^{-1})$ | $k_t(m^{-1})$ | $\mu_0$ | $\mu_f$ | $\mu_r$ | $\mu_{c0}$ | A | $A_0$ | B(m) | C | $e_{cr}$ | $u_t^f(mm)$ | t(mm) |
|---|---|---|---|---|---|---|---|---|---|---|---|---|
| $0.24\times10^5$ | $0.14\times10^5$ | 0 | 0.45 | 0.45 | 0.37 | 0.0003 | - | 0.04 | 1.2 | 0.84 | - | 2 |

**Table 5.** Constitutive parameters used for the numerical study of interface test between Fontainebleau sand and rough metal plate ($R_n = 0.25$) carried out on a ring simple shear box

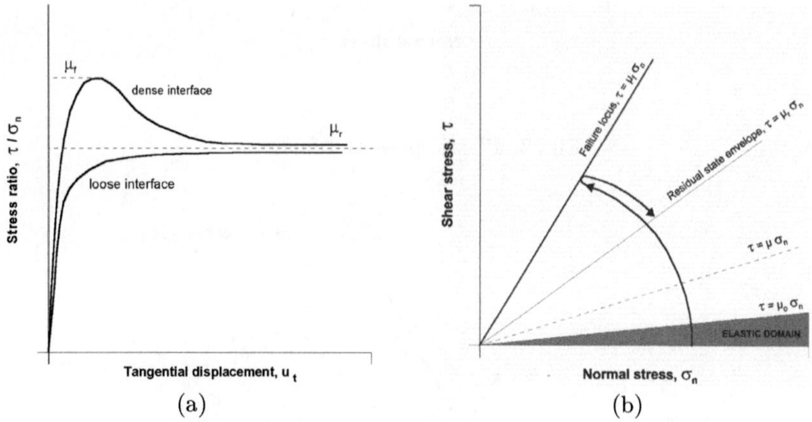


**Fig. 1.** Yield and failure of the interface: (a) typical mechanical response of interface during loading; (b) deviatoric hardening/softening mechanisms of the yield surface

## References


1. Atkinson J.H. and Bransby P.L. 1978. *The mechanics of soils. An introduction to critical state soil mechanics.* McGraw-Hill University Series in Civil Engineering, McGraw-Hill Book Company (UK) Limited.
2. Aubry D., Modaressi A. and Modaressi H. 1990. A consitutive model fro cyclic behavior of interfaces with variable dilatancy. *Computers and Geotechnics,* 9: 47-58.
3. Bencheikh B. 1991. Interaction sol-structure: modélisation et résolution numérique. *Thèse de Doctorat de l'Université des Sciences et Techniques de Lille,* Lille, France.

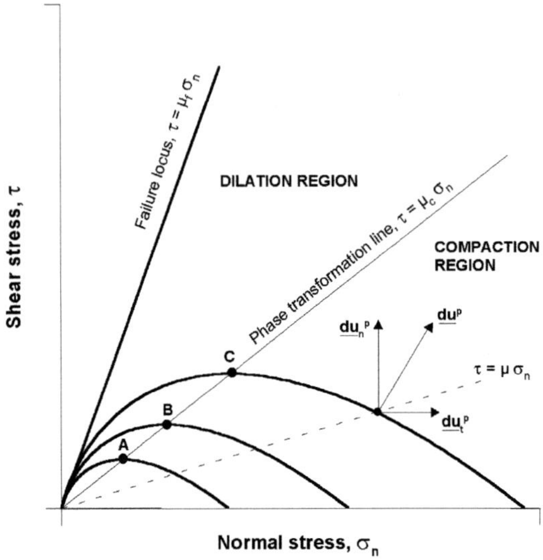


**Fig. 2.** Plastic potential function

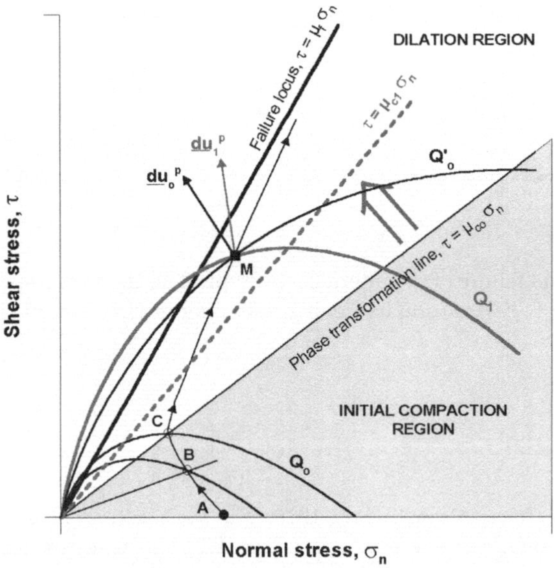


**Fig. 3.** Evolution of the plastic potential function in the $\sigma_n$-$\tau$ stress plane

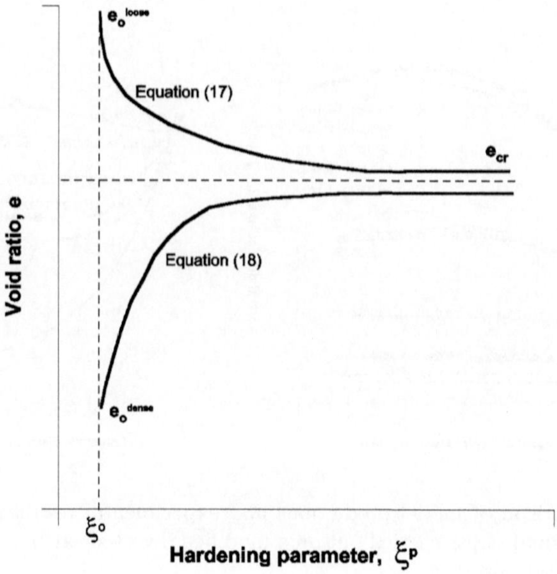


**Fig. 4.** Evolution of the void ratio versus parameter $\xi^p$

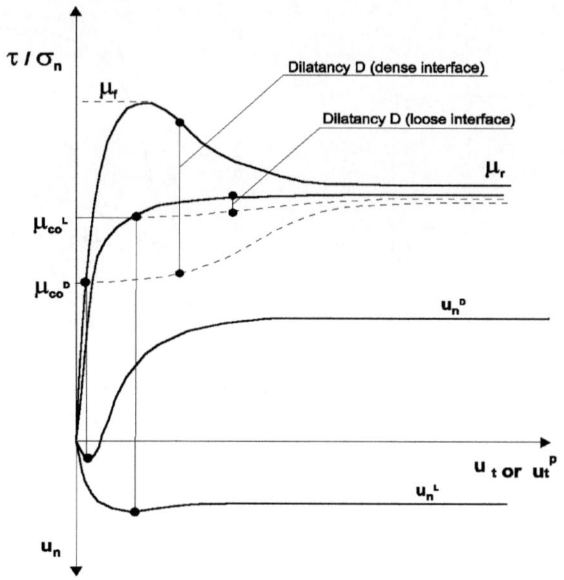


**Fig. 5.** Evolution of $\mu$, $\mu_c$ and D versus parameter $\xi^p$

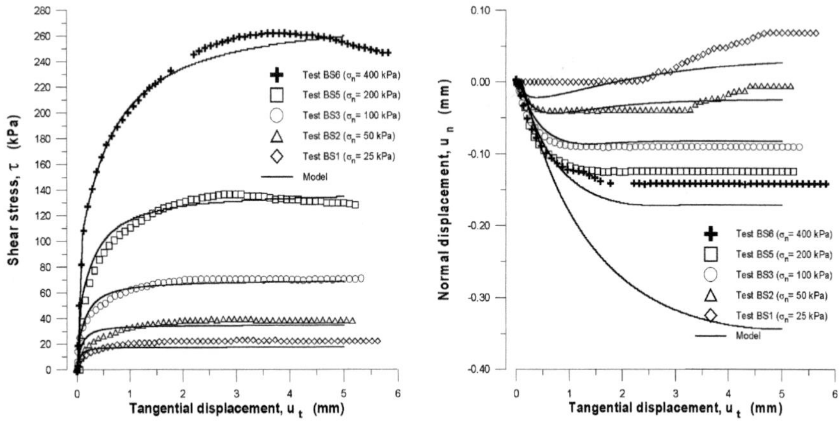


**Fig. 6.** Comparison of model predictions and experimental results: interface tests at constant normal stress carried out on a modified direct shear box (Fontainebleau sand-rough metal plate)

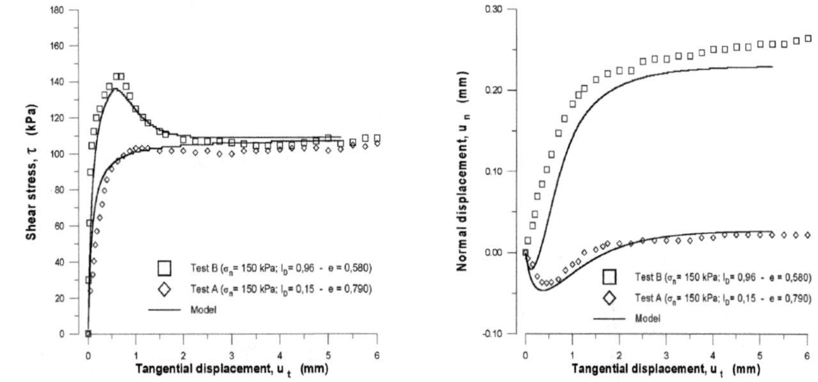


**Fig. 7.** Comparison of model predictions and experimental results: interface tests at constant normal stress carried out on a modified direct shear box (Sydney sand-rough metal plate, experimental data from Tabucanon and Airey 1992

4. Boulon M. 1988. Contribution à la mécanique des interfaces sol-structures. Application au frottement latéral des pieux. *Mémoire pour l'habilitation à diriger des recherches, Université Joseph Fourier – Grenoble I*, Grenoble, France.
5. Boulon M., Hoteit N. and Marchina P. 1988. A complete constitutive law for soil structure interfaces. *Numerical Methods in Geomechanics*, Innsbruck, Swoboda (eds), Balkema, Rotterdam: 311-316.
6. Boulon M. and Nova R. 1990. Modelling of soil-structure interface behaviour, a comparison between elasto-plastic and rate type laws. *Computers and Geotechnics,* 9: 21-46.

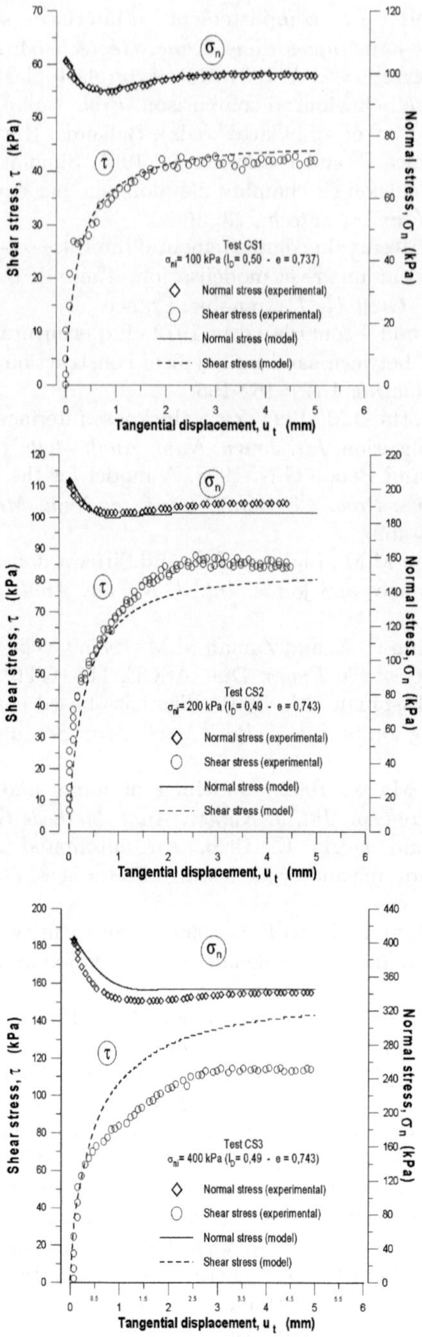


**Fig. 8.** Comparison of model predictions and experimental results: interface tests at constant volume carried out on a ring simple shear box (Fontainebleau sand-rough metal plate)

7. Boulon M. 1991. Le comportement d'interface sol-structure: aspects expérimentaux et numériques. *Rev. Franç. Géotech.* 54: 27-37.
8. Boulon M. and Jarzebowski A. 1991. Rate-type elastoplastic approaches for soil-structure interface behaviour: a comparison. *Proc. Comp. Methods and Adv. in Geomech.*, Beer, Booker and Carter (eds), Balkema, Rotterdam: 305-310.
9. Boulon M., Garnica P. and Eissautier M. 1995. Simulation numérique 3D du frottement sol-inclusion en chambre d'étalonnage par équations intégrales aux frontières. *Rev. Franç. Géotech.*, 73: 36-52.
10. Boutrif A. 1993. Mesure du comportement d'interface sol-structure à la boîte de cisaillement direct annulaire et modélisation. *Thèse de Doctorat de l'Université Joseph Fourier – Grenoble I*, Grenoble, France.
11. Brumund W.F. and Leonards G.A. 1973. Experimental study of static and dynamic friction between sand and typical construction materials. *Journal of Testing and Evaluation* 1 (2): 162-165.
12. Day R.A. and Potts D.M. 1994. Zero thickness interface elements. Numerical stability and application. *Int. Journ. Num. Anal. Meth. Geomech.*, 18: 689-708.
13. De Gennaro V. and Pande G.N. 1998. A model for the behaviour of the soil-structure interface. *Proc. 4th Europ. Conf. on Num. Meth. in Geotech. Eng.*, Udine, Italy: 453-462.
14. Desai C.S., Zaman M.M., Lightner J.G. and Sirirwardane H.J. 1984. Thin-layer element for interfaces and joints. *Int. J. Numer. Anal. Methods Geomech.*, 8: 19-43.
15. Desai C.S., Drumm C. E. and Zaman M.M. 1985. Cyclic testing and modelling of interfaces. *J. Geotech. Engrg. Div.*, ASCE, 111 (GT6): 793-815.
16. Desai C.S. and Fishman K.L. 1991. Plasticity-based constitutive model with associated testing joints. *Int. J. Rock Mech. Min. Sci. and Geomech. Abstr.* 28 (1): 15-26.
17. Desai C.S. and Ma Y. 1992. Modelling of joints and interfaces using the disturbed-state concept. *Int. J. Numer. Anal. Methods Geomech.*, 16: 623-653.
18. Fakharian K. and Evgin E. 1996. An automated apparatus for three-dimensional monotonic and cyclic testing of interfaces. *Geot. Testing J.* ASTM, 19 (1): 22-31.
19. Gens A., Carol I. and Alonso E.E. 1990. A constitutive model for rock joints: formulation and numerical implementation. *Computers and Geotechnics* 9: 3-20.
20. Ghaboussi J., Wilson E.L. and Isenberg J. 1973. Finite element for rock joints and interfaces. *J. of Soils Mech. and Found. Div.*, ASCE, 99 (SM10): 833-848.
21. Goodman R.E., Taylor R.L. and Brekke T.L. 1968. A model for the mechanics of jointed rocks. *J. of Soils Mech. and Found. Div.*, ASCE, 94 (SM3): 637-659.
22. Goodman R.E. and Dubois J. 1972. Duplication of dilatancy in analysis of jointed rocks. *J. of Soils Mech. and Found. Div.*, ASCE, 98 (SM4): 399-422.
23. Hassan A.H. 1995. Etude expérimentale et numérique du comportement local et global d'une interface sol granulaire-structure. *Thèse de Doctorat de l'Université Joseph Fourier – Grenoble I*, Grenoble, France.
24. Hoteit N. 1990. Contribution à l'étude du comportement d'interface sable-inclusion et application au frottement apparent. *Thèse de Doctorat de l'Institut National Polytechnique de Grenoble,* Grenoble, France.
25. Ladanyi B. and Archambault G. 1970. Simulation of shear behaviour of a jointed rock mass. *Proc. 11th Int. Symposium on Rock Mech.*, AIME, New York: 249-260.

26. Ladanyi B. and Archambault G. 1972. Evaluation de la résistance au cisaillement d'un massif rocheux fragmente. *Proc. 24$^{th}$ Int. Congress on Rock Mech.*, Sec. BD, Montreal: 249-260.
27. Leong E.C. and Randolph M.F. 1991. Modelling of sliding behaviour at rock interfaces. *Proc. Comp. Methods and Adv. in Geomech.*, Beer, Booker and Carter (eds), Balkema, Rotterdam: 365-369.
28. Leong E.C. and Randolph M.F. 1994. Finite element modelling of rock-socketed piles. *Int. J. Numer. Anal. Methods Geomech.* 18: 25-47.
29. Lerat P. 1996. Etude de l'interface sol-structure dans les milieux granulaires à l'aide d'un nouvel appareil de cisaillement annulaire. *Thèse de Doctorat de l'Ecole Nationale des Ponts et Chaussées*, Paris, France.
30. Lerat P., Schlosser F. and Vardoulakis I. 1997. Nouvel appareil de cisaillement annulaire pour l'étude des interfaces matériau granulaire-structure. *Proc. 14$^{th}$ Int. Conf. Soils Mech. and Found. Eng.*, Hambourg, vol. 2: 363-366.
31. Patton F.D. 1966. Multiple modes of shear failure in rock. *Proc. 1$^{st}$ Int. Conference of Rock Mechanics*, vol. 1, Lisbon: 509-513.
32. Pietruszczak S. and Niu X. 1992. Numerical evaluation of bearing capacity of a foundation in strain softening soil. *Computers and Geotechnics* 13: 187-198.
33. Pietruszczak S. and Niu X. 1993. On the description of localized deformation. *Int. J. Numer. Anal. Meth. Geomech.*, 17: 791-805.
34. Plytas C. 1985. Contribution à l'étude expérimentale et numérique des interfaces sols granulaires-structures. Application à la prévision du frottement latéral des pieux. *Thèse de Doctorat de l'Institut National Polytechnique de Grenoble*, Grenoble, France.
35. Poorooshasb H.B. and Pietruszczak S. 1985. On yielding and flow of sand; a generalized two-surface model. *Computers and Geotechnics*, 1: 33-58.
36. Potyondy J.G. 1961. Skin friction between various soils and construction materials. *Géotechnique* 11 (4): 339-353.
37. Rouainia M., Boulon M. and Garnica P. 1992. Modelling and validation of a non linear soil-structure interface behaviour for pile problems. *Proc. Int. Symp. Num. Models in Geomech. (NUMOG IV)*, Swansea, UK: 223-230.
38. Sadrnejad S.A. and Pande G.N. 1989. A multilaminate model for sands. *Proc. Int. Symp. Num. Models in Geomech. (NUMOG III)*, Niagara Falls, Canada: 17-27.
39. Sharma K.G. and Desai C.S. 1992. Analysis and implementation of thin-layer element for interfaces and joints. *J. of Engin. Mechanics*, ASCE 118 (12): 2442-2462.
40. Shahrour I. and Rezaie F. 1997. An elastoplastic constitutive relation for soil-structure interface under cyclic loading. *Computers and Geotechnics*, 21: 21-39.
41. Tabucanon J. T. and Airey D. W. 1992. Interface tests to investigate pile skin friction in sands. *Research Report No. R662*, University of Sydney: pp. 14.
42. Tabucanon J. T., Airey D. W. and Poulos H. G. 1995. Pile skin friction in sands from constant normal stiffness tests. *Geot. Testing J.* ASTM, 18 (3): 350-364.
43. Taylor D.W. 1948. *Foundamentals of soil mechanics.* John Wiley and Sons, New York.
44. Tejchman A. and Tejchman J. 1990. Scale effect in pile model tests. *Archiwum Hydrotechnicki* XXXVII: 97-126.
45. Tejchman J. and Wu W. 1995. Experimental and numerical study of sand-steel interfaces. *Int. Journ. Num. Anal. Meth.Geomech.*, 19: 513-536.

46. Uesugi M. and Kishida H. 1986. Influential factors of friction between steel and dry sands. *Soils and Foundations*, 26 (2): 33-46.
47. Uesugi M., Kishida H. and Tsubakihara Y. 1988. Behavior of sand particles in sand-steel friction. *Soils and Foundations*, 28 (1): 107-118.
48. Wernick E. 1978. Skin friction of cylindrical anchors in non-cohesive soils. *Symposium on Soi Reinforcing and Stabilising Techniques,* Sydney, Australia: 201-219.
49. Yoshimi Y. and Kishida T. 1981. A ring torsion apparatus for evaluating friction between soil and metal surface. *Geot. Testing J.* ASTM, 4 (4): 145-152.
50. Zong-Ze Y., Hong Z. and Guo-Hua X. 1995. A study of deformation in the interface between soil and concrete. *Computers and Geotechnics*, 17: 75-92.

# Neural-Fuzzy Modelling to Analyse Complex Geotechnical Systems

Paola Provenzano

Dipartimento di Ingegneria Civile
Università di Roma "Tor Vergata"
00133 Roma, Italia

**Abstract.** Engineers have always been aware of uncertainties in analysis of geotechnical system complexity, related to soil inherent variability, site conditions, construction tolerance, and failure mechanisms. Geotechnical and structure design deal with these uncertainties by heuristic and expertise knowledge using input data that fall in the category of non-statistical uncertainties. In this article, an expert system based on an Artificial Neural Network with fuzzy input vectors is suggested for settlement prediction of shallow foundations in cohesionless soils.

## 1 Introduction

In many applications, the geotechnical systems under study are too complex to be described by classical mechanical model. On the other hand, very complex models cannot be proposed, for the difficult to assign correct values to each parameter. Deterministic procedures adopt usually conservative design values, but for most structures, even for critical facilities, use of this values cannot be economically justified. The probabilistic approach, in which all uncertainties are considered random, has a sophistication that is often in stark contrast to the crudeness of data which engineers must work with. An alternative approach is given by the adaptive inference systems that derive relations between the system variables by a learning process from available information, numerical and linguistic. In this way, they work to do resembles that is in an experimental study. It means that the single analysis is a simulation of the reality and the problem to be solved concerns with the exploiting of the maximum of information from these trials [2]. Promising inference systems are suggested by the fuzzy logic theory, able to deal with non-random uncertain variables and by the Artificial Neural Network theory that, it is well known [5], can approximate any non linear function arbitrary well over a compact interval.

## 2 Fuzzy sets and operations

As a crisp set A is defined by a *characteristic function,* declaring which elements $x$ of a given universal set $X$ are members of A and which are not, a fuzzy set $\tilde{A}$ is defined by a *membership function* defining the membership grade of each element in the set. Such a function maps the elements of X into real numbers in [0,1].

$$\tilde{A} : X \to [0, 1] \tag{1}$$

The value $\mu_A$(x)=0 means that an object x is not a member of the set A, and the value 1 indicates that x entirely belongs to A.

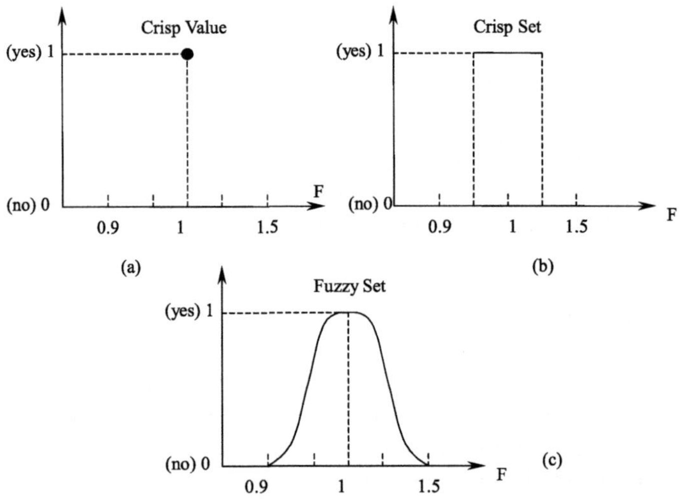


**Fig. 1.** Crisp versus fuzzy definition of slope instability

Figures (1) illustrate the difference between fuzzy set and "crisp" set membership, using different way to define the safety factor of a natural slope. It is the ratio $\eta = t_f/t_m$ of shearing resistance to the shearing mobilized by the disturbing forces. In limit equilibrium condition, with no uncertain input, is $\eta = 1$ (fig.1.a). An ordinary crisp set (fig.1.b) is able to describe the sensitivity of the slope behavior for interval inputs, while a fuzzy set (fig.1.c) is more useful to describe the stability condition for uncertain non random parameters. Note that an ordinary crisp set can be viewed as special case of fuzzy set with only two allowed membership values$\mu_A$(x)=0 or $\mu_A$(x)=1.

If $\tilde{A}$ and $\tilde{B}$ are two normalized fuzzy sets, extension principle [4] allows to calculate any operation (*):

$$(\tilde{A} * \tilde{B})(z) = \sup_{z=x*y} \min\,[A(x), B(y)] \tag{2}$$

In this paper, the equation (2) has been traduced by Vertex Method [5]. Given a fuzzy set $\tilde{A}$ on X and any number $\alpha \in [0,1]$ (Fig. 2.A), the $\alpha$-cut is ${}^{\alpha}\tilde{A} = [a, b]$. In general, if y=f(x) is a continuous and monotonic function on ${}^{\alpha}A$=[a, b], the interval representing $\tilde{B}$ at a particular value of $\alpha$, ${}^{\alpha}$B, can be obtained by:

$$ {}^{\alpha}B = f\left({}^{\alpha}A\right) = \left[\min\left(f\left(a\right), f\left(b\right)\right), \max\left(f\left(a\right), f\left(b\right)\right)\right] \tag{3} $$

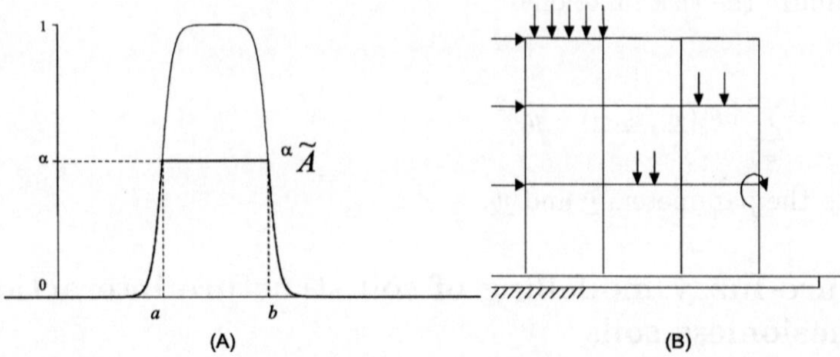


**Fig. 2.** (A) $\alpha$-level of a fuzzy number ; (B) Type of Frame structure modelled

## 3 Artificial neural network

Artificial Neural Network (ANN) is a computational mechanism able to acquire, represent and compute a mapping $f(\underline{x}, \omega)$ from a multivariate space of information to another, given a set of data $(\underline{x})$ representing that mapping. It consists of simple computational units, called *neurons,* which are highly interconnected. Each interconnection has a strength that is expressed by a number referred to as a *weight* $(\varpi)$. A Multilayer Feed-Forward Neural Network has been proposed here [5]. It consists of an input layer, one or multiple hidden layers and an output layer. The units in the hidden layer play a key role in the internal representation of the input patterns. The input-output flow of the network model is determined by the strength of the connections and the operation function of the neurons. The operation of a single neuron consists in a weighted sum of the incoming signals and a bias term, fed through an *activation function*, F, resulting in the output value of the neuron. In an ANN with an hidden layer the output value is mathematically expressed as

$$y_i = F\left(\sum_{r=1}^{n} w_{ir} F\left(\sum_{j=1}^{m} v_{rj} x_j + v_{0r}\right) + w_{0i}\right), \quad i = 1, ..., l \tag{4}$$

where $\underline{x}$ is the $m \times 1$ input vector, $\underline{y}$ is the $l \times 1$ output vector, $n$is the number of neurons in the hidden layer, v and w are the weight factors and $v_{0r}$ and $w_{0i}$ the bias values of the neurons in the hidden layer and in the output layer, respectively. A sigmoidal activation function have been used in this research.

Considering input and output samples, $x_i$ and $y_i$, respectively, a stochastic approximation procedure (or training), called backpropagation [4], is used, that minimize the risk functional

$$R_{emp} = \sum_{i=1}^{l} \left(F\left(\underline{x}_i, \underline{w}, \underline{v}\right) - y_i\right)^2 \tag{5}$$

respect to the parameters $\underline{v}$ and $\underline{w}$.

## 4 Neuro-fuzzy modelling of soil structure interaction cohesionless soils

In cohesionless soils, soil-structure interaction analysis has motivated primarily by the need to limit differential settlements within buildings to avoid structural or architectural damage. In this research, a Fuzzy-Neural Network method to predict the behaviour of structures built on complex cohesionless soils is proposed. The method is based on an ANN to model the soil-foundation interaction. The learning process analyses over 200 records of building foundations, tanks and embankments settlements on sand and gravel. Once validated, ANN is introduced in soil-foundation-sovrastructure interaction model. Using fuzzy sets to define vague and ambiguous variables, the Fuzzy-Neural Network method predicts the system behaviour and quantifies the uncertainty of its response.

The presented soil-structural interaction analysis involves frame structures, divided into elements connected by nodes, founded in granular soils (Fig.2.B). Let $\tilde{Q}$a loading vector with fuzzy coefficients $Q_i=\{\mu_Q(x)/x, x\in \Re\}$ that includes uncertain loads and boundary conditions. In the same way, any uncertain parameter of the analytical problem can be expressed by a fuzzy number, from the available information. By the extension principle, for a linear elastic structure, the solving equation system is:

$$\begin{aligned} \underline{\underline{\tilde{K}_s}}\underline{\tilde{D}}_s &= \underline{\tilde{Q}} + \underline{\tilde{R}}_{sf} \\ \underline{\underline{\tilde{K}_f}}\underline{\tilde{D}}_f &= \underline{\tilde{R}}_{fs} + \underline{\tilde{R}}_{fs'} \\ \underline{\tilde{D}}_t &= \underline{\tilde{D}}_t\left(\underline{\tilde{R}}_{fs'}\right) \end{aligned} \tag{6}$$

where (fig.3) $\underline{\underline{\tilde{K}}}_s$and $\underline{\underline{\tilde{K}}}_f$ are the fuzzy stiffness matrices of superstructure and foundation, $\underline{\tilde{Q}}$ is the fuzzy load vector, $\underline{\tilde{R}}_{sf}$, $\underline{\tilde{R}}_{fs}$ and $\underline{\tilde{R}}_{fs'}$, $\underline{\tilde{R}}_{s'f}$ are the fuzzy interaction strength vectors between superstructure ad foundation and between foundation and soil, respectively. $\underline{\tilde{D}}_s$, $\underline{\tilde{D}}_f$ and $\underline{\tilde{D}}_{s'}$ are unknown fuzzy displacements of the system nodes.

The third equation of the system (4) is the foundation settlement-predicting model: since the soil-structure interaction analysis is to be considered within serviceability limit states, soils exhibit non-linear behaviour even at small strain. Different levels of fuzziness can be used combining th uncertain in the model and in the parameters [5].

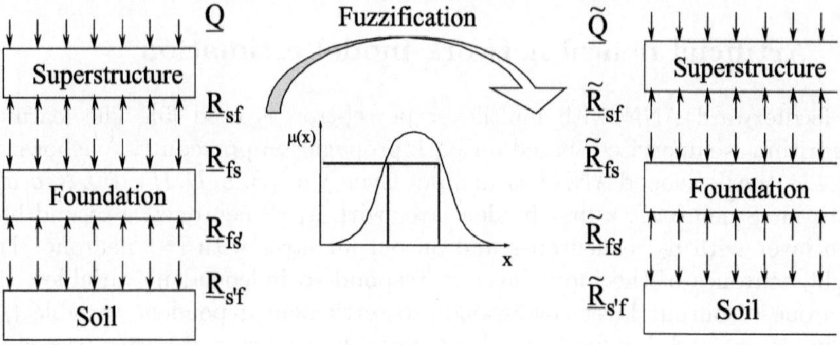


**Fig. 3.** Interaction of components

## 5 Statement of the problem and hypothesis formulation

Soil-foundation interaction analysis assumes the foundation more rigid than the superstructure and known applied loads. The foundation geometry, the thickness of granular deposit, the blow-counts of Standard Penetration Test (N) along borehole and the hydraulic boundary conditions are all known variables. Eight variables are chosen to characterise the system (Fig. 4.A): breadth (B, [m]), length (L, [m]) and depth (D, [m]) of the foundation, depth of water table beneath ground level ($H_w$, [m]), thickness of sand or gravel layer ($H_s$, [m]), average SPT blow-count over the depth of influence (N), gross pressure applied at foundation level (q, [KPa]) and maximum previous effective overburden pressure ($\sigma'_{v0}$, [KPa]). System response is expressed by means of the medium foundation settlement value ($\rho$, [mm]). Crisp intervals and linguistic variables are used to represent imprecise and uncertain knowledge that arises because of soil complexity.

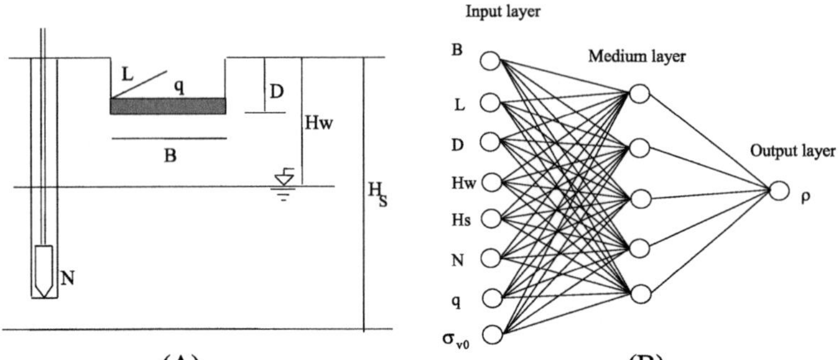


**Fig. 4.** (A) Geometry of foundation; (B) Neural Network Architecture

## 6 Artificial neural network model estimation

A feedforward ANN with multilayer perceptron is used [5]. The learning algorithm is supervised, based on back-propagation procedure. As shown in fig.4.B, the developed ANN has an input layer $\underline{X} = [B, L, D, H_w, H_s, N, q, \sigma'_{v0}]$ with m=8 neurones, a first hidden layer with $n_1$=8 neurones, a second hidden layer with $n_2$=5 neurones and an output layer with l=1 neurone. The eight neurones of the input layer correspond to independent variables, the neurone of output layer corresponds to settlement dependent variable ($\rho$). Neurones of each hidden layers have been chosen by an iterative procedure [7]. A summary of measured settlement and of its estimate made by using the mentioned ANN is given in the figure (3). The results are compared with those obtained by empirical [3] and analytical methods [1], assuming as a basic variable R=$\rho_s/\rho_m$, that is the ratio between simulated and measured settlement.

## 7 Example

This case history was presented in Maugeri et al. [4]. The analyzed building is part of a residential area located in Fondachello, a seaside village in Mascali (Catania, Italy). A laboratory and in situ testing program was performed to evaluate the soil proprieties. The results of standard penetration tests (N) are pictured in fig. 6.a.

In order to consider the uncertain compressibility of soil, it is possible define the membership function of $\tilde{N}$ variable (Fig.6.b). For the input layer, with fuzzy variable ($\tilde{N}$) and crisp variables (B, L, D Hw), the ANN is able to predict medium settlements, and their variation intervals, for different membership grades (tab.1).

The results are compared with that obtained by current methods (Fig.7).

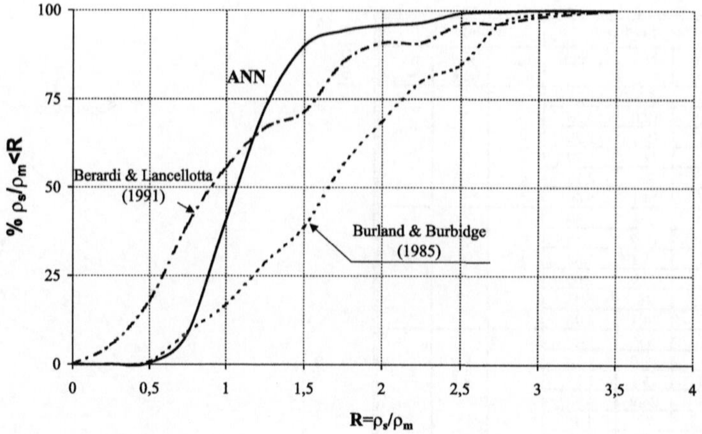


**Fig. 5.** Cumulative frequency curves of ANN, Burland-Burbidge and Berardi-Lancellotta methods

# 8 Conclusions

The aim of this research is to focus the role of an expert system, based on a fuzzy-neural network, to forecast displacement of shallow foundations in granular soils, keeping into account non-linear effects. Using of fuzzy numbers is effective to deal with soil parameter uncertainty and to calculate the extreme values of a structure response to all possible boundary condition combinations. A procedure to analyze soil-structure interaction, based on this Fuzzy-Neural Model is able to be extended also to loading uncertainties.

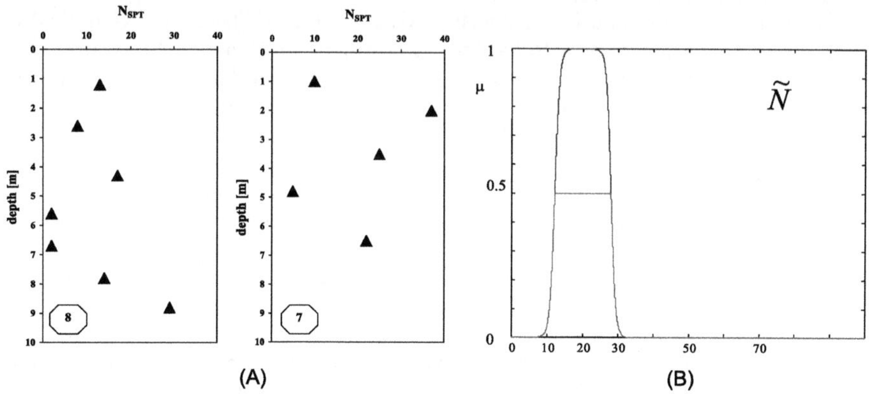


**Fig. 6.** (A) $N_{SPT}$ profiles; (B) Membership function of N variable

| Neural-fuzzy prediction | | | | Monitoring | | |
|---|---|---|---|---|---|---|
| q' | ?? | N | $w_s$ [mm] | $w_m$ min | $w_m$ max | $w_m$ average |
| | 0 | 5.9 | 1.21 | | | |
| | 0.5 | 12.0 | 1.28 | | | |
| 16.24 | 1 | 15.1 | 1.33 | 0.8 | 2 | 1.48 |
| | 1 | 24.9 | 1.79 | | | |
| | 0.5 | 28 | 2.19 | | | |
| | 1 | 34.1 | 5.09 | | | |
| | 0 | 5.9 | 1.36 | | | |
| | 0.5 | 12.0 | 1.50 | | | |
| 25.82 | 1 | 15.1 | 1.62 | 1.5 | 3 | 1.94 |
| | 1 | 24.9 | 2.60 | | | |
| | 0.5 | 28 | 3.43 | | | |
| | 1 | 34.1 | 8.78 | | | |
| | 0 | 5.9 | 1.49 | | | |
| | 0.5 | 12.0 | 1.70 | | | |
| 32.82 | 1 | 15.1 | 1.88 | 2 | 3.4 | 2.35 |
| | 1 | 24.9 | 3.32 | | | |
| | 0.5 | 28 | 4.45 | | | |
| | 1 | 34.1 | 6.15 | | | |
| | 0 | 5.9 | 1.69 | | | |
| | 0.5 | 12.0 | 2.03 | | | |
| 42.59 | 1 | 15.1 | 2.29 | 2.8 | 4 | 3 |
| | 1 | 24.9 | 4.45 | | | |
| | 0.5 | 28 | 6.15 | | | |
| | 1 | 34.1 | 15.18 | | | |

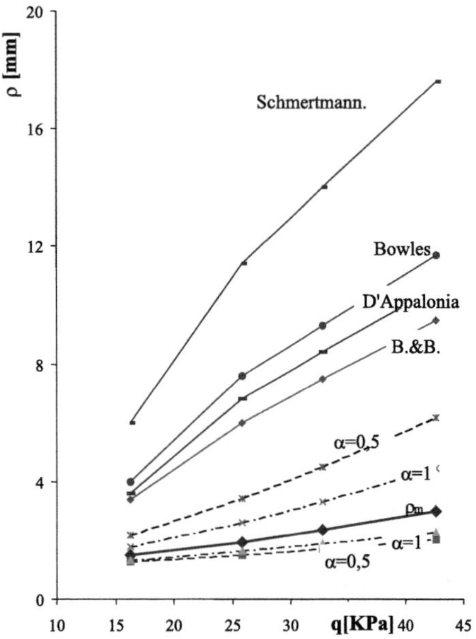


**Fig. 7.** Settlements predicted by fuzzy-ANN method and some classical methods

# References


1. Berardi, R. & Lancellotta, R. 1991. Stiffness of granular soil from field performance. Geotechnique 41, No 1, 149-157
2. Biondini, F., Bontempi, F. & Malerba P.G. 2000. Fuzzy Theory and Genetically-Driven Simulation in the Reliability Assessment of Concrete Structures. $8^{th}$ ASCE Joint Specialty Conference on Probabilistic Mechanics and Structural Reliability, July 24-26, 2000, University of Notre Dame, Indiana.
3. Burland, J.B. & Burbidge, M.C. 1985. Settlement of foundation on sand and gravel. Proc. Instn. Civ. Engrs, Part 1 78, 1325-1381
4. Maugeri, M., Castelli, F., Massimino, Maria M.& Verona, G.1998. Observed and Computer Settlements of two Shallow foundations on sand. *J.of Geot. and Geoenv. Eng.*, Vol.124, No.7, July
5. Provenzano, P. 2000. Modelli di Inferenza neurali e fuzzy nell'analisi di sistemi geotecnici complessi. Phd. Thesis. University of Palermo. Palermo. Italy
6. Provenzano, P. & Bontempi, F. 2000. Impostazione dell'analisi strutturale in presenza di informazioni incerte attraverso Logica Fuzzy. STUDI E RICERCHE – Vol. 21, 2000 Scuola di Specializzazione in Costruzioni in C.A. – Fratelli Pesenti. Politecnico di Milano, Italia
7. P. Provenzano (2001)"*Soil-structure interaction: a Neural-Fuzzy model to deal with uncertainties in footing settlement prediction*" ISEC-01, First Int. Str. Eng. and Constr. Conf., Honolulu, Hawaii, January 24-26

# Speckle and Infrared Images Processing Applied to the Analysis of the Thermomechanical Behaviour of a Semicrystalline Polymer

Jean-Michel Muracciole, Bertrand Wattrisse, André Chrysochoos

Laboratoire de Mécanique et Génie Civil,
cc 081, Université Montpellier II,
F-34095 Montpellier Cedex 5, France

## 1 Introduction

The mechanical behaviour of thermoplastics with glassy amorphous regions is characterised by a strong dependence in temperature and strain-rate. It seems thus relevant to take a thermomechanical point of view into account. As these materials may also exhibit both tough dissipation and strong strain localisation, we have decided to use two complementary local techniques to study their response.

Infrared thermography can constitute a highly reliable means to monitor temperature maps during the deformation process of flat samples. The temperature and heat distribution within the specimen obeys the diffusion equation and the heat sources can be locally estimated thanks to the processing of these thermal data.

This determination is a first step, using energy balances, to set up constitutive thermomechanical equations. To achieve such evaluations, the evolution of mechanical data (strain, strain-rate, stress, ... ) is also to be detected. During tensile tests, these variables are often supposed to remain homogeneous and their evolution is estimated through rather "global" measurements. In order to analyse the kinematics fields on the same level of space resolution than the thermal ones, we have chosen to develop a digital image correlation technique [9].

Results obtained on a semicrystalline thermoplastic polymer (PolyAmide) are presented in this paper. Quasi-static tensile tests, at constant stroke velocity, were performed at room temperature (*i.e.* below the glass transition temperature) on a MTS hydraulic machine. The conventional stress-strain curve is characterised by a yield point and a fall in load At this stage a neck has formed and its shoulders propagate along the gauge part of the sample. Two optical sensors with different characteristics (infrared – visible) are set by each side of the machine. We first describe the image processing techniques in the infrared spectrum (temperature charts) and for the visible wavelengths (mechanical fields). And, when the deformation process becomes heterogeneous, the analysis of the material behaviour is proceeded by the simultaneous use of these two imaging techniques.

## 2 Thermal and calorimetric analysis

### 2.1 Experimental set-up

The infrared thermography set-up involves a mono-detector camera (Agema 880 SWB, short wave) which records thermal images (256 × 180 pixels) of the flat sample during the loading.

Using the heat equation, the specific heat sources induced by the straining are related to temperature evolutions. For quasi-static tests, assuming an isotropic conduction and neglecting the terms of convection, this relation can be written as [3]:

$$\rho C \left( \frac{\partial \theta}{\partial t} + \frac{\theta}{\tau_{th}} \right) - k \Delta_2 \theta = w'_{ch} \tag{1}$$

where $\rho$, $C$, $\tau_{th}$ and $k$ stand respectively for the density, the heat capacity, a time constant characterising the lateral heat exchanges, and the heat conductance of the material. The variable $\theta$ represents the deviation from the temperature of the reference state. The specific heat sources ($w'_{ch}$ ) can then be determined by applying differential operators on the temperature fields (time derivative: $\partial/\partial t$ and spatial derivative: $\Delta_2 = \partial^2/\partial x^2 + \partial^2/\partial y^2$ where $x$ is the sample loading axis, $y$ the width axis).

### 2.2 Infrared data processing

Before any differentiation, the discrete and noisy thermal data are to be filtered. Several methods can be implemented [4]. Hereafter, a local least square fitting technique has been chosen. The degree of the polynomial is taken equal to 2, and the regression reflects that, in the neighbourhood of each pixel, heat exchanges by conduction remains locally constant. The Approximation Zone size ($AZ$) affects evidently the data fitting, and is set to optimise the filtering efficiency.

### 2.3 Validation of the infrared processing

Computer generated thermal images related to a given heat sources distribution were created in order to validate the infrared data processing. According to our experimental observations, a gaussian white noise ($0.1K$ amplitude), was superimposed to the virtual data. The processing is then applied to these images. For $15 \times 15$ square pixels approximation zones, the speed of the computation and the accuracy of the measurement are balanced, as demonstrated in Figure 1.

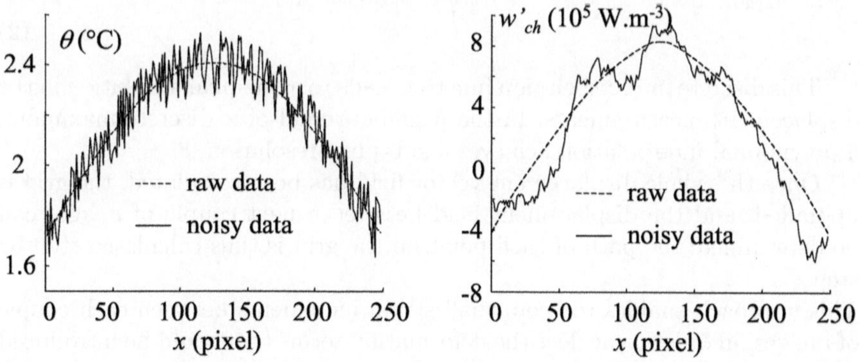


**Fig. 1.** Validation of thermal data processing. Temperature profile $\theta(x, 0, t_0)$ and related heat sources $w'_{ch}(x, 0, t_0)$

## 3 Strain analysis

### 3.1 Experimental set-up

A C.C.D. camera (Panasonic WV-CP410G) is rooted to the frame of the machine during the test. The image processing is performed afterwards in two steps. The displacement field is first estimated, then the strain and strain-rate fields are deduced by differentiation.

### 3.2 Speckle data processing

**Displacement field measurement** The displacement vector of each point is resolved into in-plane and out-of-plane components. The first one is reached by a direct correlation computation. A point, say $M$, is defined by its neighbourhood "optical signature", so-called the Correlation Zone ($CZ$). The position of $M(i_0, j_0)$ in the "reference" image $I_1$ is sought inside a Research

Zone ($RZ$) in the "deformed" image $I_2$. Its displacement is associated with the couple $(k,l)$ which maximises the intercorrelation function $\varphi$ defined by [1]:

$$\varphi(k,l)_{(k,l)\in[\frac{-ZR}{2},\frac{ZR}{2}]} = \frac{\sum_{i=-CZ/2}^{CZ/2}\sum_{j=-CZ/2}^{CZ/2} I_1(i_0+i,j_0+j).I_2(i_0+i+k,j_0+j+l)}{\sqrt{\sum_{i=-CZ/2}^{CZ/2}\sum_{j=-CZ/2}^{CZ/2} I_1^2(i_0+i,j_0+j)}.\sqrt{\sum_{i=-CZ/2}^{CZ/2}\sum_{j=-CZ/2}^{CZ/2} I_2^2(i_0+i+k,j_0+j+l)}} \quad (2)$$

This discrete intercorrelation function leads to a one-pixel resolution in the displacement measurements. In the neighbourhood of $\varphi$ discrete maximum, a polynomial interpolation achieves a sub-pixel resolution [8].

Once the whole displacement vector field has been calculated, the grid is upgraded, and the displacement field between a next couple of images can be determined: the path of each point on the grid is thus calculated step by step.

This approach enables to keep small strain increments between each couple of images, in order to neglect the deformation terms that could be introduced following [7,2].

**Strain and strain-rate computation** The differentiation of the displacement field allows to compute a whole range of kinematics fields (speed, acceleration, strain $\epsilon$ , strain-rate $\dot{\epsilon}$ , ... ) [9]. As with the calorimetric processing, the noisy experimental data are locally filtered using the same kind of polynomial fitting.

### 3.3 Validation of the processing

The algorithms of correlation and differentiation were both tested on several analytic examples based on numerically deformed virtual images, pictured in classical shooting conditions. As illustrated in Figure 2, the use of $20 \times 20$ square pixels Correlation Zones and $25 \times 25$ square pixels Approximation Zones leads to an accurate estimation of strong and weak strain gradients.

## 4 Coupling of thermal and kinematics measurements

The results obtained with these two different techniques in very similar conditions [5] – specimen, speed, room temperature, ... – motivate their combined use during the same tests as temperature maps and the related material

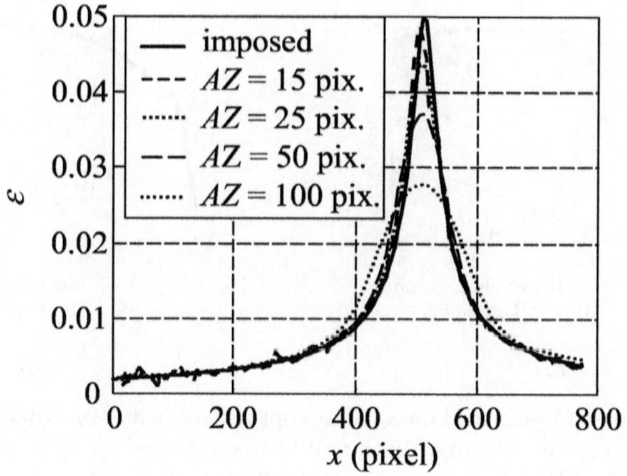


**Fig. 2.** Dependence of the strain measurement with the Approximation Zone dimensions in the case of a heterogeneous strain profile $\epsilon\,(x, 0, t_0)$, (CZ=20 square pixels)

flow are both required to compute the term of convection in the heat equation. Stretching materials with an important heat conduction as metals, at strain-rates about $10^{-2}$ $s^{-1}$, this term was found to be negligible [6]. But, in polymers where the heat conduction is poor, important temperature gradients may appear during necking (up to 100 $K.m^{-1}$ in PolyAmide 12 for strain-rates as low as $10^{-4}$ $s^{-1}$). Moreover as shown in Figure 3, the shoulder is roughly the only area still straining.

The tests were performed on standard ISO R527 samples with initial gauge length $L_0$ of 60mm and initial cross-section $S_0$ of 40mm$^2$. In the neck, the term of convection is up to 20% of the time derivative. The first results presented hereafter illustrate the heterogeneity of the thermomechanical fields during PolyAmide stretching.

### 4.1 Local analysis of necking

As the gradients remain much more acute along the tensile axis ($x$) than along the sample width ($y$), only $x$-distributions are taken into account. As shown in Figure 3, those profiles $-w'_{ch}(x, 0, t)$ and $\dot{\epsilon}_{xx}(x, 0, t)$ – are piled up in chronological order with magnitudes represented using a greyscale: the horizontal axis stands for time (or homogeneous strain) and the vertical for space ($x$). The engineering stress-strain curve is also plotted onto the map to connect the global mechanical response with the local one. The abscissa then represents the conventional strain ($\epsilon_C = v_T t / L_0$ ), and the ordinate the conventional stress ($\sigma_C = F/S_0$ , where $F$ is the applied load).

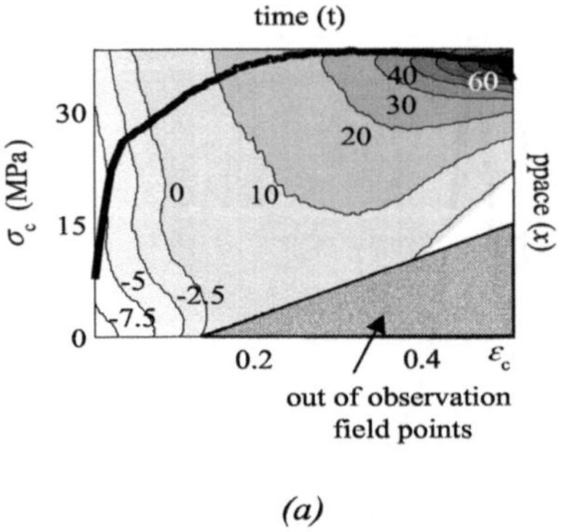


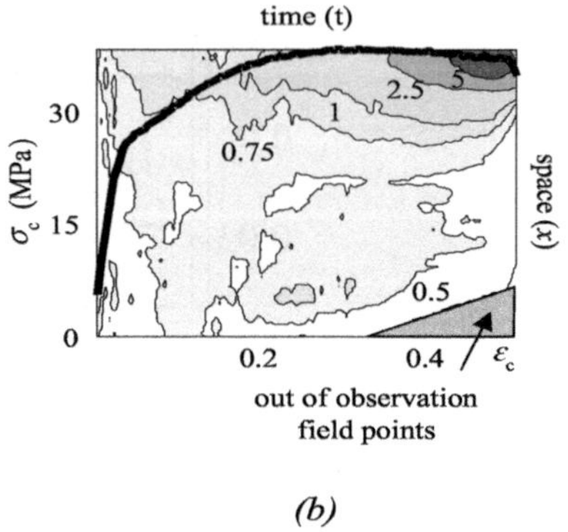


**Fig. 3.** Tensile test performed on a PA at constant velocity ($v_T = 0.05\text{mm.s}^{-1}$) and at room temperature – (a) distribution of heat sources $w'_{ch}(x, 0, t)$ in $10^3$ W.m$^{-3}$ – (b) distribution of Hencky strain-rate $\dot{\epsilon}_{xx}(x, 0, t)$ in $10^{-3}$ s$^{-1}$

At first, the material response is linear. As the stress rises further, it becomes more or less anelastic until yielding occurs. Then, the nominal stress drops somewhat and extension proceeds easily, with the formation of a neck that continuously undergoes growing.

Those three stages can be noticed on the evolution of heat sources (Figure 3.a): they are first negative due to the thermoelastic effect, then they increase as heat is dissipated. And the heat sources concentrate in the necking region when yield occurs. At that stage, strain-rates are also localised (Figure 3.b). The heat sources and the strain-rate are negligible outside the necking zone. The material behaviour can hardly be withdrawn from the engineering curve exhibiting such structural effects: the material local response has to be investigated.

## 4.2 Stress analysis

The main difficulty is here to estimate locally the stress tensor components. A method, based on the exploitation of the strain data and the equilibrium equations is presented in [10]. Provided the strain remains homogeneous in each cross-section, and the volume constant, the "true" stress might be written as:

$$\sigma_{xx}(x, t) = \frac{F(t)}{S_0} \exp(\varepsilon_{xx}(x, t)) \tag{3}$$

The three stress–strain curves shown in Figure 4 illustrate three different processing of data monitored during the same test. The first one ($G$), reflects the macroscopic response. The others exhibit the "local" responses of two virtual extensometers: *A* is located around the point where necking first appears, and *B* by the point reached by the shoulders of the neck when the test is stopped. Natural strain-rates are also plotted on the same graph.

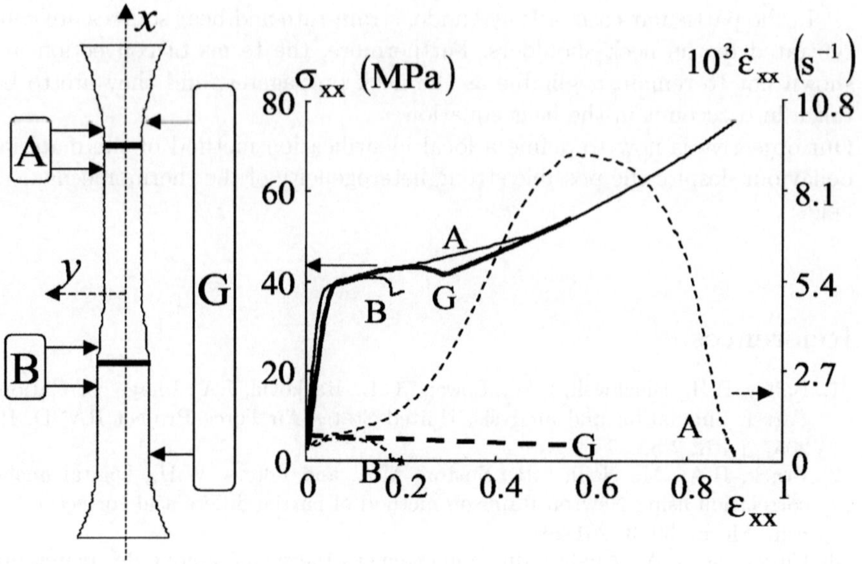


**Fig. 4.** $\sigma_{xx} - \epsilon_{xx}$ (full line) and $\dot{\epsilon}_{xx} - \epsilon_{xx}$ (dotted line) relations for a macroscopic strain G from 0 to 55%

It is well known that considering the true stress-strain curve in spite of the conventional one increases strain hardening (see curve *G*). This is intensified handling local data (curve *A*): the stress increases because the load decrease is always balanced by the section reduction. Between the neck inception and its subsequent propagation the strain-rate increases tenfold. When the neck travels out of the gauge of extensometer *A*, the strain-rate vanishes.

The response of extensometer *B* is similar until the neck onset. Then the material unloads and it won't strain anymore until the shoulder reaches it. Every other point has a similar behaviour, which is more or less delayed by the neck displacement time. So the whole straining process is concentrated in the vicinity of the neck shoulders.

## 5 Conclusions

This work represents the very first step to proceed energy balances during the deformation of materials undergoing necking. The local material behaviour could be more accurately analysed by the combined use of infrared and speckle techniques, which will be helpful to deduce the nature of the different mechanisms induced by the material deformation.

In the particular case of PolyAmide, strain-rate and heat sources are concentrated in the neck shoulders. Furthermore, the terms of convection are shown not to remain negligible as the neck propagates, and they are to be taken into account in the heat equation.

Our objective is now to define a local identification method of the material behaviour despite the possible strong heterogeneity of the thermomechanical fields.

## References

1. Bailey, H.H., Blackwell, F.W., Lowery, C.L., Ratkovic, J.A., Image correlation: Part I. Simulation and analysis, United States Air Force Project RAND, R-2057/1-PR, 7-8.
2. Bruck, H.A., Mc Neill, S.R., Sutton, M.A. and Peters, W.H., Digital image correlation using Newton-Raphson method of partial differential correction, J. Exp. Mech., 39, 3, 261-267.
3. Chrysochoos, A., Analyse du comportement thermomécanique des matériaux par thermographie infrarouge; Photomécanique 95, 203-211, Ed. Eyrolles.
4. Chrysochoos A., Louche H., An infrared image processing to analyse the calorific effects accompanying strain localisation, Int. J. Eng. Sc., 38,1759-1788.
5. Chrysochoos, A., Louche, H., Muracciole, J.-M., Saurel, J.-L., Némoz-Gaillard, M., Wattrisse, B., Experimental analysis of localization mechanisms in mild steels by infrared and speckle image processing, IUTAM Symposium "Advanced optical methods and applications in solid mechanics", Poitiers 98, 313-320, Ed. Kluwers.
6. Chrysochoos, A., Muracciole, J.-M., Wattrisse, B., Experimental analysis of strain and damage localization, Symposium on continuous damage and fracture, Ed. A. Benallal, 41-51.
7. Chu, T.C., Ranson, W.F., Sutton, M.A. and Peters, W.H., Application of Digital-Image-Correlation Techniques to Experimental Mechanics, J. Exp. Mech., 25, 232-244.
8. Oulamara, A., Tribillon, G. and Duvernoy, J., Subpixel speckle displacement measurement using a digital processing technique, J. Mod. Opt., 35.
9. Wattrisse, B., Chrysochoos, A., Muracciole, J.-M., Némoz-Gaillard, M., Analysis of strain localisation during tensile test by digital image correlation, J. Exp. Mech., 41,1,29-38.
10. Wattrisse, B., Chrysochoos, A., Muracciole, J.-M., Némoz-Gaillard, M., Kinematic manifestations of localisation phenomena by image intercorrelation, Eur. J. Mech., A/solids, (to be printed).

# A Delamination Model

Nelly Point[1], Elio Sacco[2]

[1] Département de Mathématiques
Conservatoire National des arts et Métiers
F-75141 Paris Cedex 03, France
[2] DiMSAT
Università di Cassino
03043 Cassino, Italia

## Abstract

A delamination model for laminated composites is proposed. A damage variable takes into account the degradation of the adhesive properties of the interfaces. It is described by a nonsmooth, nonconvex functional [1].

It must be emphasized that this model permits to recover, mathematically, the fracture mechanics theory and, furthermore, to generate various regularized models, each of them with different mechanical meanings. These regularized models correspond to approaches based on specific interface constitutive laws.

Numerical results for the regularized problem applied to a laminated beam and to the drilling of a composite laminated plate are presented. They are obtained using finite elements and a elastic predictor-damage corrector method.

The results for the beam shows a good agreement with the analytical solution [1] [2].

Both models, the nonsmooth one and the regularized one, are used to study analytically the classical double cantilever beam specimen. This permits to predict the external loads leading to delamination in pure mode I or pure mode II [2].

From a mathematical point of view, proof of the existence of a solution of both problems is given [3].

## References

1. Point N., Sacco E. (1996) A Delamination Model for Laminated Composites. International Journal of Solids and Structures, **33**, 4: 483-509
2. Point N., Sacco E. (1996) Delamination of beams : a method for the evaluation of strain energy release rates of DCB specimen. International Journal of Fracture, **79**: 225-247
3. Point N., Sacco E. (1998) Mathematical Properties of a Delamination Model. Mathematical and Computer Modelling, 28/4-8 (aug-oct), 359-371

# Granular Decohesion Thermal-Damage in Marble Monuments

Gianni Royer Carfagni

Dipartimento di Ingegneria Civile
Università di Parma
43100 Parma, Italia

**Abstract.** We present some recent tests on marbles performed using a high sensitivity dilatometer. Marble permanent dilatation consequent to thermal cycles is correlated with a microstructural damage, usually referred to as granular decohesion, observed in marble monuments exposed to rigid climates, of which Arval Aalto's Finland Hall in Helsinki is a paradigmatic example.

## 1 Introduction and practice

My personal experience, corroborated by various contribution discovered in the Literature, is that marble tends to deteriorate when exposed to rigid climates. To this respect, paradigmatic is the case of Arval Aalto's Finland Hall in Helsinki, one of the symbols of Finland. Just a few years after the building completion, the white marble panels of the façade showed a tendency towards concave bowing so that, just before the cladding was replaced in 1998, the borders protruded outwards up to 10 *cm* (Figure 1a).

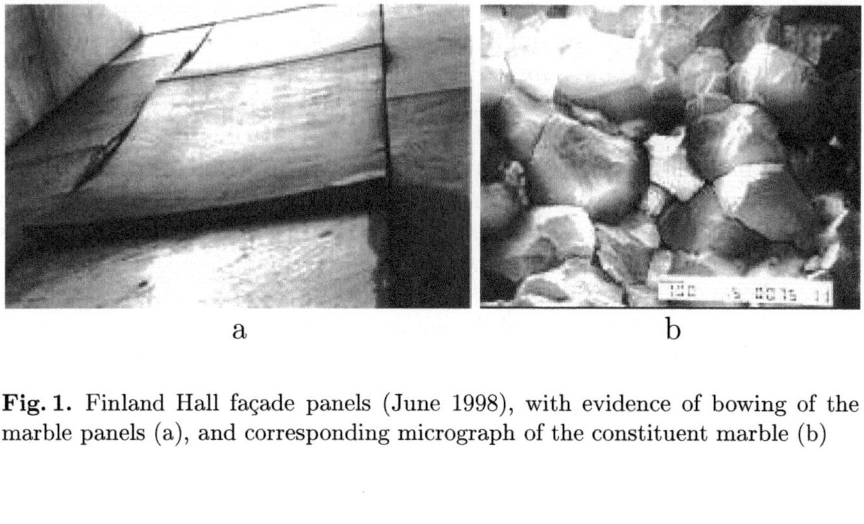

a b

**Fig. 1.** Finland Hall façade panels (June 1998), with evidence of bowing of the marble panels (a), and corresponding micrograph of the constituent marble (b)

An accurate analysis with a scanning electron microscope was performed in order to investigate, at the microscopic level, the causes of bowing [4]. One of the main conclusions of this research was that chemical attack was responsible only of a superficial degradation, causing the formation of a patina and the consequent loss of brightness of the marble panels. X-ray examinations confirmed that significant traces of chemical agents were not present at a few tenths of millimeter depth from the outer surface.

The microscopic appearance of the inner portion of the marble slab is represented in Figure 1b. Marble may be considered a sintered composite of calcite crystal granules and in Finland Hall's façade they appeared almost completely detached one from the other. Hardly any traces of transgranular cleavage fracture could be found, but it simply looked as if the grain cementing material had been gradually destroying. Because of this peculiarity, this type of damaged is usually referred to as *granular decohesion.* It represents the common denominator to other similar cases of marble-panel bowing, occurred also in recent times.

The reason for such a particular deterioration are not completely clear yet. Possible explanations in the Literature essentially call for three different possible causes: *i*) the chemical-physical attack of sulfates and chlorides [5]; *ii*) the mechanical action of soluble salts which, penetrating in a solution, expand when the solvent evaporates and open existing cracks [1]; *iii*) thermal variations, which may be responsible of internal stress [3]. Particular hygrometic conditions may have a synergetic effect with the other causes.

My personal opinion is that *i*) and *ii*) are of importance only at the superficial level, whereas *iii*) seems to be the leading factor. Reasons for this conclusion can be explained by an elementary model. Calcite crystallizes in the rhombohedral form and, as a result of this asymmetry, not only its elasticity but also its thermal expansion is strongly anisotropic in character. For example, the coefficient of thermal expansions varies from $25{\cdot}10^{-6}\ {}^\circ C^{-1}$ in the direction of the crystallographic $c$-axis to $-5{\cdot}10^{-6}\ {}^\circ C^{-1}$ (a negative value) at right angle to that. Consequently, temperature variations, even when uniformly distributed in the specimen, may provoke an incongruent deformation when, as in marbles, the calcite grains are randomly oriented. It results that even a few degrees' temperature increase can produce a self-equilibrated state of stress that can overcome the resistance of the grain bonding. The results of a qualitative model [3], corroborated by more sophisticated FEM simulations assuming a cohesive softening constitutive law for the interface contact forces [2], confirm that the most dangerous states of stress occur at the grain borders, especially in proximity of the texture nodes. This explains the intergranular rather than transgranular character of fractures, even not invoking the fact that the strength of the interface bonding is, presumably, less than the strength of calcite itself.

Of course, granular decohesion is always accompanied by volume increase. Marble bowing can thus be explained assuming a gradient of the decohesion level through the slab thickness.

## 2 Thermal expansion of marble

The reasons why the granular-decohesion level should vary through the slab thickness may be numerous and different in type. For example, as confirmed by careful measurements, temperature may vary between the inner and outer surface of the façade due to the building internal heating; differences in the humidity level, also confirmed by measurements, might as well play an important role.

In order to assess whether the measured difference in temperature and moisture might be sufficient to provoke bowing, a series of tests was performed using a high-sensitivity dilatometer, able to monitor the expansion of marble while varying its temperature. Since granular decohesion implies volume increase, the permanent dilatation may represent an index of the amount of damage produced.

Different qualities of marbles were considered. Figure 2 shows the results for an A-type marble (for convenience, here commercial names have been substituted by letters), the same used for the replacement of the façade. Specimens were subjected to quasi-static thermal cycles varying from approximately $-50°\,C$ to $+55°\,C$ (such values are not too far from the temperature excursion measured at the top of the roof of Finland Hall, during seasonal changes) and their dilatation continuously recorded by the experimental apparatus.

Five different phases can be recognized in the material response. The first one, referred to as (a), corresponds to the first temperature increment, from $t$=+20° $C$ (room temperature) to $t$=+55° $C$. What should be noticed is the superlinear growth of thermal strain, which should be compared with the slope of the straight line also drawn in the Figure, corresponding to the theoretical average dilation of pure calcite ($\alpha$=12·10$^{-6}$ °$C^{-1}$). The greater expansion of the composite with respect to the component can be attributed to the development of cracks. SEM investigations show that these are transgranular in type. Moreover, the analysis of the simple model discussed in [2], shows that the transgranular cracking is very likely to be accompanied by the relative sliding of the grain borders, in order to accommodates the incongruent deformation of the grains themselves.

The cooling stage can be conveniently distinguished into three branches: phase (b), from $t$=+55° $C$ to $t$=+30° $C$; phase (c), from $t$=+30° $C$ to $t$=−40° $C$; phase (d), from $t$=−40° $C$ to $t$=−50° $C$. The main difference between (b) and (c) is that the graph in (b) is pseudo-horizontal, whereas the beginning of (c) is indicated by a sudden change in the slope. This can be explained assuming that the frictional contact among the grains prevents an instantaneous re-

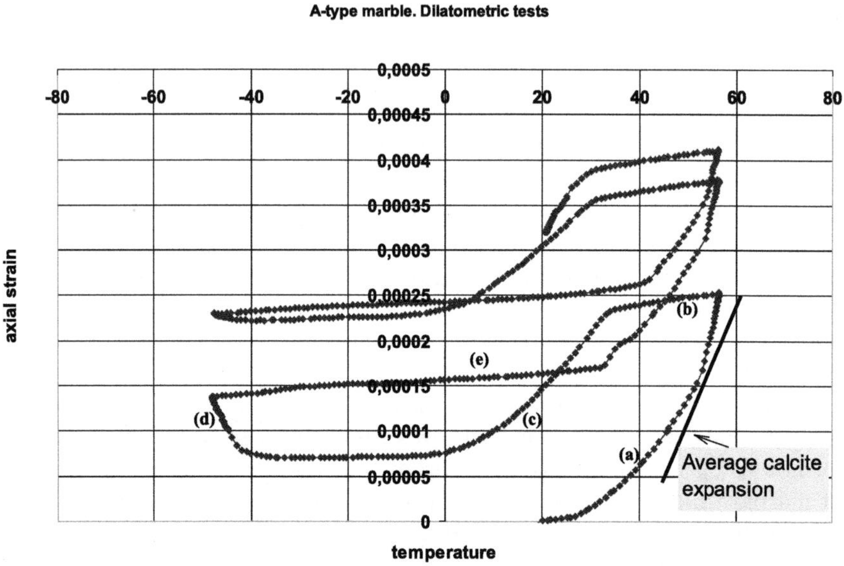


**Fig. 2.** Average strain as a function of temperature for A-type Marble. First and second cycle

covery of the strain. Reverse sliding starts only at the beginning of phase (c) and then gradually slows down, as logically expected, when the temperature is further decreased. This may be a confirmation that the dilatation accumulated during phase (a) should be, at least partially, attributed to sliding of the grains. The distinguishing character of phase (d) consists in the fact that, lowering the temperature beyond a certain limit, the material expands rather than shrinking further. This phase can be correlated with the breaking of intergranular bonds subsequent to the incongruent granular contraction at low temperatures. We observe in passing that the aforementioned model of [2] suggests that *tensile* stresses at the grain borders are greater for negative (temperature decrease) rather than for positive temperature variations, for the same temperature interval. This may explain why marble deterioration is more pronounced in cold rather than warm climates.

Finally, if at this point temperature is increased again, the material starts to expand once more, entering phase (e). The distinguishing character of (e) is that the slope of the corresponding graph is almost parallel to that of phase (b), an evident sign, in our interpretation, of the role played by friction. Phase (e) ends at $t \cong +30° C$. It is important to notice that if the cycle had stopped at room temperature ($t = +20° C$), marble would have shown a significant permanent dilatation.

In further cycles the strain follows a similar characteristic trend, where all the five aforementioned phases can again be recognized. A few comments are however necessary. First of all, the slope of the graphs referring to phases (b) and (e) goes slowly increasing, an indication that the friction coefficient diminishes with ongoing cycles, probably do to the smoothening out of the contact surfaces. Secondly, the beginning of phases (a), characterized by superlinear growth of strain with temperature, is shifted rightwards with ongoing cycles. For example, it was around $33°\,C$ in the $2^{nd}$ cycle; it becomes around $42°\,C$ at the $9^{th}$ cycle. What is even more important is that, as it is clear also from Figure 2, the volume increase occurring in phases (d) diminishes with cycling. This is a characteristic of ongoing damage. The higher is the number of thermal cycles the specimen has undergone, the lower is the number of bonds that can be broken if a new cycle is performed. In fact, roughly speaking, if the weakest bonds have already been broken in the previous history, the strongest bonds are the ones that remain active.

One-sided thermal variations may as well produce a permanent dilatation. Figure 3 represents, for example, the response of the same A-type marble when its temperature is varied from $+20°\,C$ to $-35°\,C$ (the two pictures refer to cycles 1-2 and 4-5 respectively). Again, similarly to phase (d) in Figure 2, we notice a characteristic expansion when the temperature is decreased beyond a certain limit. However, such contributions decreases very quickly at each cycle, as Figure 3b confirms. These experiments show that, in general, two-sided thermal conditioning may be more dangerous than one-sided variations.

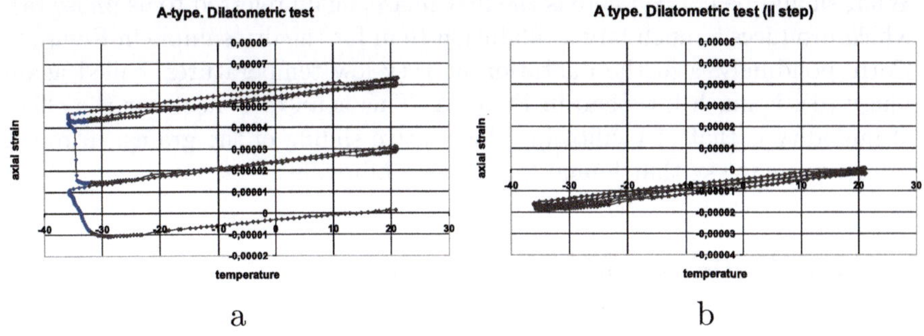


**Fig. 3.** Average strain as a function of average temperature for A-type marble. Thermal cycles between $t = +20°C$ and $t = -35°C$. Cycles n. 1-3 (a) and n. 4-6(b)

It is also important to mention that the marble response may show remarkable anisotropy. Figure 4 shows, for example, the behavior of the same quality of marble when the average strain is measured in a direction at right angle with respect to the one considered before. Despite there are similarities with Figure 2, now the distinction among the aforementioned five phases (a)-(e) is less clear than before. Such a noteworthy difference is due to the fact that, in general, there are directions where the orientation of the grains is statistically more pronounced so that also the marble response is influenced at the macroscopic level.

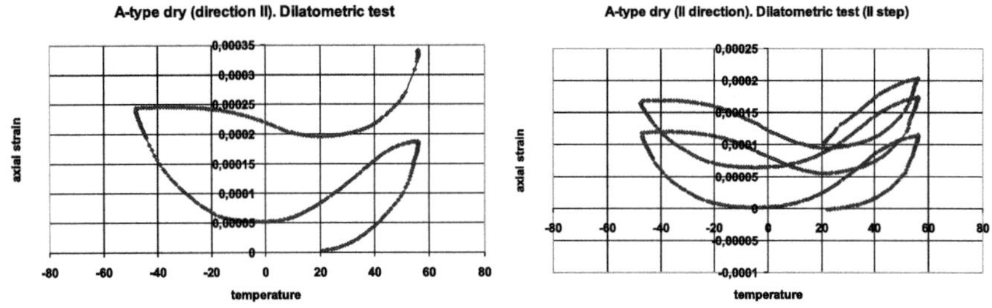


a b

**Fig. 4.** Average strain vs. temperature for A-type Marble. Same material as in Figures 2 and 3 but specimens rotated 90°. Thermal cycles between $t = -50°C$ and $t = +55°C$. Cycles n. 1-2 (a) and n. 4-5 (b)

In order to investigate the influence of moisture, wet sample were also tested. The diagrams for the same A-type marble, with specimens carved in the very same direction as in Figures 2 and 3, are reported in Figure 5. What should be noticed here is the first phase, again referred to as phase (a), which produces a much larger dilatation than for the dry sample in Figure 2. Correspondingly, also the dilatation at very low temperatures, called again phase (d), is much less than in Figure 2. This is an evidence that the effect of humidity is that of facilitating the relative sliding of the grains. In rough words, we may say that humidity acts as a lubricant.

## 3 Applications and conclusions

The forthcoming deductions are necessarily conjectural at this time, but it is expected that at least some of the questions mentioned will be clarified by further studies. In any case, what seems to be confirmed is that thermal cycles may produce the permanent dilation of marble. Figure 6a shows a SEM photograph of an inner portion of marble having undergone thermal cycles. In this picture it is evident the intergranular nature of fracture and the similarity with Figure 1b.

In summary, we can affirm that thermal variations can produce granular decohesion similar to that observed in Finland Hall façade. This is a very

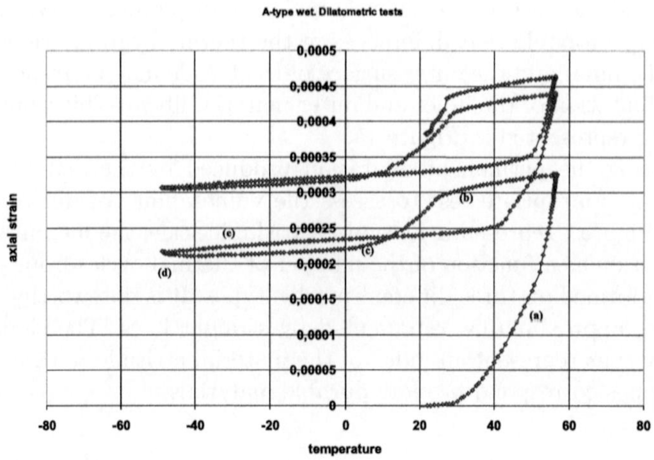


**Fig. 5.** Average strain vs. temperature. A-type Marble in wet conditions. First and second cycle

**Fig. 6.** Intergranular fractures in marble having undergone thermal cycles (a). Transgranular fractures in marble broken in bending (b)

important distinguishing character. If, for example, marble is broken in bending, the resulting fracture surfaces appear as in Figure 6b, i.e. transgranular fracture is dominating. This is because bending produces tensions on a wide portion of the specimen, whereas tensile stresses caused by thermal variations are concentrated at very small portions of the grain, and are the most dangerous at the interfaces.

In this study, we have also evidenced the anisotropy of marble response (Figure 4) and we have discovered the effects of humidity, which acts as a lubricant and may accelerate the accumulation of permanent strain (Figure 5). It is very probable that differences in the thermo-hygrometric conditions between the inner and the outer surface of the façade may be responsible of a different dilatation of the inner and outer material fibers. This would explain the bowing represented in Figure 1a.

Measuring the permanent dilatation produced by thermal cycling may also be a an appropriate test to assess the vulnerability of different marble qualities. Figure 7 represents, in semi-logarithmic scale, the measured permanent dilatation as a function of the number of thermal cycles from $-50° C$ to $+55° C$ performed on three different marble types. It is noteworthy that such graphs are, approximately, represented by straight lines. The slope of each graph may thus represent an index of the material attitude towards damage: smaller slopes correspond to more durable materials.

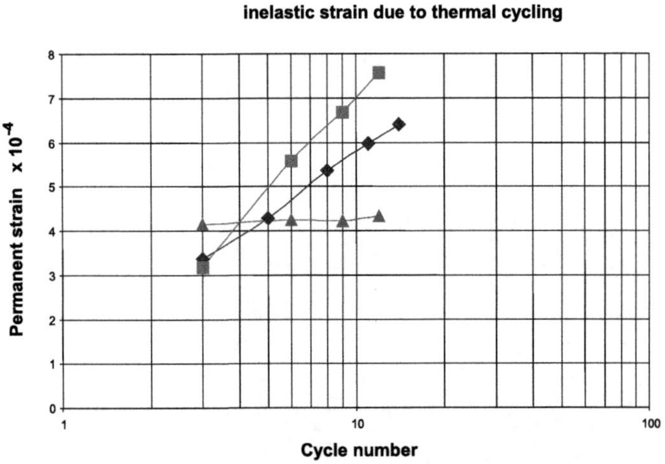


**Fig. 7.** Permanent dilatation vs. number of cycles (semi-log scale) for three different qualities of marble. Cycles performed between $-50°C$ and $+55°C$

The reason why different qualities behave differently consists in the microscopic arrangement of the crystals. Any marble is composed for more than 99% of pure calcite, whereas the other accessory components have practical no influence on the mechanical properties even if they are responsible of chromatic effects, likewise the veining. The main difference in the examined marble types consists, at the microscopic level, in the textural arrangement of the constituent calcite grains, which may range between the two extreme cases termed "homoblastic" and "xenoblastic", schematically represented by thin sections in Figures 8a-b, respectively. The first is composed of regular-shaped grains, with straight or gently curving boundaries; the second is character-

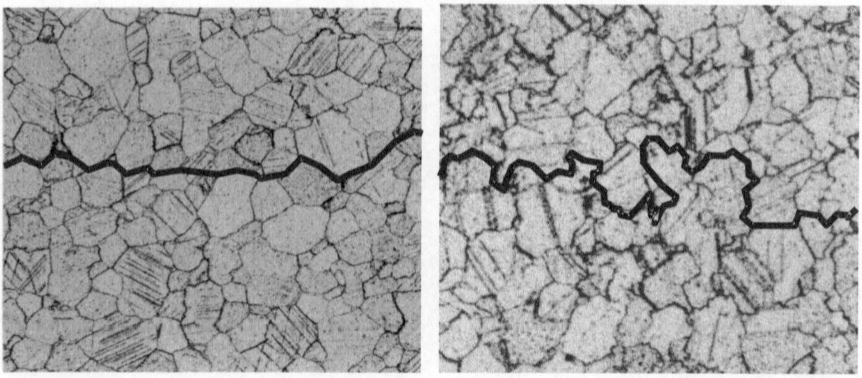

a b

**Fig. 8.** Typical homoblastic (a) and xenoblastic (b) textures for two different marble types

ized by the interlacing of irregular crystals closely fitting along their wavy contours. Of course, due to the higher imbrications, granular decohesion is certainly more favored in homoblastic than in xenoblastic marbles. Indeed, the curve with lowest slope in Figures 7 corresponds to xenoblastic marbles. Therefore, this study suggests that the grain texture may be a major qualifying characteristic for selecting marbles able to endure over the years.

**Acknowledgement.** This research has been supported jointly by the City of Helsinki and the European Community and is part of the research program "Developing long term durability of marble façades" (MARA project). The Experiments referred to in Section 2 have been performed in the VTT laboratories, Espoo, Finland.

## References

1. Franzini M. (1995) Stones in Monuments: natural and anthropogenic deterioration of marble artifacts. Eur. Journ. of Mineral. **16**, 735-743
2. Giacomini A, Iori I, Royer-Carfagni G. (2001) A micromechanical model for the granular decohesion of marble. In: Särkkä P, Eloranta P. (eds.): Rock Mechanics - A Challenge for Society, Proceedings of the International Symposium Eurock 2001, Espoo, Finland. Balkema, Rotterdam, 71–76
3. Royer-Carfagni G. (1999) On the thermal Degradation of Marble. Int. J. Rock Mech. Min. Sc. **36**, 119-126
4. Royer-Carfagni G. (2000) Some Considerations on the Warping of Marble Façades. The example of Alvar Aalto's Finland Hall in Helsinki. Constr. Build. Mat. **13**, 449-457
5. Winkler E. M. (1994) Stone in Architecture: Properties, durability. Springer, Berlin

# Study of Thermal and Kinematic Phenomena Associated with Quasi-static Deformation and Damage Process of some Concretes

Vincent Huon, Bruno Cousin, Olivier Maisonneuve

Laboratoire de Mécanique et Génie Civil,
cc 048, Université Montpellier II,
F-34095 Montpellier Cedex 5, France

## 1 Introduction

Since many years, the use of optical methods for studying behaviour of materials has made great strides. These techniques give information to improve macroscopic behaviour laws and allow observing initiation and propagation of localization phenomena. In the field of civil engineering materials and structures, optical measurements of the displacement field are more and more used (stereophotogrammetry [3], for example). On the other hand, infrared thermography is not very used to study the behaviour of concretes. For dynamical solicitation, it is common to refer of M.P Luong's works [6].

This work shows what we are able to obtain in the framework of quasi-static mechanical solicitations imposed on concretes by using digital speckle images correlation in whiting light mode and infrared thermography. In this paper, we study both Reactive Powder Concrete (RPC) and damaged freeze /thaw plain concrete.

## 2 Experimental arrangement and studied materials

### 2.1 Experimental device

The experimental set-up uses a tension-compression testing machine (either a DARTEC ± 100 kN or a MFL UED40 400 kN) coupled with an infrared camera or a CCD video camera. The thermal measures are realized either with a camera AGEMA 880 sw (NETD = 0.2°C at 30°C) or with a camera CEDIP JADE 3 mw (NETD = 0.025°C at 35°C). An image processing is performed in order to filter noisy data provided by the camera.
For speckle images, we used a CCD video camera PANASONIC WC-CP410. The method of measure of the displacement is based on digital images correlation techniques. The differentiation of the displacement field is done by a local approximation method [10].

### 2.2 Studied materials

RPC are ultra-high-strength concretes reinforced with steel fibres [9]. The mechanical properties are a compressive strength of 200 MPa, a flexural strength of 35 MPa and Young modulus around 65 GPa.

The damaged freeze/thaw concrete is a plain concrete with a classic mixture: a granular class of 0-4 mm with water/cement and cement/sand ratios of 0.5 and 0.3, respectively. The specimens have been aged by rapid freeze/thaw cycles at the Norwegian Building Research Institute [3]. The procedure was the ASTM C666. Fourteen cycles were applied with a cooling rate approximately 12°C/h in the limit +5/-18°C.

## 3 Strain localization phenomena occurring during tensile test on RPC 200 samples

The sample is shown in the figure 1a. The grip setting of the specimen, used to assign the tensile loading, is described in the figure 1b. The experimental results are obtained during monotone tensile tests performed at constant cross-head velocity ($V_{CH} = 10^{-2}$ mm.s$^{-1}$).

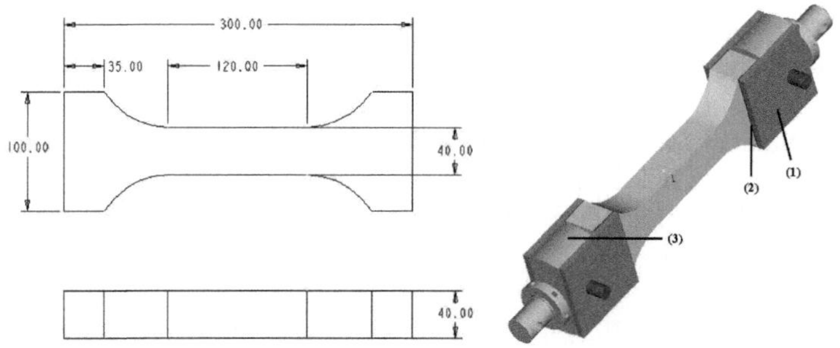


**Fig. 1. a** Direct tensile specimen – **b** Grip system

### 3.1 Loading/unloading tensile tests

To represent the entire test in the same figure, we choose to plot the axial distribution of the tension component of the strain tensor. Figure 2 presents the tension component of the strain tensor of the points located along the longitudinal axis (i.e. $x_1$ axis) for a given loading state (black curve) as a function of time (image number). In a first observation, during the first loading cycle, the RPC behaviour exhibits two stages. The first, until image 45, is due to the elastic behaviour of material with slight homogeneous strain.

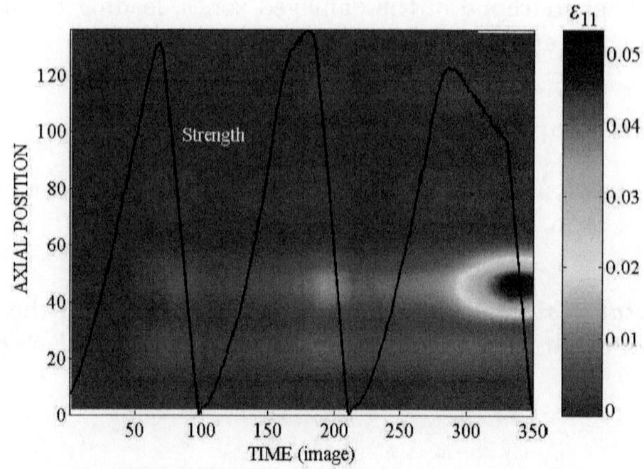


**Fig. 2.** Lengthwise profiles of the longitudinal component of the strain tensor rate ($F_{max} = 26kN$, acquisition 1 $i/s$)

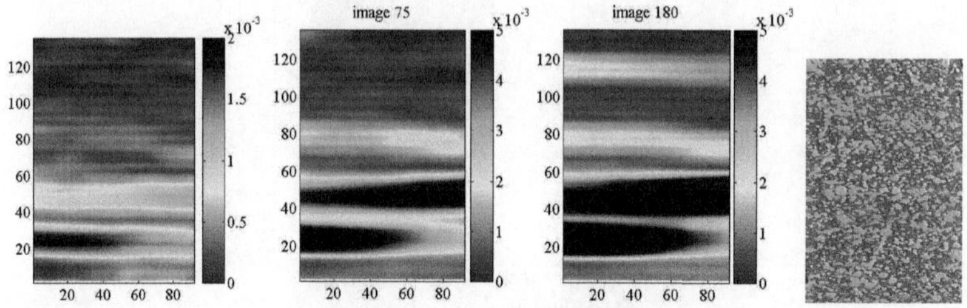


**Fig. 3.** Longitudinal strain field at the image
**a** 45 – **b** 75 – **c** 180 – **d** image 45 of the sample

In the second stage, many areas of strain localization ("multi localization") appear which may lead to the development of macro cracks.

Although we observe two strain localization areas, on figure 3a, no crack is visible on the sample. At this time, it seems that the macro crack will appear in the localization area A (fig. 3a). In fact, this is the area B that leads to the creation of macro crack (fig. 2).

During the loading, we observe (fig. 2) that residual strain are more or less important depending on the level of localization. Contrary to the area C, the areas A and B are the seats of residual strain.

The second loading cycle shows that the localization areas are irreversible (permanent). From the outset of loading, we observe an accumulation of deformation in the areas already identified. These areas are due to a state of damage more important than in the other parts of the sample. This second cycle also illustrates the essential role of the fibres in the transmission of stress around the neighbourhood of the damaged zones, leading to the apparition of a new damaged area D, for example.

### 3.2 Tensile-compressive tests

An outstanding characteristic of the concrete behaviour is the unilateral effect. How is it for a RPC?

***Experimental results*** The figure 4 gives the evolution of the longitudinal component of the strain tensor for a given loading state (black curve).

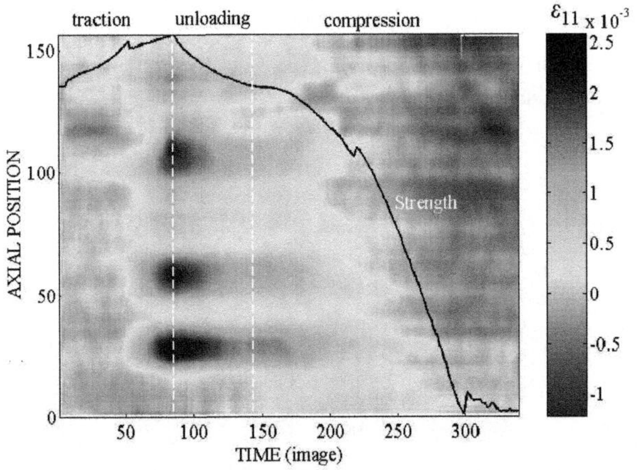


**Fig. 4.** Lengthwise profiles of the longitudinal component of the strain tensor rate ($F_{max} = 9.5kN$ in traction and –60 kN in compression, acquisition 1 i/s)

We distinguish on this figure three phases according to the loading state: (1) the traction which leads to "multilocalisation" of the strain; (2) the unloading associated with residual strain distributed on the entire sample but more important where there were strain localizations; (3) during the compression the residual strains progressively come out. At the end of the compression ($\sigma = 40$ MPa) we have a quasi-homogeneous strain state.

***Stress-strain relationship analysis*** For a plain concrete, a compressive loading applied to the damaged materials induces a recovering of the stiffness (unilateral character). Let us consider the case of RPC?

With the kinematic measures, we can plot the stress-strain diagram for a gauge length of 30 mm at the centre of the sample.

On figure 5a, we observe damage during the traction and a recovering of the stiffness at the beginning of the compressive loading. We can see a difference of the stiffness between an undamaged sample and a damaged sample in the figure 5b. It seems there is not a total recovering of stiffness due to the fibres that interfere with crack closure.

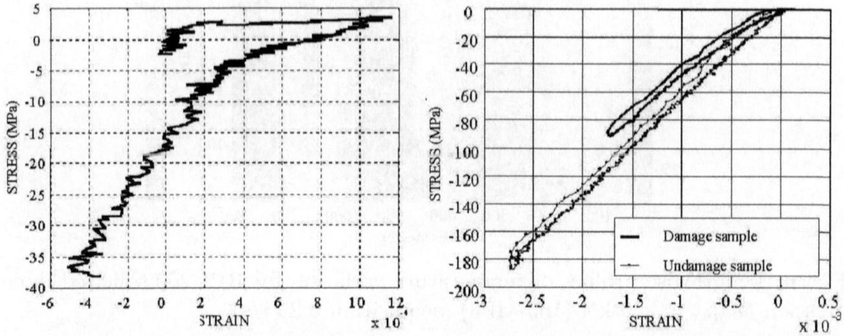


**Fig. 5.** Stress-strain relationship of a RPC sample – **a** Undamaged solicited in traction-compression – **b** Undamaged and damaged solicited in compression

## 4 Thermal effects associated to deformation of RPC200

Observations by the infrared thermography have been achieved on the undamaged sample during three loading/unloading cycles in compression. The experimental results are obtained during test performed at constant cross-head velocity ($V_{CH} = 5.10^{-2}$ mm.s$^{-1}$).

### 4.1 Experimental results

The evolution of the temperature variation of the median axis of the sample versus time is represented figure 6. In overprinting, the strength evolution is plotted (white curve).

During the test, the temperature is uniform in the sample and we observe positive evolution of the temperature associated with the loading evolution. The amplitude of the temperature variation is around 0.3°C for a maximal stress of 160 MPa. At first sight, this effect could be attributed to a thermoelastic coupling that we would like study.

### 4.2 Valuation of the thermoelastic coupling

We start our analysis with the linearized heat conduction. The second member only includes the thermoelastic coupling as heat source term. The linearization is justified since the thermal effect is negligible.

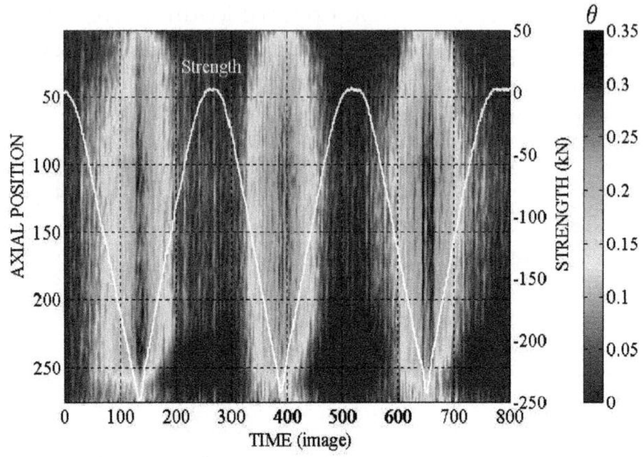


**Fig. 6.** Lengthwise profiles of temperature variation for RPC200 solicited in compression ($F_{max}$ = 250 kN (160 MPa), acquisition 6.25 i/s)

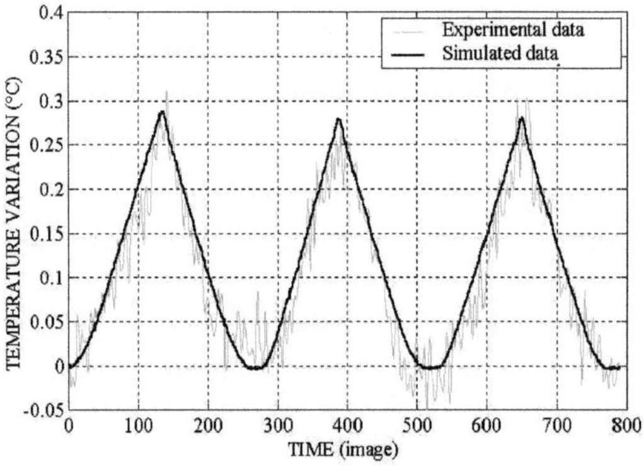


**Fig. 7.** Comparison simulation-experiment

The experimental observations (temperature variation in phase with the loading) and the thermal features of the material (low conduction) enable us to do the quasi-adiabatic working hypothesis during the test. After the corresponding simplification, we can write:

$$\rho C \frac{\partial \theta}{\partial t} = -\left(\theta + T_0\right) E \alpha \dot{\epsilon} \tag{1}$$

where $\rho$, E et $\ddot{\epsilon}$ have the usual significance and C and $\alpha$ denote respectively the specific heat capacity and the isotropic dilatation coefficient. While integrating with the initial condition $\theta$ = 0 at t = 0, $\theta$ being always little opposite $T_0$, we get:

$$\theta = -\frac{T_0 \alpha \sigma}{\rho C} \tag{2}$$

where $\sigma$ is the stress in the material at the considered moment.

Figure 7, we note that simulation results are consistent with experimental $\theta$ evolution with $T_0 = 298$ K, $\rho = 2{,}5.10^3$ kg.m$^{-3}$, C = 838 J.kg$^{-1}$.K$^{-1}$ [4] and $\alpha = 12.10^{-6}$ K$^{-1}$, reference value given for concrete with silica granular (BPEL 91). It confirms the interpretation given in § 4.2.

# 5 Compressive loading tests on freeze/thaw damaged plain concrete

Three freeze/thaw damaged samples have been subjected to a compressive test of 20 loading-unloading cycles. We have previously checked the damaged status with a preliminary loading-unloading compressive test. The experimental results are obtained during a test performed at constant cross-head velocity ($V_{CH} = 5.10^{-2}$ mm.s$^{-1}$). The strength evolution is quasi-identical for the three samples.

The thermal effects are studied with the camera CEDIP. For reasons of a storage capacity, we only made several shots, 3 or 4, covering over 3 or 5 loading cycles.

## 5.1 Experimental results

They are displayed for the sample 1563F6 (the more damaged). The evolution of the temperature variation of the median axis of the sample during the test is represented figure 8. In overprinting, the strength evolution is plotted (white curve).

We observe two phenomena corresponding, the first, to the temperature variation in phase with the loading cycles and, the second, to a progressive warming-up of the sample during the test. The same observations (different quantitatively) have been made with the sample 1563F2 and 1563F10 (the less damage).The first phenomenon has already been identified as thermoelastic effect (see §4-RPC). In the following, we will focus on the analysis of the second phenomenon.

## 5.2 Study of the progressive warming-up

Figure 9, we compare the evolution of the temperature variation of 3 points for each sample, during the test. The most important temperature variations are observed on the previously identified sample as the most damage. To operate the results appropriately, it is necessary to quantify what could be

due to the drift of the camera and of the ambient temperature. A sample has been filmed without any solicitation during a time equivalent to the time-test. The drift (positive) remains lower than 0.07°C. There is a small impact of camera-environment drift in the observed warming-up.
Consequently, the evolution of the temperature map of the sample 1563F10 can be allocated to the phenomenon of drift. On the other hand, it is not the same way for the samples 1563F2 and 1563F6. The increase of temperature observed at the end of the test (about 0.3°C) are much higher than what has been observed for the drift. The differences between the sample 1563F2 and 1563F6, on the one hand, and the sample 1563F10, on the other hand, must be related to the states of different damage induced by the freeze/thaw cycles. We can attribute this progressive warming-up to dissipative phenomena associated with the state of damage of the sample. Besides, we note (fig.8 and fig.9) a non-homogeneous warming-up along the sample. Unless there was a non-uniform drift of the camera in the space that we did not study, this heterogeneity of temperature could be associated with heterogeneity of the state of damage.

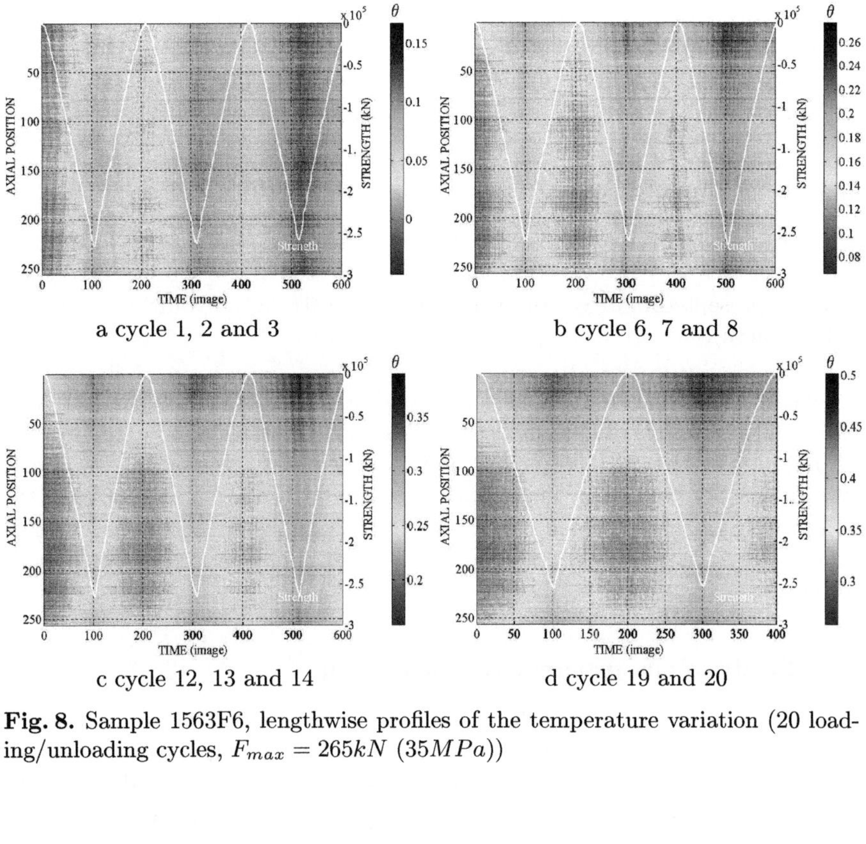


a cycle 1, 2 and 3

b cycle 6, 7 and 8

c cycle 12, 13 and 14

d cycle 19 and 20

**Fig. 8.** Sample 1563F6, lengthwise profiles of the temperature variation (20 loading/unloading cycles, $F_{max} = 265kN$ $(35MPa)$)

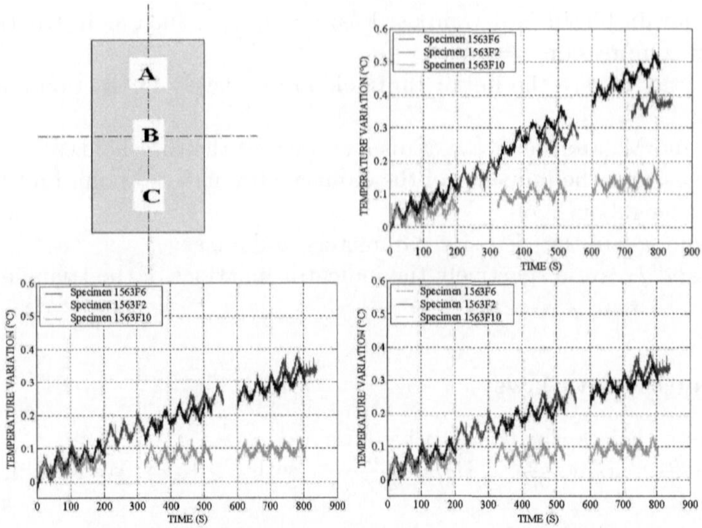


**Fig. 9.** Comparison of the temperature variation

# 6 Modelling of the damage thermoelastic behaviour of a plain concrete

To take into account thermal phenomena associated with the deformation of concrete, which damage it, we propose to expand the M. Fremond and B. Nedjar's model [8]. In the modelling, the temperature variable is taken into account through thermal phenomena associated with thermal dilatation.

## 6.1 Free energy and pseudo-potential of dissipation

We are led to choose the free energy, $\psi$, and the pseudo-potential of dissipation $\phi$ as:

$$\Psi = \beta_t \left[ \tfrac{1}{2} \left( 2\mu \, tr \left[ \epsilon_t . \epsilon_t \right] + \lambda \left( \left( \langle tr \epsilon_t \rangle^{+} \right)^2 \right) \right) - (3\lambda + 2\mu) \, \alpha \, tr \epsilon_t \right] +$$
$$\beta_c \left[ \tfrac{1}{2} \left( 2\mu \, tr \left[ \epsilon_c . \epsilon_c \right] + \lambda \left( \left( \langle tr \epsilon_c \rangle^{-} \right)^2 \right) \right) + (3\lambda + 2\mu) \, \alpha \, tr \epsilon_c \right] +$$
$$W_t \left(1 - \beta_t\right) + W_c \left(1 - \beta_c\right) + \tfrac{k}{2} \left[ \left(grad \beta_t\right)^2 + \left(grad \beta_c\right)^2 \right] - \tfrac{\rho C}{2 T_0} \theta^2 + I_c \left(\beta_t, \beta_c\right)$$
$$\Phi = \tfrac{1}{2} \left\{ c_t \dot{\beta}_t^2 + c_c \dot{\beta}_c^2 \right\} - \tfrac{1}{2} \dot{\beta}_t \left( \tfrac{1-\beta_t}{1 - M_t \beta_t} \right) \left[ 2\mu tr \left[ \epsilon_t . \epsilon_t \right] + \lambda ( (\langle tr \ \epsilon_t \rangle^{+})^2 \right] -$$
$$\tfrac{1}{2} \dot{\beta}_c \left( \tfrac{1-\beta_c}{1 - M_c \beta_c} \right) \left[ 2\mu tr \left[ \epsilon_c . \epsilon_c \right] + \lambda ( (\langle tr \ \epsilon_c \rangle^{-})^2 \right] + I_{-} \left( \dot{\beta}_t, \dot{\beta}_c \right)$$

With the notations:

- $\beta_t$ and $\beta_c$ are the damage quantity respectively in tension and compression ($\beta = 1$, undamaged material and $\beta = 0$, complete damaged material);

- the strain tensor is decomposed as: $\epsilon = \epsilon_t$ - $\epsilon_c$, such as in traction $\epsilon = \epsilon_t$ and in compression $\epsilon$ = -$\epsilon_c$;
- $W_t$ and $W_c$ are the initial thresholds respectively in traction and compression;
- $M_t$ and $M_c$ are the factor of displacement of the thresholds $W_t$ and $W_c$;
- K measures the influence of the damage at a material point on the damage of its neighbourhood;
- $c_t$ and $c_c$ are the viscosity parameters of damage;
- $I_C$ and $I_-$ are respectively the indicator functions of the triangle $C$ and of the set $]-\infty,0] \times ]-\infty,0]$.

### 6.2 Constitutive laws

With these choices, the constitutive laws are:

$\sigma^r = \frac{\partial \Psi}{\partial \epsilon}$ , $B_i^r \in \frac{\partial \Psi}{\partial \beta_i}$ , $H_i^r = \frac{\partial \Psi}{\partial grad \beta_i}$ , with $i \in \{t,c\}$. We obtain:

$$\sigma^r = -\beta_c\left(2\mu\epsilon_c + \lambda\left(\langle tr\epsilon_c\rangle^-\right)Id + (3\lambda+2\mu)\,\alpha\theta Id\right)$$
$$+\beta_t\left(2\mu\epsilon_t + \lambda\left(\langle tr\epsilon_t\rangle^+\right)Id - (3\lambda+2\mu)\,\alpha\theta Id\right) \quad (3)$$

$$B_i^r \in \tfrac{1}{2}\left(2\mu tr\left[\epsilon_i.\epsilon_i\right] + \lambda\left(\langle tr\epsilon_i\rangle^{sgn(\eta)}\right)^2\right) + \eta\,(3\lambda+2\mu)\,\alpha\theta tr\epsilon_i - W_i + \partial I_c(\beta_i)$$

with $\eta = 1$ if $\epsilon_i = \epsilon_t$, $\eta$ =-1 if $\epsilon_i = \epsilon_c$. We let $H_c = H_c{}^r + H_c{}^{ir}$, $H_t = H_t{}^r + H_t{}^{ir}$, $B_c = B_c{}^r + B_c{}^{ir}$ et $B_t = B_t{}^r + B_t{}^{ir}$, which in addition $\sigma = \sigma^r + \sigma^{ir}$, according to take into account in the power of the internal forces associated with damage and its gradient.

### 6.3 Equations of damage evolution

Like [8], the principle of virtual power is rewritten to take into account the power of the internal forces associated with microscopic motions.

$$\rightarrow c_i\dot{\beta}_i - k\Delta\beta_i + \partial I_-(\dot{\beta}_i) + \partial I_c(\beta_i)$$
$$\ni -\frac{1}{2}\left(1 - \frac{1-\beta_i}{1-M_i\beta_i}\right)\left(2\mu tr\left[\epsilon_i.\epsilon_i\right] + \lambda\left(\langle tr\epsilon_i\rangle^{sgn(\eta)}\right)^2\right) +$$
$$W_i + \eta\,(3\lambda+2\mu)\,\alpha\theta tr\epsilon_i \quad in\ \Omega \quad (4)$$

$$\rightarrow \quad k\frac{\partial\beta_i}{\partial n} = 0 \quad on\ \partial\Omega$$

### 6.4 Heat conduction equation

While adopting the assumptions leading to the classic form of the heat conduction equation:

$$\rho C\dot{T} - r + divq = D_1 + \rho T\frac{\partial^2\psi_1}{\partial T\partial\alpha_k} : \dot{\alpha}_k = w_{ch} \quad (k = 1, ..., n) \tag{5}$$

We obtain after linearisation:

$$\rho C\left(\frac{\partial\theta}{\partial t} + \frac{\theta}{\tau_{th}}\right) = w_{ch} \tag{6}$$

where the $\alpha_k$ are the n state variables (except the temperature), $W_{ch}$ is the volume density of heat sources, D1 is the intrinsic dissipation, r symbolizes the external heat supply and q the heat influx vector. $\tau_{th}$ is a characteristic constant of the local thermal looses. The subsequent assumptions to the linearization are the following: (1) the heat influx vector q is written $q = -k_{th}gradT$ where $k_{th}$ is the thermal conductivity coefficient assumed constant and independent of the temperature; (2) the external heat supply r is time independent; (3) for quasi-static processes, the convective terms of the material time derivative are negligible; (4) the thermal looses are linear [1].

The expression of $W_{ch}$ is, while assuming k, $W_t$, $W_c$, grad$\beta_t$, grad$\beta_c$, $\beta_t$, and $\beta_c$ independent of the temperature:

$$w_{ch} = D_1 + (3\lambda + 2\mu)\,\alpha tr\epsilon_c T\dot{\beta}_c - (3\lambda + 2\mu)\,\alpha tr\epsilon_t T\dot{\beta}_t - \tag{7}$$

$$(3\lambda + 2\mu)\,\alpha\beta_c Ttr\dot{\epsilon}_c - (3\lambda + 2\mu)\,\alpha\beta_t Ttr\dot{\epsilon}_t$$

with

$$D_1 \in c_c\dot{\beta}_c^2 - \tfrac{1}{2}\dot{\beta}_c\left(\tfrac{1-\beta_c}{1-M_c\beta_c}\right)\left(2\mu tr\left[\epsilon_c.\epsilon_c\right] + \lambda\left(\left(\langle tr\epsilon_c\rangle^-\right)^2\right)\right) + \dot{\beta}_c\partial I_-\left(\dot{\beta}_c\right)$$

$$+c_t\dot{\beta}_t^2 - \frac{1}{2}\dot{\beta}_t\left(\frac{1-\beta_t}{1-M_t\beta_t}\right)\left(2\mu tr\left[\epsilon_c.\epsilon_t\right] + \lambda\left(\left(\langle tr\epsilon_t\rangle^+\right)^2\right)\right) + \dot{\beta}_t\partial I_-\left(\dot{\beta}_t\right) \tag{8}$$

and for the following terms in the order: the terms of temperature - damage coupling (respectively in compression and in traction) and the terms of isentropic sources.

### 6.5 Numerical simulation of a compressive test

Six load-unloading cycles in compression piloted in deformation have been simulated. The state of stress is considered like uniform in the space, what corresponds to grad $\beta_i = 0$ and k = 0. $\tau_{th}$ is determined by a proportionality ratio of the thermal conductivity of steels to those of the concretes, $\tau_{th} =$ 10000 s. Some experimental observations confirmed this value and leads to do the assumption of quasi-adiabatic tests (elevated $\tau_{th}$). For $W_c$, $M_c$, and $c_c$, the proposed values in [8] have been taken: $W_c$ = 0,7.10$^{-2}$ MPa, $M_c$ = 0,8, $c_c$ = 0,5 MPa.s with besides E = 25 GPa, $\gamma$ = 0,2, $\rho$ = 2,2.10$^3$ kg.m$^{-3}$, $T_0$ = 298 K, $\dot{\epsilon}_0$ = -1,3.10$^{-4}$ s$^{-1}$ and taken in the literature $\alpha$ = 10.10$^{-6}$ K$^{-1}$, C = 838 J.kg$^{-1}$.K$^{-1}$.

**Mechanical behaviour** The figure 10 gives the stress-strain answer of the material and the figure 11 the evolution of the damage during the time. We recover a classic behaviour for the concrete requested in compression. To every cycle of loading the material damages it a few more (fig. 11). The consequences of the thermoelastic coupling are as well imperceptible on these curves as on the experimental curves.

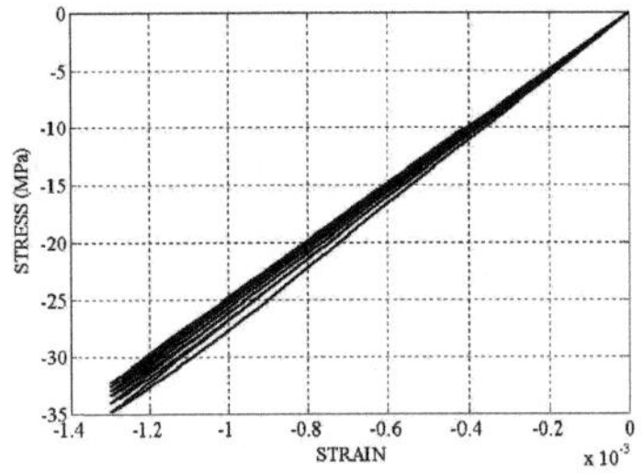


**Fig. 10.** Stress-strain relationship

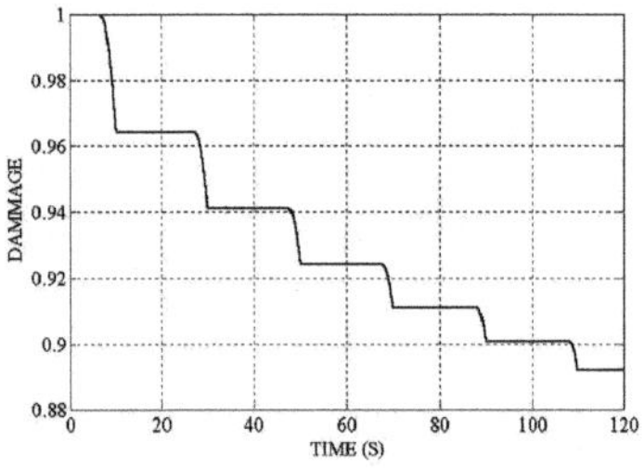


**Fig. 11.** Damage evolution

**Thermal behaviour** The thermal behaviour, resulting of the deformation and the damage of the material, is characterized by the evolution during the time of the temperature dependence, $\theta$ (fig. 16), and of the heat sources, $W_{ch}$ (fig. 15), corresponding to the intrinsic dissipation, $D_1$ (fig. 12), and to the terms of coupling temperature-damage, noted T-$\beta$ (fig. 13), and temperature-deformation, noted T-$\epsilon$ (fig. 14). During a cycle, $D_1$ (fig. 12) and T-$\beta$ (fig. 13) are nonzero if the damage grows ($\beta_c$ decreases) to the load, the threshold of damage being exceeded. The maximal values of D1 and T-$\beta$ decrease with the passing of the cycles because of the increase of the damage threshold. We note D1 is negligible in relation to the other heat sources terms. It is not major in the temperature variations generated by the processes of deformation and damage.

With regard to the terms of coupling, we note, for a given cycle, a "competition" between T-$\beta$ (fig. 13), that increases, with the damage and T-$\epsilon$ (fig. 14), that falls as and when the damage grows. The figure 15 gives, in practice, the result of this competition, D1 being negligible.

If we look at the evolution of $\theta$, under the hypothesis of quasi-adiabaticity, we note (fig. 16) the apparent superposition of two phenomena:

- the linear temperature variations, in phase with the loading-unloading

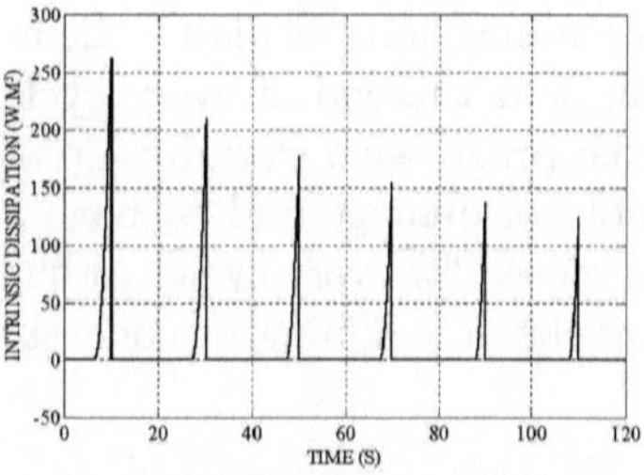


**Fig. 12.** $D_1$ evolution

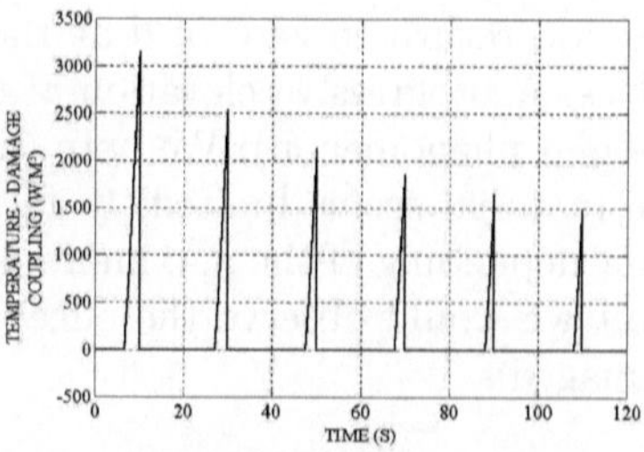


**Fig. 13.** T-$\beta$ evolution

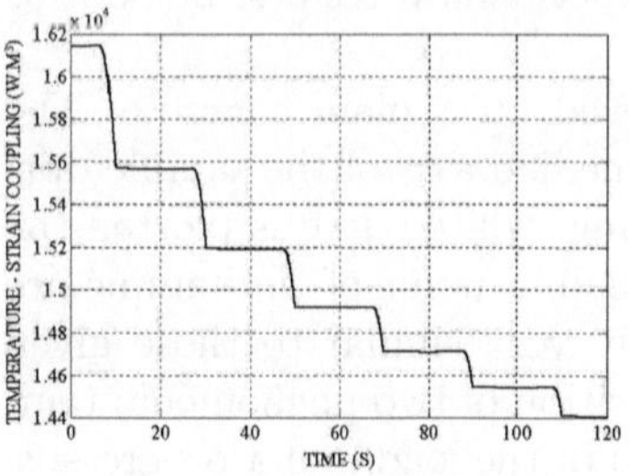


**Fig. 14.** T-$\epsilon$ evolution

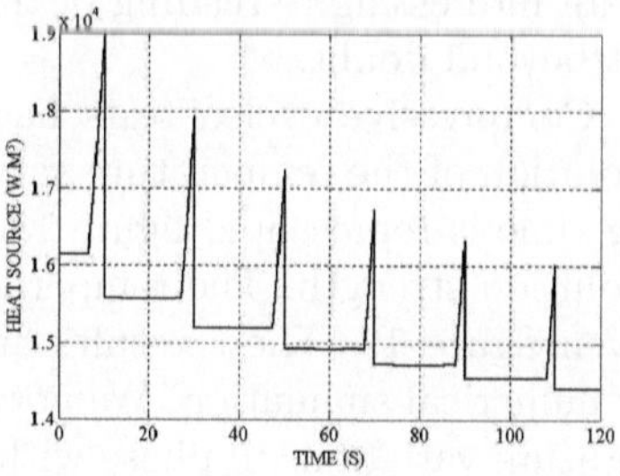


**Fig. 15.** Heat sources evolution

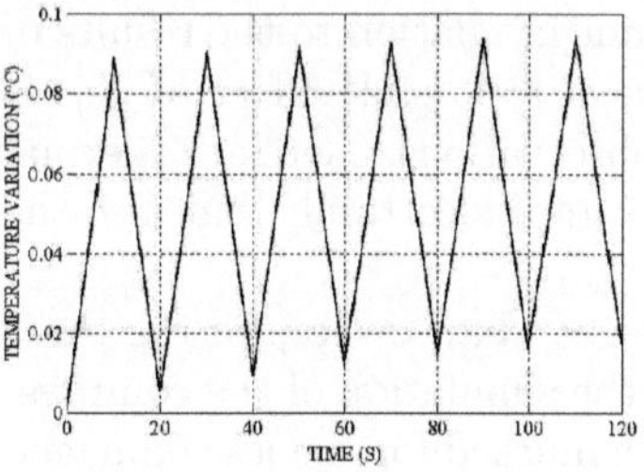


**Fig. 16.** Temperature evolution

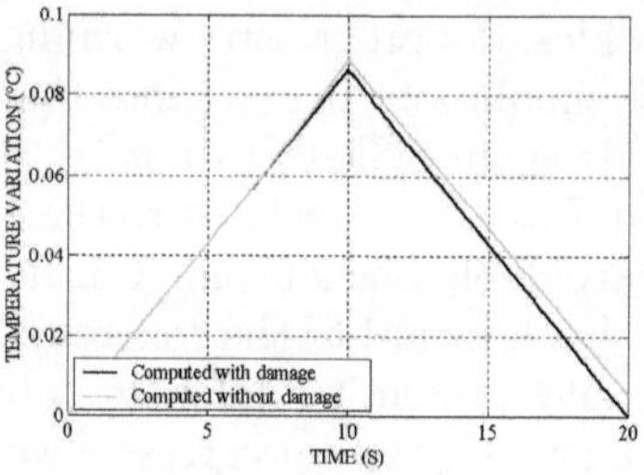


**Fig. 17.** For the first cycle, comparison with computed temperature

cycles, even if the damage evolves;

- a progressive warming-up of the material.

The first phenomenon results from the competition between T-$\beta$ and T-$\epsilon$. If we perform a zoom, on the first cycle, while making represent the evolution of the temperature of the material while assuming that they don't damage it, we note (fig. 17):

- a thermoelastic evolution before the threshold of damage (5 s);
- then an influence of T-$\beta$ and T-$\epsilon$ that offset themselves (until 7,5 s);
- then the major influence of T-$\beta$ lead to a temperature variation (nonlinear) more important than the one due to the only thermoelastic effect (linear) and, a fortiori, to the thermoelastic effect with damage (nonlinear).

If we add that the thermoelastic effect to the unloading of the damaged material ($E' < E$), cannot compensate the contribution of heat to the load, the non return to zero of $\theta$ at the end of the loading quite explains themselves. A progressive elevation of $\theta$ results from it at the end of every cycle (second phenomenon). We note (fig. 16) that the progressive warming-up of the material seems limited. In fact, the threshold of damage and $\beta_c$ have a limit depending of the maximum deformation imposed for every cycle. At the limit we would observe the effect of the thermoelastic coupling of concrete damaged.

**Confrontation with the experience** The results of the simulation lead to an interesting re-reading of the obtained experimental results, but not as yet beyond doubt.

Compressive cycled tests have been achieved on a plain concrete. The evolution of the temperature variation of the median axis of the sample during time is represented figure 18. In overprinting, white curve is plotted the evolution strength. The temperature variations in a point of the sample are given figure 19. These results are qualitatively very similar to those given by numerical simulation. We note the superposition of two phenomena: temperature variations in phase with the evolution of the load and a progressive warming-up of the sample. This last lets think about an effect of the damage. However, a doubt subsists on the decrease of the peak load, during the cycles. Besides, the progressive warming-up is important in relation to the results of the simulation. It is to note that the detection of a possible effect of T-$\beta$ is made again difficult by a noisy signal. In our observations, there are thermal manifestation associated to the damage of the samples certainly, but also the unavoidable phenomena of drift.

With regard to the concrete damaged by freeze/thaw cycles, for the three samples presenting different states of damage, the simulation of the compressive cycles gives a progressive warming-up more marked for the less damaged sample. The tests gave an inverse result and of intensity more marked (0,3°C at the end of test). It seems to indicate that other dissipative phenomena has been activated, on these pre-damaged sample, linked for example to contact rubbing between the lips of the microcracks, a priori of any orientations because non generated by an axial loading.

## 7 Conclusions

In spite of important experimental difficulties due to the studied material nature and to the thermal phenomenon weakness, it has been possible:

- to clearly show the thermoelastic coupling RPC's;
- to perform a first scanning of the temperature-damage couplings and dissipative phenomena either appearing in the damage thermoelastic behaviour of damageable concrete solicited with quasi-static way in simple compression.

Otherwise, during tests led on a RPC, the mechanical measures permitted to show precocious "multilocalisation" phenomena of the deformation.

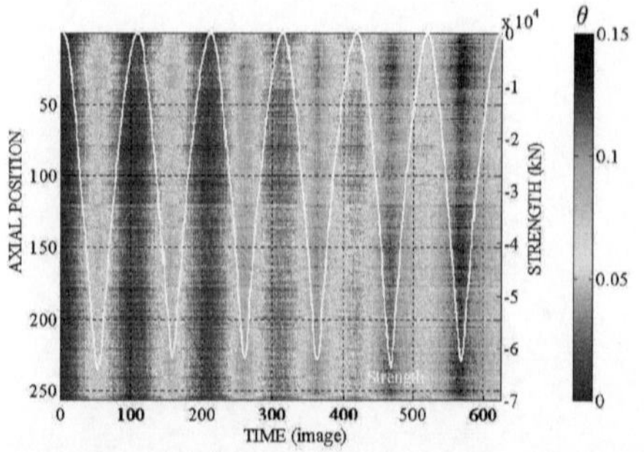


**Fig. 18** Lengthwise profiles of temperature variation for RPC200 solicited in compression ($F_{max}$ = -64 kN (40 MPa))

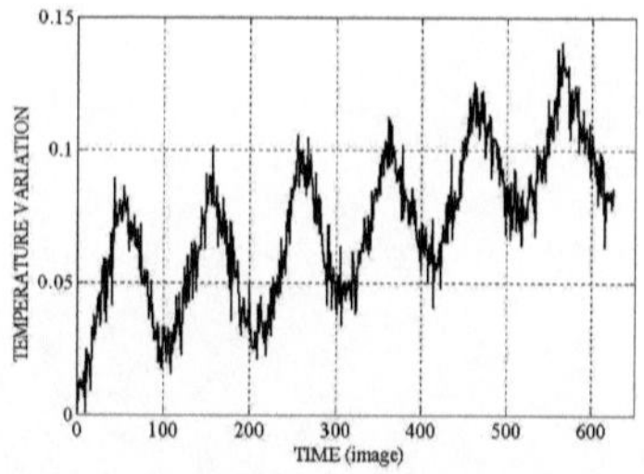


**Fig. 19** Temperature variation in one point of the sample

To go farther in the analysis of the phenomena, it will agree to define the more rigorous experimental protocols, in order to be able to do the part of that reverts to the elastic deformation and the existence, the creation or to the development of micro or macrocracks.

On the point of view of the applications, the obtained results on the freeze/thaw damaged samples permit to consider interesting repercussions in the domain of the fatigue. In any case, it will be important to really master the phenomena of drift of the camera-environment set and to be able to, as for the metallic materials, to pass from the temperature maps to the heat sources.

## References


1. Chrysochoos A (1987) Dissipation et blocage d'énergie lors d'un écrouissage en traction simple. Thèse de docteur d'état, Université Montpellier II
2. Chrysochoos A, Louche H (2000) An infrared image processing to analyze the calorific effects accompanying strain localisation, Int. J. of Eng. Sci., Vol 38, pp 1759-1788
3. Desrues J (1983) Sur l'application de la stéréophotogrammétrie à la mesure des grandes déformations, Rev. Fr. de Mécanique, Vol 3, pp 55-63
4. Jacobsen S. (1999) Freeze/thaw cracking of w/c=0.5 prisms, Byggjorsk report 9526, Norvège
5. Lemaitre J, Chaboche JL (1985) Mécanique des matériaux solides, Dunod.
6. Luong MP (1985) Vibrothermographie infrarouge d'un béton endommagé, C. R. Acad. Sci. Paris, t 301, Série II, n°7, pp 459-464
7. Mazars J, Berthaud Y (1989) Une technique expérimentale appliquée au béton pour créer un endommagement diffus et mettre en évidence son caractère unilatéral, C. R. Acad. Sci. Paris, t 308, Série II, pp 579-584
8. Frémond M, Nedjar B (1996) Damage, gradient of damage and principle of virtual power, Int. J. Solids Struct., Vol 33, pp 1083-1103
9. Richard P, Cherezy M (1995) Les Bétons de Poudres Réactives, Annales de l'ITBTP Série béton 320, n°532, pp8 5-102
10. Wattrisse B et all (2001) Analysis of strain localization during tensile tests by digital image correlation, Experimental Mech., Vol 41, n°1, pp 29-39

# Using Infrared Thermography for the Experimental Analysis of the Thermal Profile of Building Elements Subjected to Thermal and Mechanical Action

Italo Meroni, Valter Esposti

Istituto per le Tecnologie della Costruzione
Consiglio Nazionale delle Ricerche
20098 San Giuliano Milanese, Italia

**Abstract.** Infrared thermography is nowadays a well-established method of analysis for thermal irregularities in construction materials, components and systems. Many laboratory experimentations, but most of all the great number of in-field surveys, have allowed to get a good knowledge about the tools used to this end and the physical phenomena that make the basis for surveying the parameters to be used for the definition of the investigated performances. Investigations were therefore directed towards different fields characterised by physical phenomena that could be identified with available tools.
This short presentation will illustrate some meaningful analyses carried out by ITC, whose results show good possibilities of using the thermographic method especially for the verification of adhesions and detachments of plasters and the determination of structural characteristics of construction materials and components.

## 1 Adhesion and detachment of plaster

Using a thermal methodology to point out plaster detachment, implies a different behaviour with respect to heat propagation between construction materials and air; it is assumed that there is a thin air layer between plaster and masonry, due to the detachment.
The existence of this air layer in a state of rest, modifies the heat propagation mechanisms with respect to an area without any such layer, due to the additional thermal resistance it provides. By applying thermal impulses to the area of interest, any thermal irregularity caused by detachment appears on the thermogram during the cooling phase as a warmer area since the greater thermal resistance of the detached area prevents the heat from leaking out, keeping it for a longer time in the wall at the rear. The thermogram represented in Figure 1 shows an example of plaster detachment detected in-situ by applying the above illustrated principles. This thermogram clearly shows the light detachment areas and the dark adhesion areas.

The thermographic output was subsequently proved to be correct by tearing off the investigated part. Another example is given in the thermographic analysis of Figure 2 carried out in laboratory on plastered concrete specimens on which 4 detachments of plaster have been artificially executed with increasing thickness values from 1 to 4 mm.

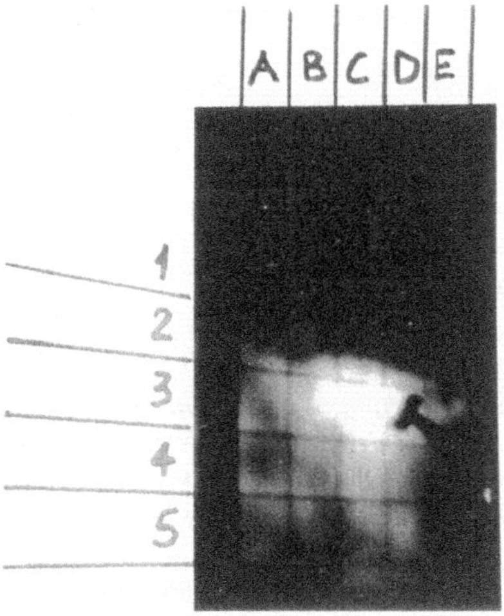


**Fig. 1.** In-situ plaster analysis

The thermogram clearly shows the proportionality between the thermal evidence of the detachment and its extent. Therefore, what has been observed in-situ can be reasonably interpreted in a similar way.

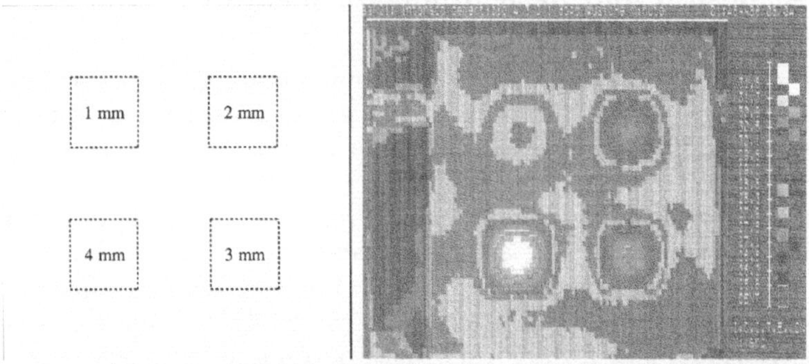


a) detachment thicknesses b) thermogram

**Fig. 2.** Laboratory plaster analysis

**Fig. 3.** Bar under tensile stress

a) b)

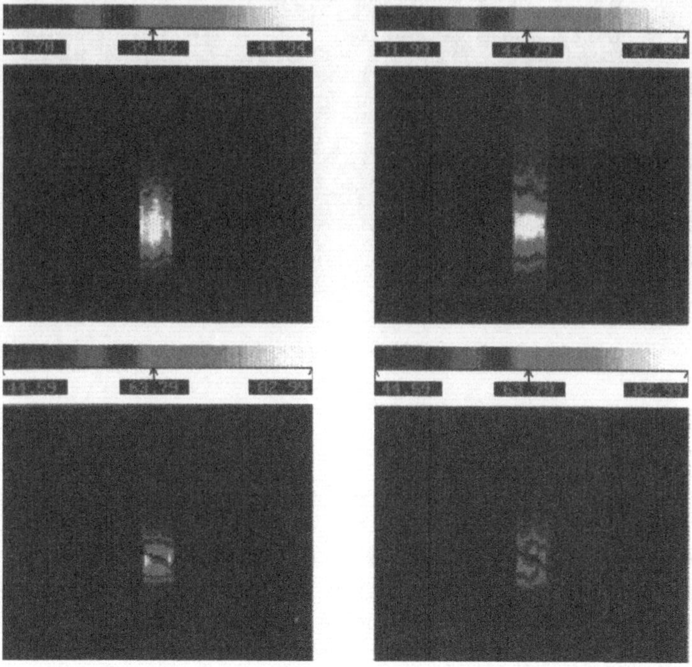

c) d)

**Fig. 4.** Thermograms' sequence of a bar subjected to tensile stress tests up to failure

## 2 Identification of stress states

Stress states may take place in materials in the form of heat dispersion in points where the material is weaker or more stressed, thus outlining areas at risk of failure before failure actually occurs. This condition was analysed by performing tensile stress tests and compressive tests on two samples consisting of steel bars and lightweight concrete prisms. The thermographic analysis highlighted these areas.

**Fig. 5.** Concrete prism under compressive stress

### 2.1 Tensile stress test

The tensile stress test was performed on bars subjected to an increasing stress starting from an initial value up to yield point and failure. The thermograms' sequence illustrated in Figure 4 shows the thermal behaviour of the bar during the phases previous to failure. In Figure 4a it is evident how the stress state is producing a temperature variation in the area of the bar where failure will later occur. In Figure 4b, the area with the highest temperature clearly becomes outlined as yield begins. Figure 4c shows the beginning of the failure phase; it can be clearly noticed how the area where temperature has increased is progressively reducing since all the stress applied is absorbed by the creep and detachment of the material. Finally, in Figure 4d the bar breaks.

## 2.2 Compressive test

The compressive test was executed on lightweight concrete samples subjected to increasing loading and unloading cycles during which thermographies were carried out up to the failure of the sample. Thermographies of Figure 6 illustrate the thermal trend of the sample's surface, starting from conditions of thermal balance represented in the thermogram of Figure 6a where it is possible to notice a slight thermal gradient which increases with height and two slight thermal bridges due to the backing plates of the press. The thermogram of Figure 6b shows the reversal of this gradient and how the section where the temperature increases to a greater extent moves towards the lower part of the specimen. As it may be remarked by examining the thermogram of Figure 6c, the area with the temperature rise goes up towards mid-height and it is precisely in that section that the specimen yields, as shown in the thermogram of Figure 6d.

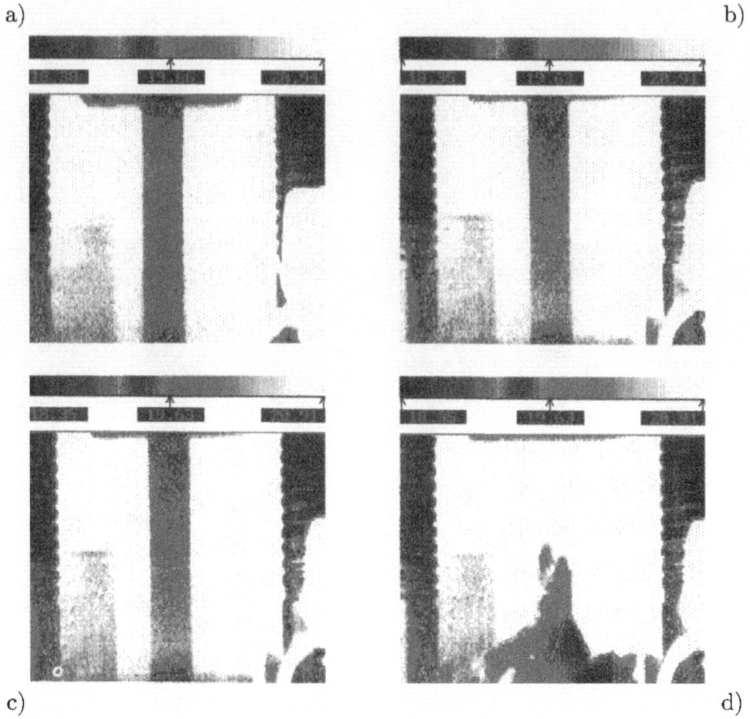


**Fig. 6.** Thermograms' sequence of the sample subjected to cyclic compressive stress with increasing intensity up to failure

## 3 Conclusions

The purpose of the executed analyses was to observe thermal dynamics in the area of plaster detachments and as a result of stress states applied to a series of meaningful samples.

As observed, the analyses provided meaningful responses which help to clearly understand the trend of the investigated phenomena and the sectors where the application of this method may prove to be meaningful.

In particular, it was demonstrated that the possible presence of detachments in plasters could be detected and it may be assumed that, by applying repeated loads on structures, it could be possible to underline the areas that are particularly sensitive to stress, which are therefore potentially exposed to damage or failure risks.

The future development of this method is expected to be based on a deeper knowledge of the minimum energy levels obtained by means of dynamic stress, able to thermally highlight the greatest-stress areas in some typologies of construction components.

# Strains and Stresses Induced by Water Transport in Hygroscopic Elastic Media

Jean Claude Bénet, Ibrahim Mrani, Gilles Fras

Laboratoire de Mécanique et Génie Civil,
cc 048, Université Montpellier II,
F-34095 Montpellier Cedex 5, France

**Abstract.** We propose to analyse the phenomenological equation of water transport in a biconstituent medium with stresses. The work is limited to the case of elastic, isotropic behaviour and an isothermal transfer process.

## 1 Introduction

Water transport in certain heterogeneous media such as gels and biopolymers may be accompanied by considerable deformation causing internal mechanical stresses which interact with water transport mechanisms. In previous work (Auria R and Bénet JC, 1990) on transfer of water in natural rubber, the pattern of variation of the transport coefficient with the water content remained unexplained; the hypothesis of interaction between transfer and deformation was mentioned. It has been shown that internal strain occurring during transfer affect the expression of local diffusion flux (Larché FC and Cahn JW, 1982). Following these publications, we propose to analyse the phenomenological equation of water transport in a biconstituent medium with stresses. The work is limited to the case of elastic, isotropic behaviour and an isothermal transfer process.

## 2 Phenomenological transport equation

### 2.1 General equation, elastic case

The thermodynamics of irreversible linear processes gives the following transport equation in a biconstituent medium during an isothermal process in which evolution is assumed to be quasi-static (Mrani I, 1993) :

$$\boldsymbol{J} = \rho_e(\boldsymbol{u} - \boldsymbol{\vartheta}) = -\frac{L}{T}\,\boldsymbol{grad}\,\mu_e \qquad (1)$$

Gravity and accelerations have been ignored in this equation. $\boldsymbol{J}$ is the transport flux defined from velocities of water $\boldsymbol{u}$ and the solid component $\boldsymbol{\vartheta}$, $\rho_e$ is the apparent bulk mass density of water, $L$ is the phenomenological

transport coefficient, $T$ is the temperature and $\mu_e$ is the chemical mass potential of the liquid.
The state variables used to describe the local state of the system are water content ($w$) and the strain tensor ($\varepsilon_{ij}$). In the neighbourhood of a reference state ($w^\circ$, $\varepsilon_{ij}$), the increment of the chemical potential of water $\mu_e$ is related to the infinitesimal increments in state variables (Coussy O, 1993) by the equation:

$$\rho_s \Delta\mu_e = d_w \Delta w - 3K\beta\Delta tr\varepsilon \tag{2}$$

where $\rho_s$ is the apparent bulk mass density of the solid phase. $d_w$ is defined by $d_w = \rho_s\ (\partial\mu_e/\partial w)$ with constant deformation. K is the coefficient of compressibility: $3K = E/(1 - 2\ \nu)$, where $E$ is Young's modulus and $\nu$ is Poisson's ratio. $\beta$ is the hygrometric expansion coefficient; $tr\varepsilon$ is the trace of the strain tensor.
Within the framework of infinitesimal changes around the reference state, the elastic mechanical behaviour of the medium accounts for the incremental variation in stress tensor $\sigma_{ij}$ according to increments in state variables (Coussy O, 1991) :

$$\Delta\sigma_{ij} = \lambda\Delta tr\varepsilon\delta_{ij} + 2\mu\Delta\varepsilon_{ij} - 3K\beta\Delta w\delta_{ij} \tag{3}$$

where $\delta_{ij}$ is the unit tensor and $\lambda$ and $\mu$ are Lamé's coefficient. According to (3), coefficient $\beta$ is defined by: $3K\ \beta$ = -$(\partial\sigma_{ij}/\partial w)\ \delta_{ij}$, with constant deformation.
Using (2), transport equation (1) becomes:

$$J = -D_w \mathbf{grad} w + D_\varepsilon \mathbf{grad} tr\varepsilon \tag{4}$$

Transport coefficients $D_w$ and $D_\varepsilon$ are related to the phenomenological coefficient $L$ by the following equations:

$$D_w = L\frac{dw}{T\rho_s}; \qquad D_\varepsilon = L\frac{3K\beta}{T\rho_s} \tag{5}$$

## 2.2 Transport equation for a plate

In general, equation (4) is difficult to analyse experimentally. In the case of a plate, the increment of the trace of the stress tensor depends solely on the water content increment (Mrani I. 1993):

$$\Delta\mu_e = \frac{1}{\rho_s}[d_w^* + \frac{2E\beta^2}{1-\nu}]\Delta w \tag{6}$$

According to (6), equation (1) becomes:

$$J = -D^* \frac{\partial w}{\partial x} \tag{7}$$

This equation is similar to Fick's law. Apparent transport coefficient $D^*$ depends on the hygroscopic and mechanical parameters of the material:

$$D^* = \frac{L}{T\rho_s}[d_w^* + \frac{2E\beta^2}{1-\nu}] \tag{8}$$

With :

$$d_w^* = d_w - 9K\beta^2 \tag{9}$$

# 3 Experimental study of the phenomenological equation

## 3.1 Experimental verification

Validation of transport law (7) and deduction of transport coefficient values $D^*$ are proposed. The experimental study concerned water transport in agar gel plates initially 15 mm thick and with a uniform water content of 10. The samples were placed in a polyethylene glycol (PEG) solution with controlled temperature (30°C). The difference in concentration at the interface between the sample and the solution caused movement of water from the material to the solution, accompanied by shrinkage of the material. Transfer was one-directional along axis $ox$ perpendicular to the facets of the agar gel plate. Experimental conditions are those of the conditions of application of equation (7).

Samples were taken at intervals of time and cut into fine slices 0.6 mm thick at right angles to the transfer direction. Water contents were measured in the slices. This makes it possible to draw water content profiles agains time and to deduce the flux J and the water content gradient.

Figure 1 shows the variation in transport flux according to water content gradient for different values of $w$. This figure shows that the transport flux can be considered to be proportional to the water content gradient. This validates equation (7) in the water content range studied.

## 3.2 Apparent transport coefficient D*

Apparent transport coefficients were calculated using the slopes of the lines in Fig 1. Figure 2 shows the variation in coefficient $D^*$ according to the water content for an interval $1.5 < w < 8$. It can be seen that the coefficient depends strongly on water content. At $w > 6$, the transport coefficient fell rapidly to a minimum of about $w = 6$. Similar movements of coefficient $D^*$ to those in Fig 2 were observed in the case of biphase media (Auria R, 1988, Do Amaral S.P.J, 1992, Gehrmann D and Kast W, 1978) and triphase media (Constock GL, 1963, Crausse P et al., 1984, Ketelaars AAJ, 1992)

In triphase media, the rise in transport coefficient can be attributed to the preponderance of transfers in the gas phase with low water contents. This explanation cannot be used here because of the biphase nature of agar gel. Interpretation based on the equation for coefficient $D^*$ (8) is proposed here. This requires analysis of the term related to contraction of the medium: $2E\beta^2/(1 - \nu)$.

The Young's modulus $E$ of agar gel was determined by one-directional compression tests (Mrani I, 1993). Its variation according to water content reveals considerable rigidification of the material at below $w$=400%. Ultrasonic measurements of Poisson's ratio $\nu$ (Mrani I, 1993) show that this coefficient is practically constant at 0.5 in the water content range investigated. In the hypothesis of a biphase medium, coefficient $\beta$ can be expressed as follows (Mrani I, 1993): $\beta = 1/\ 3(\alpha + w)$; the coefficient increases when $w$ decreases. In the case of agar gel, these variations cause very strong increase in the term $2E\beta^2/(1 - \nu)$ at low water contents, which may account for the increase in coefficient $D^*$ in Figure2.

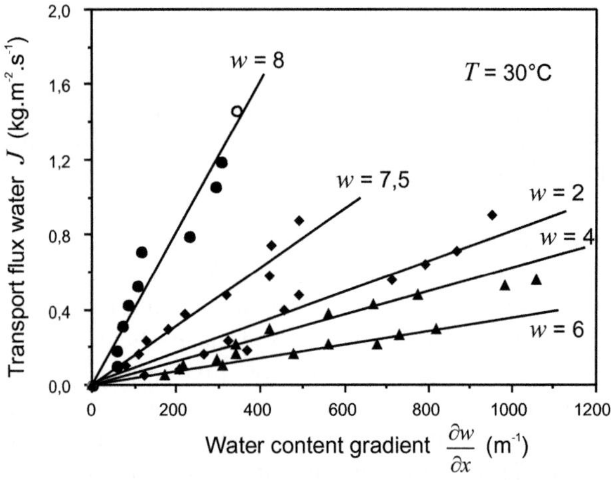


**Fig. 1.** Variation of water transport flux coefficient according to water content gradient for different values of $w$

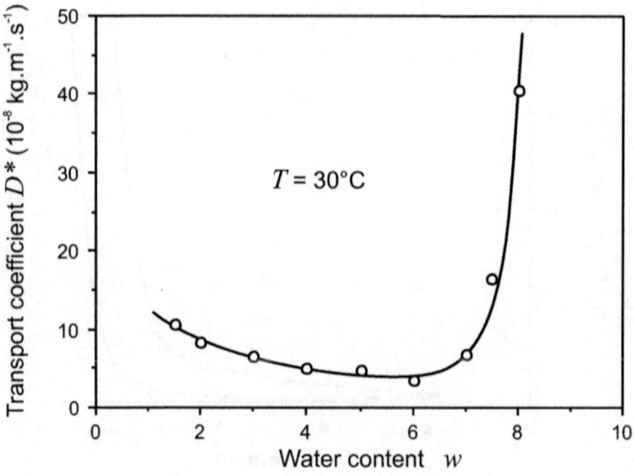


**Fig. 2.** Variation of transport coefficient of water in agar gel

### 3.3 Identification of intrinsic transport coefficients

Unlike the transport coefficient $D^*$ defined in the case of a plate, coefficients $D_w$ and $D_\varepsilon$ are intrinsic to the material. It is thus necessary to determine these coefficients to describe transfers for any geometrical case.
Using expressions (5) and (8), $D_w$ and $D_\varepsilon$ are given by:

$$D_w = \frac{3K\beta}{d_w^* + \frac{2E\beta^2}{1-\nu}} D^* \quad ; D_\varepsilon = \frac{d_w^* + 9K\beta^2}{d_w^* + \frac{2E\beta^2}{1-\nu}} D^* \tag{10}$$

Coefficient $d_w^*$ is determined using the desorption isotherm (Mrani I, 1993) in the case of agar gel. Variation of $D_w$ and $D_\varepsilon$ according to water content are shown in Figure 3 This figure shows that unlike the transport coefficient $D^*$, coefficients $D_w$ and $D_\varepsilon$ display monotonic decreasing on the water content and approach zero when the material is dry. Rise in coefficient $D^*$ (Fig 2) with $w < 6$ did not affect the rate of variation of coefficients $D_w$ et $D_\varepsilon$, given the intrinsic nature of the two coefficients.

## 4 Numerical simulation of the dehydration of a sphere and a cylinder of agar gel

A sphere and a cylinder of gel are placed in a solution of polyethylen glycol (PEG). In these conditions, the water is eliminated of the gel by effet of concentration difference between the sample and the solution. The flux of water imposed as condition to the limit has been calculated from the dehydration kinetic. A numerical model gives water content and stresses fields and deformation of the sphere and the cylinder.

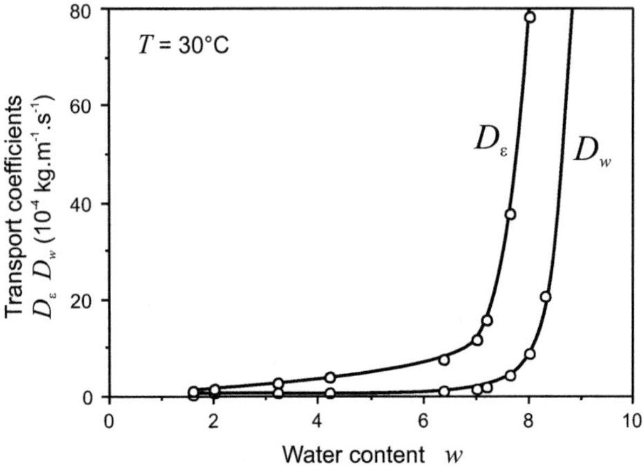


**Fig. 3.** Variation of transport coefficients $D_w$ and $D_\varepsilon$ in agar gel according to water content

Figure 4 gives an exemple of comparison between experimental and numerical water content profiles in the case of a sphere presenting initialy a radius of 2 cm. The model make it possible to analyse the evolution of stresses profiles; figure 5 shows the evolution of the circunferential stress at different times in the case of the sphere. Two phases can be seen is the figure. At the begenning of the dehydration (t<5h) the circunferential stress was positive at the surface of the sphere (traction) and negative in the material (compression). If the dehydration is to fast, this may cause cracks at the surface. In a second phase (t>7h) the circunferential stress become mainly in compression at the surface and traction in the centre. Figure 6 represents the final deformation of the cylinder. There is a good agreement between experience and theory

## 5 Conclusion

The study shows the influence of mechanical actions on diffusion fluxes. Transfer is not governed solely by water content gradient but also by the deformation gradient. It has been shown that in an infinite plate with elastic, isotropic behaviour, transport is governed by an equation similar to Fick's law: water flux is proportional to the water content gradient. In this case, the transport coefficient expression contains a part related to the hygroscopic feature of the medium and a part related to its rheology.

Experimental study of water content profiles in an agar gel plate justified the use of a transport law of the Fick type for the water content range investigated. The apparent transport coefficient decreases with the water content and reaches a minimum at $w = 6$ and then increases with low water contents. The theoretical expression proposed makes it possible to attribute the

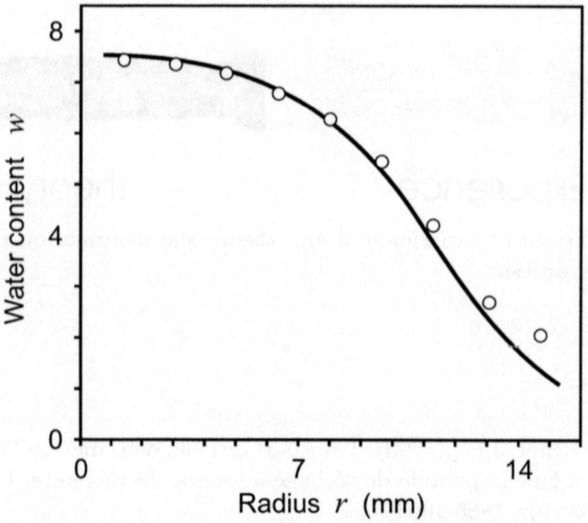


**Fig. 4.** Theoretical and experimental water content profiles in the sphere at $t = 4h$

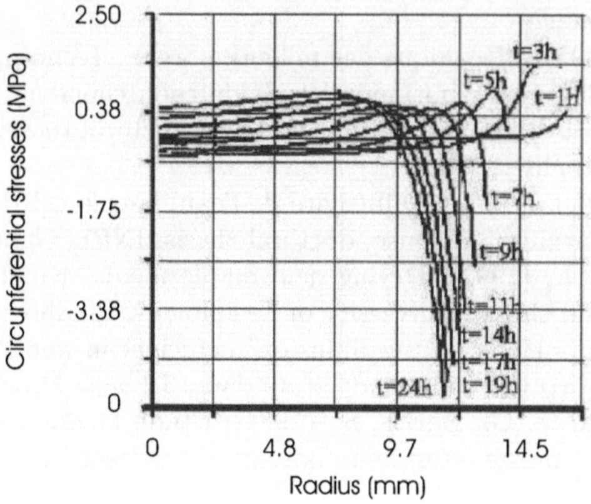


**Fig. 5.** Circumferential stresses in the sphere at different times

increase in apparent transport coefficient to variations in elastic and hygrometric coefficients.

Use of the variation of the apparent transport coefficient enabled determination of the two coefficients related to the water content and deformation gradients. These coefficients are intrinsic to the material and make it possible to describe water transport in agar gel for any geometrical layout.

The numerical results show good agreement between experience and theory in the case of dehydration of a sphere and a cylinder concerning water content profiles, stresses and deformation.

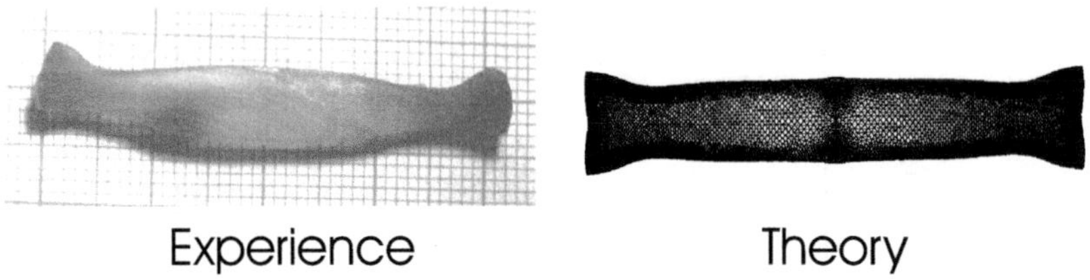


**Fig. 6.** Comparison of experimental and theoretical deformation of a cylinder at the end of dehydration

## References

1. Auria, R., Bénet, J.C., (1990): Transport de l'eau dans une feuille de caoutchouc naturel pendant la période de séchage à vitesse décroissante, Int J Heat Mass Transfer 33 (9) , 1885-1894
2. Larché, F.C., Cahn, J.W., (1982): The effect of self-stress on diffusion in solids, Acta Metall, 30, 1835-1845
3. Mrani, I., (1993): Transport de matière en milieu biconstituant élastique. Application au séchage du gel d'Agar, doctoral thesis, Université Montpellier II, Montpellier, France
4. Coussy, O., (1993): Mécanique des milieux poreux, Technip, Paris, France
5. Auria, R., (1988): Contribution à l'étude du caoutchouc naturel : structure interne et transfert d'humidité lors du séchage, doctoral thesis, Université Montpellier II, Montpellier, France
6. Do Amaral, S.P.J., (1992): Diffusivité de l'eau dans le gel Polyacrylamide-Eau autour de la transition vitreuse, doctoral thesis, INPL, Nancy, France
7. Ketelaars, A.A.J., (1992): Drying deformable media. Kinetics, shrinkage and stresses, doctoral thesis, University of Technology, Eindhoven, Netherlands
8. Comstock, G.L., (1963): Water diffusion coefficient in wood as calculated from adsorption, desorption and steady state data, J Forest Products, 97-103
9. Crausse, P., Bacon, G., Bories, S., (1984): Etude fondamentale des transferts couplés chaleur- masse en milieux poreux, Int J Heat Mass Transfer, 24, 991-1004
10. Gehrmann, D., Kast. W., ((1984): Drying of gels, in proceedings of the 1th International Drying Symposium, Montreal, Canada, Mujumdar (ed), Science Press, Princeton, New Jersey, 239-246.

# Numerical Algorithm for a Three-dimensional Stress-induced Solid-phase-transformation Model

Ferdinando Auricchio, Lorenza Petrini

Dipartimento di Meccanica Strutturale
Università di Pavia
27100 Pavia, Italia

**Abstract.** This work focuses on a three-dimensional thermomechanical model, recently proposed by Souza et al. [4] for stress-induced solid phase transformations and able to reproduce the main features of shape-memory materials in the realm of a small-deformation regime. A careful reading of the cited work highlightes how the proposed integration algorithm is only partially developed. Henceforth, we focus on the development of a effectively robust integration algorithm to be adopted in a numerical scheme, such as a finite-element framework.

## 1 Introduction

The good mechanical performances of *shape-memory alloys* (SMA) are a consequence of the fact that the material may in general present two different crystallographic structures, one characterized by a more ordered unit cell and indicated in the following as *austenite* (A), the other characterized by a less ordered unit cell and indicated in the following as *martensite* (M).

From a micro-mechanical point of view, the presence of two different crystallographic structures is the base for a reversible austenite-martensite phase transformation, while from a macro-mechanical point of view, the reversible solid-solid phase transformation results in two unique effects, i.e. the *pseudoelasticity* (PE) and the *shape memory effect* (SME).

The more and more frequent use of these unusual effects in commercially valuable applications has stimulated a vivid interest in the development of constitutive models able to reproduce the basic material response.

Hence, this work focuses on a three-dimensional thermomechanical model, recently proposed by Souza et al. [4] for stress-induced solid phase transformations and able to reproduce the main features of shape-memory materials in the realm of a small-deformation regime. Souza et al. [4] discuss also a return-map integration algorithm, to be used for the numerical solution of the model, considering some simple one-dimensional problems as well as a more complex three-dimensional non-proportional problem and showing a good performance of the model in comparison with experimental results.

However, a careful reading of the cited work highlightes how the proposed integration algorithm is only partially developed. Henceforth, we focus on the development of a effectively robust integration algorithm to be adopted in a numerical scheme, such as a finite-element framework.

## 2 The model

Cast within the framework of classical irreversible thermodynamics [2,1], the model proposed by Souza et al. [4] is developed using as control variable the strain $\boldsymbol{\epsilon}$, and the temperature $T$, and as internal variable a second-order tensor, $\mathbf{e}^{tr}$, indicated as *transformation strain.*

Assumed to be traceless, $\mathbf{e}^{tr}$ is a measure of the strain associated to the phase transformation and, in particular, to the convertion from austenite or multiple-variante martensite to single-variant martensite. Accordingly, the norm of $\mathbf{e}^{tr}$ should be bounded between zero – for the case of a material without oriented martensite – and a maximum value $\epsilon_L$ – for the case in which the material is fully transformed in oriented (single-variant) martensite.

For brevity, in the following we directly discuss the time-discrete version of the model, as proposed by Souza et al. [4], concentrating only on the deviatoric part of the model, since the volumetric component is purely elastic. Moreover, to minimize the appearance of subscripts, we find convenient to indicate with the subscript $n$ a quantity that is evaluated at time $t_n$, and with no subscript a quantity that is evaluated at time $t_{n+1}$.

The time-discrete constitutive model is described by the equations:

$$\begin{cases} \mathbf{s} = 2G\left(\mathbf{e} - \mathbf{e}^{tr}\right) \\ \mathbf{e}^{tr} = \mathbf{e}^{tr}_n + \Delta\zeta \dfrac{\mathbf{X}}{\|\mathbf{X}\|} \\ \|\mathbf{e}^{tr}\| \leq \epsilon_L \\ \mathbf{X} = \mathbf{s} - \boldsymbol{\alpha} \\ \boldsymbol{\alpha} = \left[\tau_M(T) + h\|\mathbf{e}^{tr}\| + \gamma\right] \dfrac{\mathbf{e}^{tr}}{\|\mathbf{e}^{tr}\|} \\ \gamma \geq 0 \\ F(\mathbf{X}) = \|\mathbf{X}\| - R \leq 0 \\ \Delta\zeta \geq 0 \quad \Delta\zeta F(\mathbf{X}) = 0 \end{cases} \tag{1}$$

where:

- Equation $(1_1)$ is the linear elastic relation between the deviatoric stress, $\mathbf{s}$, the deviatoric total strain $\mathbf{e}$, and the transformation strain $\mathbf{e}^{tr}$, with $G$ the shear modulus.
- Equation $(1_2)$ is the evolution equation for the transformation strain, constrained by Equation $(1_3)$, which is relative to the physical interpretation given to the internal variable $\mathbf{e}^{tr}$, with $\|\cdot\|$ indicating the Euclidean norm.

- Equation ($1_4$) defines the *transformation stress*, $\mathbf{X}$; the quantity $\boldsymbol{\alpha}$ plays a role similar to the so-called back-stress in classical plasticity and, accordingly, $\mathbf{X}$ can be identified as a relative stress.
- Equation ($1_5$) defines the *back-stress* $\boldsymbol{\alpha}$, where $\tau_M$ is a positive and monotonically increasing function of the temperature, defined as $\tau_M =< \beta(T - T_0) >$, with $< \cdot >$ the positive part, $\beta$ a material parameter, $T$ the room temperature and $T_0$ the temperature below which no twinned martensite is observed. In particular, we note that $\boldsymbol{\alpha}$ is defined only for the case $\|\mathbf{e}^{tr}\| > 0$ while $\gamma$ is constrained by Equation ($1_6$).
- Equation ($1_7$) defines the limit function, with $R$ the radius of the region with no phase transformation.
- Equation ($1_8$) are the classical Kuhn-Tucker conditions.

Given the time-discrete model, the stress history is computed from the strain history by means of a two-step elastic-predictor inelastic-corrector procedure, known as a *return-map* [3]. The algorithm proposed in the present work is detailed in Tables 1, 2 and 3, while in the following we simply comment on some significative aspects.

**1. Compute trial state**

$$\begin{cases} \mathbf{e}^{tr,TR} = \mathbf{e}_n^{tr} \\ \mathbf{s}^{TR} = 2G\left(\mathbf{e} - \mathbf{e}^{tr,TR}\right) \\ \mathbf{n}^{TR} = \dfrac{\mathbf{e}^{tr,TR}}{\|\mathbf{e}^{tr,TR}\|} \end{cases}$$

**2. Check material state**

See Table 2

**3. Update material state**

See Table 3

**Table 1.** Time-discrete model: solution algorithm.

As any return-map, the algorithm starts with the computation of an elastic trial state. Then, if the limit equation is not violated (i.e. $F \leq 0$), the trial state represents the solution at the current time step. Otherwise, if the limit equation is violated (i.e. $F > 0$), an inelastic correction should be performed, requiring the satisfaction of the limit condition ($F = 0$).

An iterative Newton method employing the trial step as initial condition is used to solve Equation 1 in the case on inelastic correction. In particular, as summarized in Table 3, we distinguish between two different situations,

**STATE AND BRANCH DETECTION**

if $\|\mathbf{e}_n^{tr}\| = 0$ then

$$\begin{cases} \text{compute:} \quad F = \|\mathbf{s}^{TR}\| - [\tau_M(T) + R] & (2) \\ \text{check } F\text{:} \quad \begin{cases} \text{if } F < 0 \text{ then} & \textbf{elastic step } (EL) \\ \text{else} & \textbf{active p.t. } (PT) \\ \text{end if} & \end{cases} \end{cases}$$

else

$$\begin{cases} \text{compute:} \quad \begin{cases} \boldsymbol{\alpha}^{TR} = \left[\tau_M(T) + h\|\mathbf{e}^{tr}\|\right] \mathbf{n}^{TR} \\ \hat{\mathbf{X}} = \mathbf{s}^{TR} - \boldsymbol{\alpha}^{TR} \\ F = \|\hat{\mathbf{X}}\| - R \end{cases} & (3) \\ \text{check } F\text{:} \quad \begin{cases} \text{if } F < 0 \text{ then} & \textbf{elastic step } (EL) \\ \text{else} & \textbf{active p.t. } (PT) \\ \text{end if} & \end{cases} \end{cases}$$

end if

**Table 2.** Time-discrete model: detection of the material state and of the evolution branch through the limit function $F$.

described by either one of the two following set of equations:

$$\tilde{\mathbf{R}}^7(\mathbf{X}, \Delta\zeta) = \begin{cases} \mathbf{X} - \mathbf{s}^{TR} + 2G\Delta\zeta \dfrac{\mathbf{X}}{\|\mathbf{X}\|} \\ \qquad + [\tau_M(T) + h\|\mathbf{e}^{tr}\|] \dfrac{\mathbf{e}^{tr}}{\|\mathbf{e}^{tr}\|} = \mathbf{0} \\ \|\mathbf{X}\| - R = 0 \end{cases} \tag{5}$$

$$\tilde{\mathbf{R}}^8(\mathbf{X}, \Delta\zeta, \gamma) = \begin{cases} \mathbf{X} - \mathbf{s}^{TR} + 2G\Delta\zeta \dfrac{\mathbf{X}}{\|\mathbf{X}\|} \\ \qquad + [\tau_M(T) + h\|\mathbf{e}^{tr}\| + \gamma] \dfrac{\mathbf{e}^{tr}}{\|\mathbf{e}^{tr}\|} = \mathbf{0} \\ \|\mathbf{X}\| - R = 0 \\ \|\mathbf{e}^{tr}\| - \epsilon_L = 0 \end{cases} \tag{6}$$

assuming in both cases $\mathbf{e}^{tr} = \mathbf{e}_n^{tr} + \Delta\zeta \dfrac{\mathbf{X}}{\|\mathbf{X}\|}$.

## 3 Numerical examples

To test the effectiveness of the proposed algorithm, we reproduce the numerical examples proposed in Souza et al. [4]. The material parameters are: $E =$

**BRANCH SOLUTION AND STATE UPDATE**

**Elastic step** ($EL$):

$$\begin{cases} \mathbf{e}^{tr} = \mathbf{e}^{tr,TR} \\ \mathbf{s} = 2G\left(\mathbf{e} - \mathbf{e}^{tr}\right) \\ \textbf{exit} \end{cases}$$

**Active phase transformation** ($PT$):

CASE $PT_1$ - Evolving phase transformation:

if $\|\mathbf{e}^{tr}_n\| = 0$ then

$$\|\mathbf{s}\| = \tau_M(T) + R$$

$$\mathbf{X} = R\frac{\mathbf{e}}{\|\mathbf{e}\|} \tag{4}$$

$$\mathbf{e}^{tr} = \left[\|\mathbf{e}\| - \frac{\|\mathbf{s}\|}{2G}\right]\frac{\mathbf{e}}{\|\mathbf{e}\|}$$

else

$\mathbf{X} = \hat{\mathbf{X}}$ [See Eq. 3]

end if

Find $\mathbf{e}^{tr}$ solving: $\tilde{\mathbf{R}}^7(\mathbf{X}, \Delta\zeta) = \mathbf{0}$ [See Eq. 5]

Check solution: $\begin{cases} \text{if } \|\mathbf{e}^{tr}\| < \epsilon_L \text{ then} \quad \textbf{exit} \\ \text{else} \quad \textbf{continue} \\ \text{end if} \end{cases}$

CASE $PT_2$ - Saturated phase transformation:

Find $\mathbf{e}^{tr}$, $\gamma$ solving: $\tilde{\mathbf{R}}^8(\mathbf{X}, \Delta\zeta, \gamma) = \mathbf{0}$ [See Eq. 6]

**exit**

**Table 3.** Time-discrete model: branch solution and detection of the new state in the different phases.

70000 MPa, $\nu = 0.33$, $R = 45$ MPa, $h = 500$ MPa, $\epsilon_L = 0.03$, $T_0 = 253.15$ $^oK$, $\beta = 7.5$ MPa$^oK^{-1}$.

The tests performed are: isothermal uniaxial (tension-compression) tests in the superelastic range (Figure 1); an isothermal uniaxial (loading-unloading) test, followed by a free-stress thermal cycle in the shape-memory effect range (Figure 2).

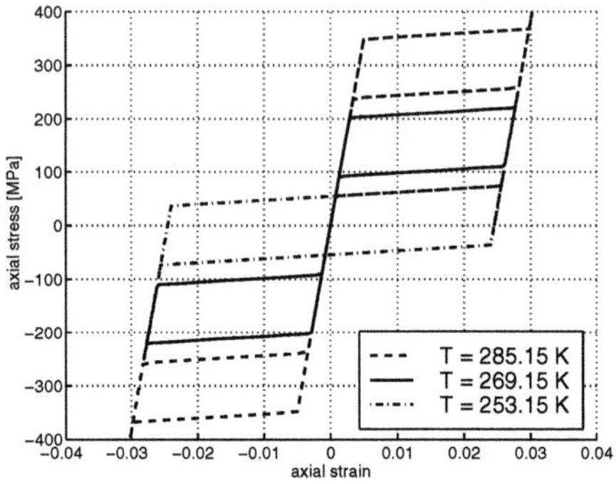


**Fig. 1.** Loops under traction-compression tests at different temperature.

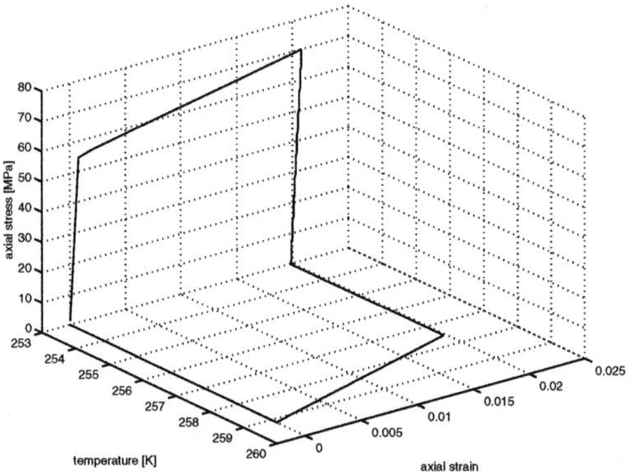


**Fig. 2.** The shape memory effect.

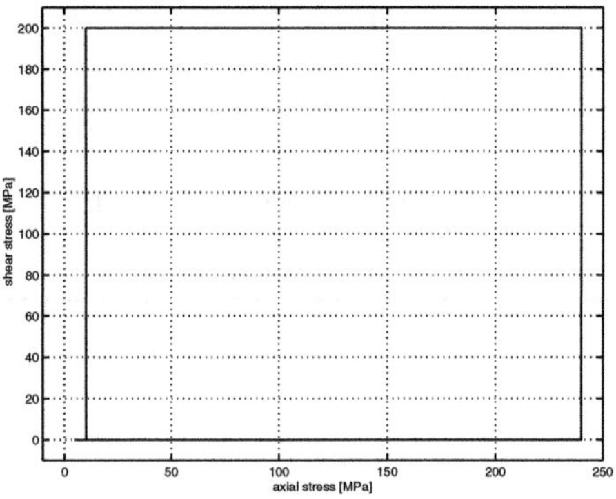


**Fig. 3.** Non proportional loading path.

Following again Reference [4], and choosing the material constants as follows: $E = 30700$ MPa $\nu = 0.36$ $R = 73.4$ MPa $h = 9230$ MPa $\epsilon_L = 0.1$ $T_0 = 253.15\ ^oK$ $\beta = 4.1$ MPa$^oK^{-1}$, we also consider a non-proportional tension torsion test (Figures 3-6).

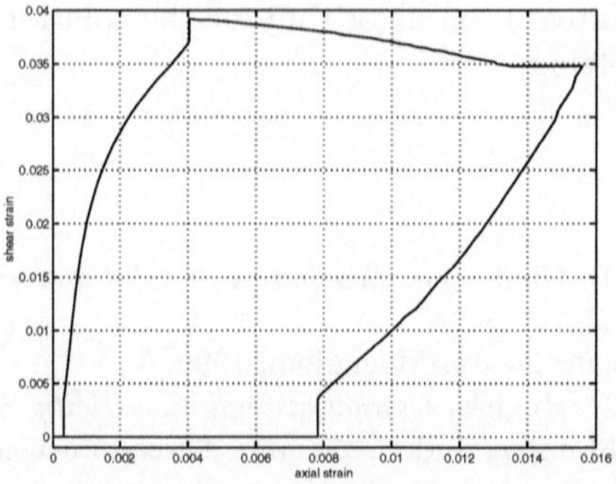


**Fig. 4.** Strain path under non proportional loading.

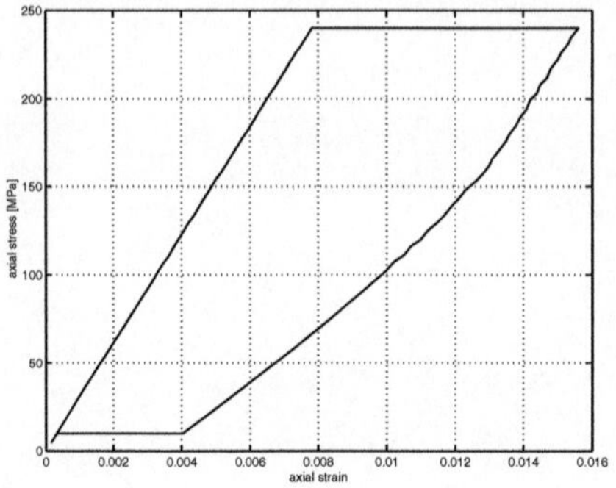


**Fig. 5.** Axial stress and strain response under non proportional loading.

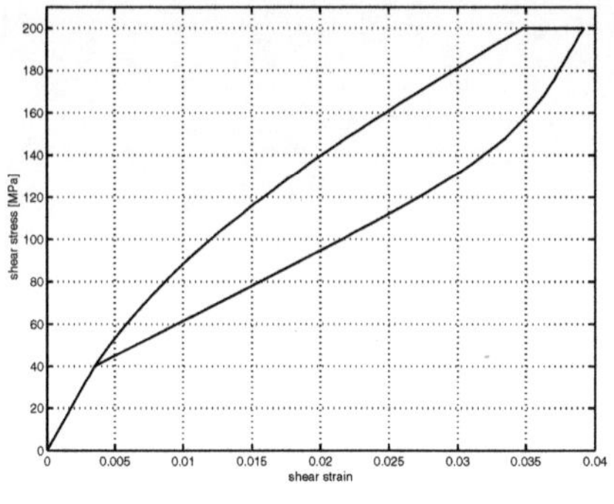


**Fig. 6.** Shear stress and strain response under non proportional loading.

## 4 Conclusions

We may conclude pointing out on one hand the qualities of the constitutive model as proposed originally by Souza et al. [4], which is to be able to catch the basic SMA features. On the other hand, the numerical solution algorithm proposed by Souza et al. has shown limits and inconsistencies. Accordingly, we report a robust integration algorithm for the solution of the cited model in a time-discrete setting.

## References

1. J. Lemaitre and J.L. Chaboche. *Mechanics of solid materials.* Cambridge University Press, 1990.
2. J. Lubliner. *Plasticity theory.* Macmillan, 1990.
3. J.C. Simo and T.J.R. Hughes. *Computational inelasticity.* Springer-Verlag, 1998.
4. A.C. Souza, E.N. Mamiya, and N. Zouain. Three-dimensional model for solids undergoing stress-induced phase transformations. *European Journal of Mechanics, A: Solids*, 17:789–806, 1998.

# Interaction Energy in some Phenomenological Modelings of Phase Transitions in Crystalline Solids

Christian Licht

Laboratoire de Mécanique et Génie Civil,
cc 048, Université Montpellier II,
F-34095 Montpellier Cedex 5, France

**Abstract.** Some phenomenological and macroscopic modelings of solid/solid phase transitions involve a so-called interaction term which may imply the nonconvexity of the strain energy function. Nevertheless, it is shown that, if this term is not too large, these modelings are valuable and account for some experimental facts.

## 1 Introduction

I will present some remarks on a category of phenomenological modelings of phase transition in crystalline solids. These modelings involve a so-called interaction term which may imply the nonconvexity of the strain energy density function of the medium. This lack of sacrosanct convexity generates some controversy, warning, or nonuse (see some comments in [2], [8], [9], [13], [19]).

I will show that, if this term is not too large, the modeling is valuable. Certainly, the stored energy density function is not convex but *quasiconvex* and, consequently, inherits a double-well like structure which may account for some experiments.

From a technical, say mathematical, standpoint, this study is nothing but a slight variant of the beautiful papers [11], [12]... Nevertheless, I hope it may interest people concerned with the modeling of shape memory alloys.

## 2 Setting the problem

First, the phenomenological modelings, that I will discuss, describe the state of the medium by a triple $(\epsilon, x, T)$ ; $\epsilon, T$ stand respectively for the linearized strain tensor and the temperature, while $0 \leq x \leq 1$ denotes a volume fraction. Since a volume fraction is put forward, note that this modeling operates at a large, say macroscopic, scale. Moreover, $x$ being a scalar, I will consider only two phases; indeed considering more phases implies some substantial difficulties (except in obvious symmetry related phases). Thus this modeling concerns the detwining of two variants of martensite (martensite may present *two* variants in the case of soda, neodynium pentaphosphate, wolfram, ... ) or is appropriate to polycristalline alloys where the variants of martensite are not told apart in the austenite-martensite mixture. Next, a free energy density function is defined by a function $W$ of the state variables. At last, the phase transitions are assumed nondissipative, which seems rather realistic, so that the medium has a thermoelastic behavior with a stored energy density function

$$\overline{W}(\epsilon, T) = Inf\,\{\, W\,(\epsilon, T, x)\ ,\ 0 \leq x \leq 1\}$$

Of course, the whole point is the choice of $W$ ; at least two approaches lead to the category of modelings that I consider. The first approach, which models the medium as a generalized standard medium [10], splits the free-energy into three terms:

$$W(\epsilon, x, T) = W^e(\epsilon - xa, T) + xW^{ch}(T) + W^{\phi}(x, T) \tag{1}$$

The first term is the thermoelastic one, $xa$ being the phase transformation part of the total strain $\epsilon$ , while the second one concerns the latent heat exchange. The last one is the stored part due to internal changes, it is affected by the interactions between phases thus it has to be minimal and equal to 0 at $x = 0$ or $x = 1$ . Obviously $x \in [0, 1] \longmapsto x(1 - x)$ is a simple and smooth function which satisfies the previous conditions, so a proposal for $W^{\phi}$ is:

$$W^{\phi}(x, T) = \phi(T)\, x\,(1 - x)\ ,\ \phi \geq 0$$

Note that in [14], [15], [16], it is reported that $x\,(1 - x)$ is proportional to the expectation value of the interfacial area, if the phases are randomly distributed over the specimen ... Moreover, this modeling implicitly assumes that the elastic coefficients of the phase are equal, so this modeling confines to detwining.

The second approach regards the medium as a mixture such that the free energy may be a $x$-weighted combination of the energies $W^i$, $i = 1, 2$, of each phase. But, referring to homogenization, subtracting a term $W^c$ , minimal and equal to 0 at $x = 0$ or $x = 1$ , yields a better estimate:

$$W(\epsilon, x, T) = x\,W^1(\epsilon, T) + (1 - x)\,W^2(\epsilon, T) - W^c(x, T) \tag{2}$$

Once again, a convenient choice is

$$W^c(x, T) = c(T)\, x\,(1 - x)\ ,\ c \geq 0$$

Note that this approach does not assume the identity of the elastic laws of the two phases. From the mixture standpoint, another approach, which will be detailed in remark 3.1, leads to an expression like (1) in the case of two linearly elastic phases with identical elastic moduli. More deeply, this last assumption, *which will stand for all the sequel*, implies the identity of the two approaches. More precisely, I assume

$$W^i(\epsilon,T) = 1/2\alpha(\varepsilon - a_i(T)).(\varepsilon - a_i(T)) + w_i \qquad , i = 1,2$$

where $a_i$ is the stress-free strain of the i-th phase, $w_i$ the associated minimum energy and $\alpha$ the tensor of elastic moduli, a positive symmetric linear map on the space $S^3$ of symmetric matrices. Then the obvious identity

$$x\, W^1(\epsilon,T) + (1-x)\, W^2(\epsilon,T) - c(T)\, x\,(1-x) \;=\; \tag{3}$$

$$1/2\; \alpha\,(\epsilon - \bar{a}(T)) \cdot (\epsilon - \bar{a}(T)) \;+\; \bar{w}(x,T) \;+\; \phi(T)\, x\,(1-x)$$

where

$$\bar{a} = x\, a_1 + (1-x)\, a_2 \quad , \quad \bar{w} = x\, w_1 + (1-x)\, w_2$$
$$\phi = 1/2\, \alpha\, a \cdot a \;-\; c \quad , \; a = a_2 - a_1$$

yields a single modeling where two equivalent expressions of the free energy are given by the two members of this identity!

It is not my purpose to list all the authors who, specifically or vaguely, proposed or discussed this modeling. Due to the many studies devoted to Shape Memory Alloys, it is impossible to be exhaustive. Generally, some information may be found in [2], [6]-[9], [13]-[17], [19] and in the references therein. My main thrust is to decide on the value of a nondissipative modeling where the free energy and the strain energy functions are :

$$W_\phi(\epsilon,x,T) = 1/2\alpha(\epsilon - \bar{a}(T)) \cdot (\epsilon - \bar{a}(T)) + \bar{w}(T) + \phi(T)x(1-x)$$

$$= xW^1(\epsilon,T) + (1-x)W^2(\epsilon,T) - c(T)x(1-x), c = 1/2\alpha a.a - \phi \tag{4}$$

and

$$\overline{W}_\phi(\,\epsilon,T\,) = Inf\,\{W_\phi\,(\epsilon,T,x)\;,\; 0 \leq x \leq 1\} \tag{5}$$

Clearly, from the first expression, $\phi$ non positive implies that $W_\phi$ is a convex function of the couple $(\epsilon, x)$ and, consequently, that $\overline{W}_\phi$ is a convex function of $\epsilon$ . But, as previously noted, $\phi$ negative does not seem plausible! ... Moreover, it is not difficult to see that $\overline{W}_\phi$ is a convex function in $\epsilon$ if $\phi$ is positive. In fact, it is not a disaster. I will show that if $\phi$ is positive, *but not too large*, the modeling is valuable... In the following technical discussions, temperature will only play a role of parameter, thus temperature dependence is suppressed hereafter.

## 3 (Quasi)convexity properties of $\overline{W}_\phi$

A mechanical interpretation of the mathematical theory of quasiconvexification [5], [18] may be : "at the macroscopic scale the behavior of a hyperelastic material is not governed by the true stored energy function but by an apparent one : its quasiconvexification, which is the infimum of those average energies that can be attained in deformations with a given average". Hence a first criterion of a good macroscopic modeling is the quasiconvexity property of the stored energy density function $\overline{W}$ which, in our case, may be read as :

$$\overline{W}(E) = Inf \left\{ 1/|U| \int_U \overline{W}(E+\epsilon(u))\,dy \; ; \; u(y) = 0 \, on \, \partial U \right\} , \; \forall E \in S^3$$

where the choice of the domain $U$ is unimportant. Note that convexity implies quasiconvexity but the converse is generally false. Here are some (quasi)convexity properties of the strain energy function $\overline{W}_\phi$ :

**Theorem 3.1.** *Let*

$$h = Min\left\{1/2\alpha(z-a)\cdot(z-a); z = k\otimes v + v\otimes k, k, v \in R^3\right\} \in [0, 1/2\alpha a\cdot a]$$

*then*
$\phi \leq 0$ *implies* $\overline{W}_\phi$ *is convex,*
$h > 0$ *and* $0 \leq \phi \leq h$ *implies* $\overline{W}_\phi$ *is not convex but quasiconvex,*
$h < \phi$ *implies* $\overline{W}_\phi$ *is not quasiconvex.*
*Moreover,* $\overline{W}_\phi$ *equals* $\overline{W}_{1/2\,\alpha a\cdot a}$ *if* $\phi \geq 1/2\,\alpha\, a\cdot a$ .

As noted in [11], $h$ differs from 0 if and only if the full transformation strain $a$ is not a symmetrized tensor product, in other words if the two stress-free strains are not compatible. Thus if $h > 0$ , it can be claimed that the modeling with $\phi \in [0, h]$ , is a "good" macroscopic modeling, whereas only $\phi = 0$ yields a good modeling if $h = 0$ !

*Proof of theorem 3.1.* Several arguments of [11] pp 201-203 are used without giving details. First, in the definition of quasiconvexity, a unit cell $Y$ and periodicity conditions may be used in place of $U$ and Dirichlet conditions. Standard results on mesurable selections imply :

$$\begin{aligned} I &:= Inf \left\{ \int_Y \overline{W}_\phi(E+\epsilon(u))\,dy \; ; \; u \in C^1_{per}(\bar{Y}) \right\} \\ &= Inf \left\{ \int_Y W_\phi(E+\epsilon(u), \theta(y))\,dy; u \in C^1_{per}(\bar{Y}), \theta \in B(Y) \right\} \end{aligned}$$

where $B(Y) := \{\theta \in L^\infty(Y); 0 \leq \theta \leq 1\}$. An elementary calculation using the second expression of $W_\phi$ in (4) gives

$$I = Inf\,\{\,m\,W^1(E) + (1-\,m)\,W^2(E) - cm\,+\,J(m)\ ;\ 0 \leq m \leq 1\,\},$$

$$J(m) = Inf\{\int_Y c\theta^2 + 1/2\,\alpha\,\epsilon(u)\cdot\epsilon(u)\,+\theta\,\alpha\,\epsilon(u)\cdot a\,;$$

$$u \in C^1_{per}(\bar{Y}), \theta \in B(Y), \int_Y \theta\,=\,m\}.$$

As in [11] p. 202, Fourier analysis yields

$$J(m) \geq Inf\{\,cm^2\,+\,(h-\phi)\,\textstyle\sum_{k\neq 0}\hat{\theta}(k)^2;$$

$$y \in Y \mapsto \theta(y)\,=\,\textstyle\sum_{k\in Z^3}\hat{\theta}(k)\,e^{2i\pi k\cdot y} \in B(Y)\}$$

which proves the quasiconvexity of $\overline{W}_\phi$ if $0 \leq \phi \leq h$.

Clearly, (4), (5) imply that $\overline{W}_{\phi_1} \geq \overline{W}_{\phi 2}$ if $\phi_1 > \phi_2$ , and that $\overline{W}_\phi = \overline{W}_{1/2\alpha a.a} = Min\,(W^1, W^2)$ if $\phi \geq 1/2\alpha a\cdot a$ . Moreover, it is shown in [11] that the quasiconvexification and the convexification of $Min\,(W^1, W^2)$ are respectively $\overline{W}_h$ and $\overline{W}_0$ . Thus, $\overline{W}_\phi$ is not convex if $0 < \phi$ and is not quasi-convex if $h < \phi$ , because quasiconvexification and convexification preserve the order.

**Remark 3.1.** A rough way to define the strain energy function of a mixture of two phases described by the energy density functions $W^i$ , in proportions $x$ and (1-$x$), is to consider

$$W^{mix}_a(\epsilon,x)\;=\;Inf\,\{\,x\;W^1(\epsilon^1)\;+\;(1-x)\;W^2(\epsilon^2)\;;\;x\,\epsilon^1\,+(1-x)\,\epsilon^2\,=\,\epsilon\,\}$$

With $W^i$ as in (3), it is shown in [11] that

$$W^{mix}_a(\epsilon,x)\;=\;W_0(\epsilon,x)\;=1/2\;\alpha\,(\epsilon-\bar{a}(x))\;\cdot(\epsilon-\bar{a}(x))\,+\,\bar{w}(x)$$

Thus, as announced previously, in order to account for interactions at a macroscopic level, some modelings ([6]-[9], [13]-[16], [19]) add a term depending on $x$, vanishing at $x$ = 0,1. Their common proposal is $\phi\,x\,(1-x)$ , hence they get $W_\phi$ as free energy density function for the medium! However, the previous definition $W^{mix}_a$ presupposes constant and compatible strains in each phase, which is open to criticism. A better definition is

$$\begin{aligned} W^{mix}_b(\varepsilon,x)\;&=\;Inf\;\{1/|Y|\textstyle\int_Y \chi_x(y)W^1(\varepsilon(u)) + (1-\chi_x(y))W^2(\varepsilon(u))dy;\\ &\chi_x\;characteristic\;function\;\textstyle\int_Y \chi_x(y)dy = x|Y|, u(y) = \varepsilon y\;on\;\partial Y\}\end{aligned}$$

It is proved in [11] that $W_b^{mix} = W_h$ . Hence, adding a macroscopical term of interaction $Ix(1-x)$ yields a free energy equal to $W_{h+I}$. Then the quasiconvexity criterion of good modeling implies that $I$ must be non positive. This sign, which is the opposite of what is widely used in the literature, corresponds exactly to what in [4] is termed, through microscopic considerations, mismatch energy.

## 4 Double-well like structure of $\overline{W}_\phi$ and experiments

From now on, I confine to the more interesting case $0 < \phi \leq h$ and will show that the lack of convexity of $\overline{W}_\phi$ accounts for some experimental results, which also guarantees a good modeling. Let

$$w = w_1 - w_2 \ , \ H_1 = \{e \in S^3; \ \ \alpha(e - a_1).a = w\}$$

Some elementary manipulations, as in [11] pp. 205-207, provide the following geometric properties of the graph of $\overline{W}_\phi$:

**Proposition 4.1.** *$\overline{W}_\phi$ is piecewise quadratic and differentiable on* $S^3$. *If $|w| > \phi$, $\overline{W}_\phi$ has only one absolute minimizer: $a_1$ (resp. $a_2$) if $w > \phi$ (resp. $w < -\phi$), and no other relative minimizers. If $|w| \leq \phi$, $\overline{W}_\phi$ has exactly two relative minimizers $a_1$ and $a_2$ : $a_1$* (resp. $a_2$) *is an absolute minimizer if and only if $w \geq 0$ (resp. $w \leq 0$).*

**Proposition 4.2.** *For each $e_1$ of $H_1$, the function $\overline{W}_\phi - \Sigma_1 \cdot$ , with $\Sigma_1 = \alpha(e_1 - a_1)$, has exactly two absolute minimizers $e_1$ and $e_2 = e_1 + a$ .*

Since $\overline{W}_\phi$ is differentiable, the previous properties can be interpreted in terms of stress and strain. First, it appears that along all straight paths $\epsilon_\Delta$ in the strain space $S^3$, $t \in R \longmapsto \epsilon_\Delta(t) = \epsilon_0 + t\,a$ , *parallel to* $a$, the derivative of the energy $\overline{W}_\phi(\epsilon_\Delta(t))$ is not monotone. In other words, "the stress-strain relation is not monotone along $\epsilon_\Delta$ ". Thus it should be of interest to do such strain controlled tests which must be *isothermal and biaxial* (in this direction see [6], [14], [15]). These hard device tests seem difficult to do, conversely it is easier to do isothermal and biaxial loading experiments [3]. Leaving hysteresis considerations aside, it seems clear that two different equilibrium configurations may occur when a specimen of shape memory alloy is subjected to certain uniform biaxial extension loadings. The present modeling may account for this fact. If $\Omega$ is a reference configuration of the specimen and $n$ the unit outward normal to $\partial\Omega$ , the total energy read as

$$I_\Sigma(v) = \int_\Omega \overline{W}_\phi(\epsilon(v))\,dx - \int_{\partial\Omega} \Sigma n.v\,ds = \int_\Omega \overline{W}_\phi(\epsilon(v)) - \Sigma \cdot \epsilon(v)\,dx$$

The previous propositions make it possible to use the arguments in [12] pp 82-83 and to prove the following *metastability* theorem :

**Theorem 5.1.** *For every $e_1 \in H_1$ , there exists $\delta > 0$ such that for every $\Sigma$ in a ball in $S^3$ of center $\Sigma_1 = \alpha(e_1 - a_1)$ and radious $\delta$ , the* *functional $I_\Sigma$ has two distinct $L^1$ -local minimizers (one of them being an absolute minimizer in $H^1(\Omega)^3$ ). As $\Sigma$ goes to $\Sigma_1$ , they converge strongly in $L^1$ towards "$x \longmapsto e_1 x$" and "$x \longmapsto (e_1 + a)\, x$" , absolute minimizers of* $I_{\Sigma_1}$ in $H^1(\Omega)^3$ .

The $L^1$ topology is rough : in $L^1$ neighborhoods the displacements are near in the mean, but their strains may be very remote. These local, but non absolute, minimizers may be observed, they correspond to metastable equilibrium configurations : experimentalists [3] report that specimen can undergo large deformations from an equilibrium configuration to another one only by slightly hitting the strings of the device or through a small disturbance in the room. A similar metastability analysis, in the more difficult framework of finite strains, may be found in [1].

## 5 Conclusion

I believe that all the foregoing arguments clearly show that the considered modeling is coherent and able to account some phenomena involved by phase transitions in crystalline solids. According to remark 3.1, a nice introduction of this phenomenological and macroscopic modeling is adding a rational micro-macro definition of the energy function of a mixture to a woolly termed function, $I\, x\,(1-x)$ , $I \leq 0$ , of the volume fraction $x$ of one phase. It should be interesting to derive this last term through a rational scale transition that, for instance, takes into account micro-scale surface interactions.

Eventually, I thank my friend C. Lexcellent who told me about some pertinent earlier works.

## References

1. Ball J M, James R D (1995) Local minimizers and phase transformations. Proceedings of the ICIAM
2. Chrysochoos A (1998) Vers une reformulation des transitions de phase du premier ordre. Rev. Roum. Sci. Techn. – Méc. Appl., Tome 43 no 3:283-293
3. Chu C (1993) Hysteresis and Microstructures: A study of Biaxial Loading on Compound Twins of Copper-Aluminium-Nickel Single Crystals. Dissertation , University of Minesota.
4. Chu C, James R D (1993) Biaxial loading experiments on Cu-Al-Ni single Crystals. In: Experiments in Smart Materials and Structure, ASME, AMD vol 181, pp 61-69
5. Dacorogna B (1989) Direct Methods in the Calculus of Variations, Applied Mathematical Sciences, vol 78, Springer, Berlin Heidelberg New York
6. Fu S, Huo Y, Müller I (1993) Thermodynamics of Pseudoelasticity – An Analytical Approach. Acta Mech. 99 1
7. Fedelich B, Zanzotto G. (1991) One-dimensional quasistatic nonisothermal evolution of shape-memory material inside the hysteresis loop. Continuum Mech. Thermodyn. 3:251-276

8. Goo B C, Lexcellent C (1997) Micromechanics-based Modeling of Two-way memory effect of a single crystalline shape-memory alloy. Acta Mater vol 45 no 2:727-737
9. Huo Y, Müller I (1993) Non-Equilibrium Thermodynamics of Pseudoelasticity. Continuum Mech. Thermodyn. 5
10. Halphen B, Nguyen Q S (1975) Sur les matériaux standards généralisés. Journal de Mécanique 14:39-63
11. Kohn R V (1991) The relaxation of a double-well energy. Continuum Mech. Thermodyn. 3:193-236
12. Kohn R V, Sternberg P (1989) Local minimizers and singular perturbations. Proceedings of the Royal Society of Edinburgh 111A:69-84
13. Lexcellent Ch, Boubakar M L (200) About the thermomechanical modelling of shape memory alloys. In Continuum Thermomechanics the Art and Science of Modelling Material Behaviour, Paul Germain's Anniversary. Kluwer, London, pp 225-236
14. Müller I (1989) On the size of hysteresis in pseudo-elasticity. Continuum Mech. Thermodyn. 1:125-142
15. Müller I, Xu H (1991) On the pseudo-elastic hysteresis. Acta Metall. 39:263-271
16. Müller I. (1996) Six lectures on Shape Memory Alloys, Technical University Berlin
17. Pagano S, Alart P (1999) Solid-solid phase transition modelling : relaxation procedures, configurational energies and thermomoechanical behaviours. International Journal of Engineering Science, vol 37 no 14:1821-1840
18. Pipkin (1986) Some examples of Crinkles. In: Ericksen J L (eds) Homogenization and Effective Moduli of Materials and Media. IMA vol Math. Appl. 1. Springer, Berlin Heidelberg New York, pp 182-195
19. Raniecki B, Lexcellent C, Tanaka K (1992) Thermodynamic model of pseudoelastic behaviour of shape memory alloys. Arch. Mech. vol 44 no3:261-284

# Thermomechanical Couplings and Scale Transitions in Mechanics of Materials

Robert Peyroux, Catherine Mabru

Laboratoire de Mécanique et Génie Civil,
cc 048, Université Montpellier II,
F-34095 Montpellier Cedex 5, France

## 1 Introduction

Mechanics of Materials has experienced this last decade a considerable expansion. The widening, to non strictly mechanic sciences, of available knowledge and data about the material behaviour has permitted to get a global view of the phenomena accompanying the deformation processes. The use of innovative experimental techniques, the writing of consistent theoretical framework and the recourse to high-performance numerical methods allow to analyse, to understand and to simulate the materials behaviour. Two research areas are particularly active these last years. On the one hand, the analysis of the microstructure of materials reveals the phenomena associated with the deformation process, and the use of scale transition techniques permits to integrate this description and to derive a more valuable macroscopic modelling. On the other hand, these phenomena often require variables that complete the classical displacements, strain or efforts of the mechanics. The taking into account of a temperature or a volumic fraction of phase in the constitutive equations implies the modelling of couplings between variables and the derivation of an adapted framework. The interest of this approach is again to get a more valuable description of the material behaviour, and also to use these additional variables as real tracers of the deformation process of the material.
In this paper, we first present the classical framework of Thermodynamics of Irreversible Processes, and more particularly the so-called Generalised Standard Materials approach, putting the emphasis on the modelling of thermomechanical couplings and their experimental estimations. Then, after a brief review of the various works carried out in the team Thermomechanics of Materials of the LMGC, we will detail two applications, one dealing with a multiscale analysis of the thermomechanical behaviour of Shape Memory Alloys, and the other connected with thermal and dissipative effects associated with fatigue of metallic alloys.

## 2 A convenient thermomechanical framework

A convenient thermodynamic framework is the Continuum Thermodynamics (Germain, 1973, Germain, 1983), that postulates the local state axiom and considers the material both as a continuum and as a thermodynamic system described by a set of state variables. More precisely, we use the formalism of Generalised Standard Materials (GSM) (Halphen, 1975) for which the constitutive equations can be derived from a thermodynamic potential and a dissipation potential.The chosen set of variables is $\{T, \varepsilon, \alpha\}$ , $T$ being the temperature, $\varepsilon$ a strain tensor and $\alpha$ standing for all the other variables. The two principles of thermodynamics give the following local equations,

$$\rho \dot{e} = \sigma : \dot{\varepsilon} + r_e - div\ q \tag{1}$$

$$\rho \dot{s} - \frac{r_e}{T} + div \left( \frac{q}{T} \right) \geq 0 \tag{2}$$

where $e, \sigma, q, s, \rho$ and $r_e$ stand for the volume internal energy, the stress tensor, the heat influx vector, the specific entropy, the mass density and the external heat supply, respectively.

Classically, the total dissipation $d$ is introduced through the Clausius-Duhem inequality (3), and is generally split into an intrinsic dissipation $d_1$ and a thermal dissipation $d_2$ . Each of these two dissipations is supposed to be non negative.

$$\sigma : \dot{\varepsilon} - \rho e,_{\varepsilon} \dot{\varepsilon} - \rho e,_{\alpha} \dot{\alpha} - \frac{q}{T} gradT = \frac{d}{T} \geq 0 \tag{3}$$

$$d_1 = \sigma : \dot{\varepsilon} - \rho e,_{\varepsilon} \dot{\varepsilon} - \rho e,_{\alpha} \dot{\alpha} \quad and \quad d_2 = -\frac{q}{T} gradT \tag{4}$$

Using the formalism of GSM, a thermodynamic potential $\psi(T, \varepsilon, \alpha)$ (specific Helmholtz free energy) and a dissipation potential $\varphi(gradT, \dot{\varepsilon}, \dot{\alpha})$ are introduced. Their derivatives with respect to the state variables and their time derivatives give the state equations (5a) and the evolution equations (5b).

$$a) \begin{cases} \text{s=-}\Psi,_T \\ \sigma_r = \rho\Psi,_{\varepsilon} \\ \text{A=}\rho\Psi,_{\alpha} \end{cases} \qquad b) \begin{cases} \text{q=}\varphi,_{grad\ T} \\ \sigma^{ir} = \varphi,_{\dot{\varepsilon}} \\ \text{A=-}\varphi,_{\dot{\alpha}} \end{cases} \tag{5}$$

In these last equations,$-A$ represents the thermodynamic force associated to the state variable $\alpha$ , such as the product $-A\dot{\alpha}$ is the part of dissipated energy corresponding to the mechanism described by $\alpha$ .

The local heat conduction equation can be derived from the preceding expressions,

$$\rho C\dot{T} + divq = d_1 + \rho T\psi,_{T\varepsilon} : \dot{\varepsilon} + \rho T\psi,_{T\alpha} : \dot{\alpha} + r_e \tag{6}$$

where $C$ denotes the specific heat capacity. Assuming a isotropic conduction law, (6) becomes

$$\rho C\dot{\theta} - k\Delta\theta = w_h \tag{7}$$

where $\theta$ is the temperature variation around the equilibrium temperature $T_0$, $(\theta = T - T_0)$, $k$ is the coefficient of thermal conduction and $w_h$ the volume heat source generated by all the terms in the right hand of equation (6).

This last term can be experimentally evaluated using temperature fields given by infrared thermography (see (Chrysochoos, 1995) fore more details). Another interesting experimental technique is the digital image correlation; it consists in correlating speckle images obtained by an optical camera to derive fields of displacement and strain. These two valuable techniques, providing thermal and mechanical fields, can be used to evaluate the various terms involved in $w_h$, and energy balances can be carried out.
This thermomechanical framework and these experimental devices have been used to investigate the behaviour of various materials: elastoplasticity in metallic alloys, thermovicoelasticity of polymers, phase change in shape memory alloys, localisation of thermomechanical fields (bands or necking), damage and unilateral effects in concrete for example. Two of these applications are detailed in this volume (Huon, 2001, and Muracciole, 2001).
We now focus on two particular studies where thermomechanical couplings have a great significance. The first one deals with Shape Memory Alloys and requires a multiscale approach, and the second one is concerned with analysing of thermal and dissipative effects associated with fatigue of metallic alloys.

## 3 Multiscale approach of the thermomechanical behaviour of SMA

SMA may undergo remarkable microstructural transformations: they may change the structure of their crystallographic lattice under mechanical and/or thermal loadings. This transformation, called martensitic transformation, is displacive in the sense that it corresponds to a collective displacement of atoms, and is considered as a first order phase transition described by a latent heat of phase change. Furthermore, the transition domain of SMA is generally close to the room temperature and thus is relatively easy to detect

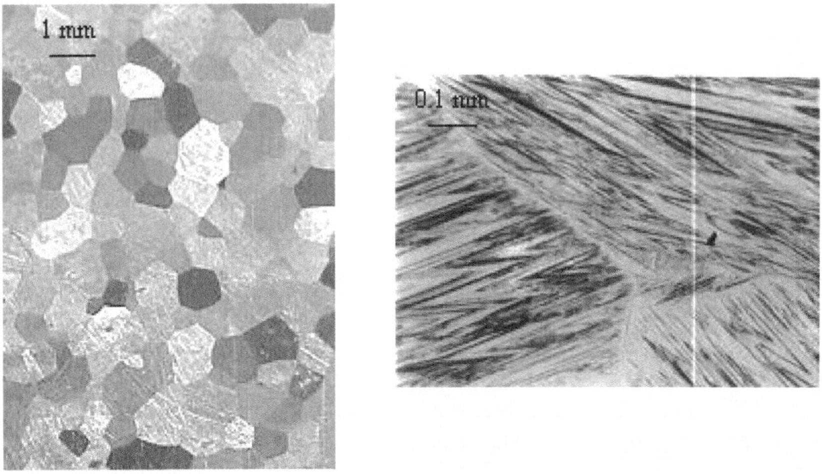


**Fig. 1.** Two characteristic views of a polycrystalline sample of CuZmAl defining the mesoscopic scale

and observe. Finally, and from a more practical point of view, the number of applications using SMA properties grows day after day.

One of the difficulties in the modelling of SMA behaviour consists in the choice of the description scale. In fact, three length scales can be introduced, starting from the crystallographic lattice level ( *microscopic* scale) up to the *macroscopic* scale of the sample, with an intermediate *mesoscopic* scale corresponding to the grain of the polycrystal or to the monocrystal (figure 1).

Some authors have systematically studied the potential microstructural arrangements (twinning) or rearrangements mechanisms due to straining (variants reorientation) when the material is in mechanical and thermodynamical equilibrium (Ball, 1987). Valuable results were obtained with good correlation with metallographic observations, but the connection with macroscopic behaviour seems difficult, especially when kinetic of phase change is to be modelled. Other authors (Auricchio, 1997, Frémond 1993, Lexcellent 1991, Patoor, 1987) were interested in the phase transition itself from a macroscopic point of view, that is pseudoelasticity, self-accommodation, one-way or two-way shape memory effects, but their works do not take into account neither microscopic considerations nor thermomechanical couplings due to phase transition. Following these works, we first present a phenomenological macroscopic model based on the analysis of thermomechanical couplings.

### 3.1 A first macroscopic analysis and modelling.

This modelling (Peyroux, 1998) is based on experimental results, obtained from a thermomechanical device, which underline the main role played by the temperature variations induced by the deformation. The detailed analysis of the associated energy balances shows that the dissipated mechanical work (i.e. intrinsic dissipated energy) remains very small (less than 2%) compared with latent heat of phase change, so that the temperature variations are essentially due to stress-induced phase transition.
On the basis of these results, a macroscopic modelling was proposed, that assumes an null intrinsic dissipation, takes account of thermomechanical cou plings and, of course, considers anisothermal deformation processes.
For the sake of simplicity, only two self-accommodating variants of martensite are considered. Thus the set of state variables is$\{T, \varepsilon, x_1, x_2\}$ , $x_1$ and $x_2$ being the volume fraction of the two variants. The strain tensor can be split into three parts, respectively due to elastic, thermal, and phase transition effects:

$$\varepsilon = \varepsilon^{el} + \alpha\theta Id + x_v\beta^v \tag{8}$$

In equation 8, $\alpha$ stands for the thermal expansion coefficient and the tensor $\beta_v$ characterises the phase change strain associated with the variant $v$ . To ensure the self-accommodation property, we suppose $\beta_1 = -\beta_2 = \beta$ , with $Tr\,(\beta) = 0$.

According to the experimental results, the dissipation potential$\varphi$ is identically equal to zero, and the thermodynamic potential is derived from the classical thermoelastic potential plus a specific part due to phase change $\psi_{ch}$.

$$\rho\psi = \rho\psi_e + \rho\psi_{ch} + I_v\,(x_1, x_2) \;\; and \; \phi = 0 \tag{9}$$

In equation 9, the indicator function $I_v$ ensures physically admissible values to $x_1$ and $x_2$ . Using the thermomechanical framework presented in section 2, we can easily derived the constitutive equations:

$$\sigma = a\,(\varepsilon \; - \; \alpha\theta \; - \; x_v\beta^v) \tag{10}$$

$$\rho C\,\dot{\theta} \; - \; k\,\Delta\theta = -\,T_0 a\,\alpha\,\dot{\varepsilon} + \rho\,L^v\dot{x}_\nu \tag{11}$$

In equation 9, $a$ stands for the elasticity modulus tensor and in equation 10, $L$ represents the latent heat of phase change. Furthermore, the nullity of the dissipation potential $\phi$ implies the nullity of the thermodynamic forces associated to $x_1$ and $x_2$ , and thus gives supplementary inequalities connecting the stress tensor, the temperature and the volume fractions of martensite. This defines the transition domain (figure 2), giving the proportion of austenite (A) and of the two variants of martensite ($M_1$ and $M_2$).

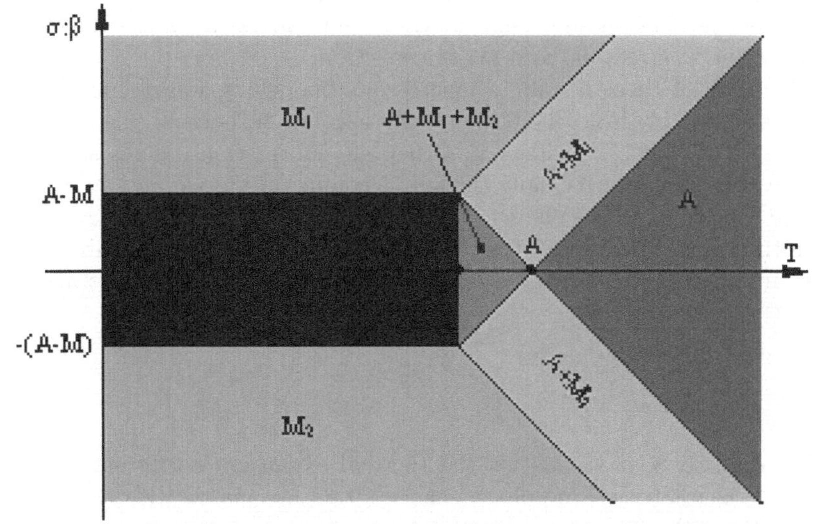


**Fig. 2.** Transition domain

This basic model gives a correct description of the main characteristic phenomena of SMA: pseudoelasticity, for high temperature behaviour, and reorientation effects for lower temperature. One of the most interesting results is in fact the prediction of temperature variations in the material. In agreement with the experimental results, these variations are found to be non negligible and of the same order of magnitude than the transition domain width. The taking into account of the anisothermal feature is then unavoidable and is the only way to predict accurately the transformation kinetic and the stress and temperature variations. As a direct consequence, the classical representation, in a stress-strain plane, of a pseudoelasticity test is not adequate, insofar as one can imagine several evolutions in the stress-strain-temperature space giving the same projection in the stress-strain plane (see figure 3).

The main drawback of this purely macroscopic modelling is that it refers to volumic fractions of martensite, characterised by their orientation tensor $\beta$ which is tightly connected to the microscopic scale. To correct this lack of scale correlation, a scale transition can be used in order to derive the macroscopic behaviour from a mesoscopic model consistent both with the crystallographic data and the thermomechanical couplings. This mesoscopic modelling is described in the next part and the transition scale aspects will be presented in the following.

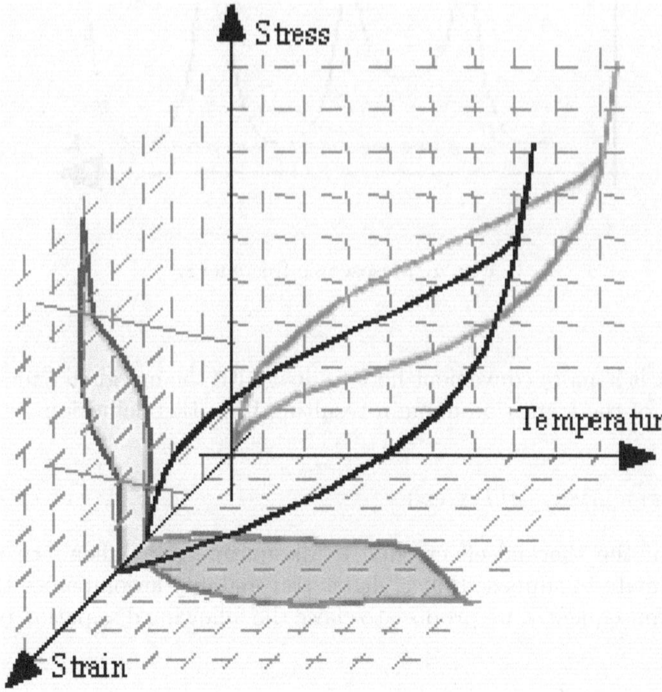


**Fig. 3.** Stress-strain-temperature evolution during a pseudoelasticity test

### 3.2 A mesoscopic thermomechanical model

In this part we aim to derive a set of constitutive equations appropriate for the behaviour of a grain of a polycrystal. To this end, we assume that each grain of the polycrystal behaves as a monocrystal and interacts with the adjoining grains. In the following, the mesoscopic model takes advantage of the crystallographic description of change phase at the microscopic level and of experimental results obtained on monocrystalline sample (Balandraud, 2000).

We consider the set of variables$\{T,\varepsilon, x_k\}$ , where $x_k$ stands for all the variants of variants of martensite required in the considered grain. The free energy $\Psi$ is chosen convex but not strictly convex with respect to $\varepsilon$ (figure 4) in order to provide a strain plateau under zero-stress (reorientation effect).

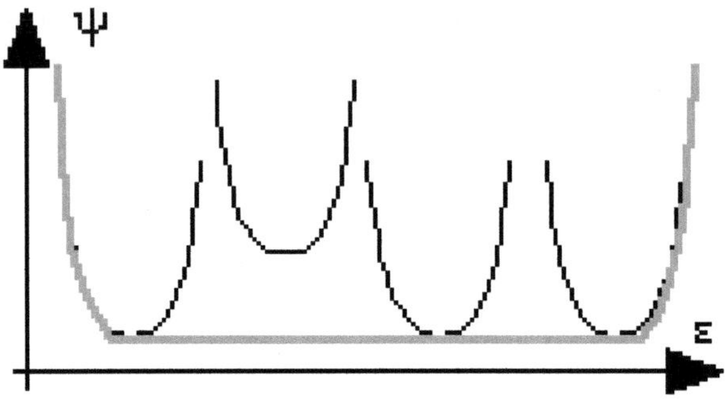


**Fig. 4.** Mesoscopic free energy

In fact it is more convenient here to deal with the marginal energy $\tilde{\psi}$, the definition of the transition domain resulting from this definition.

$$\tilde{\psi}(T,\varepsilon) = Min_{x_k}\psi(T,\varepsilon,x_k) \tag{12}$$

Even if the thermal effects due to dissipation are difficult to measure, it is legitimate to suppose that a slight irreversibility accompanies the phase change. Consequently, we propose to chose the following dissipation potential:

$$\phi = \phi(\dot{x}_k) = \frac{1}{2}K\left(\dot{x}_k^2\right) \tag{13}$$

The existence of terms in $\dot{x}_k$ in the expression of the transition diagram expresses a dependence on the sense of transformation ($M \to A$ or $A \to M$). Figure 5 shows the results of this modelling in terms of strain-strain relation and temperature evolution. The comparison with the non-dissipative case underlines that dissipation has a significant effect on the hysteresis area of the mechanical curve but a slight influence on the temperature evolutions.

Finally, this set of constitutive equations allows a correct prediction of pseudoelasticity, reorientation effect and recovery strain; it predicts phenomena such as asymmetry in tension-compression tests and time effects due to heat diffusion.

### 3.3 Towards a macroscopic model

We now focus on the macroscopic behaviour of a SMA polycrystalline aggregate. We again suppose that each grain behaves as a monocrystal and that the macroscopic behaviour can be derived from the thermomechanical

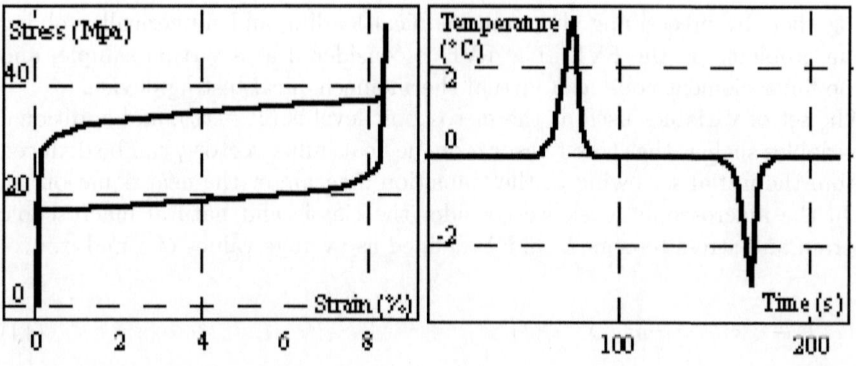


**Fig. 5.** Numerical stress-strain curve and temperature evolution in a pseudoelastic test

response of a representative volume element (RVE). The observation of a polycrystalline sample by means of electronic and optical microscopy, allows to gather data on the crystallographic texture, the shape and the statistical representation of the grains in the RVE (Figure 6). We also consider that the grains differ only by their crystallographic orientation.

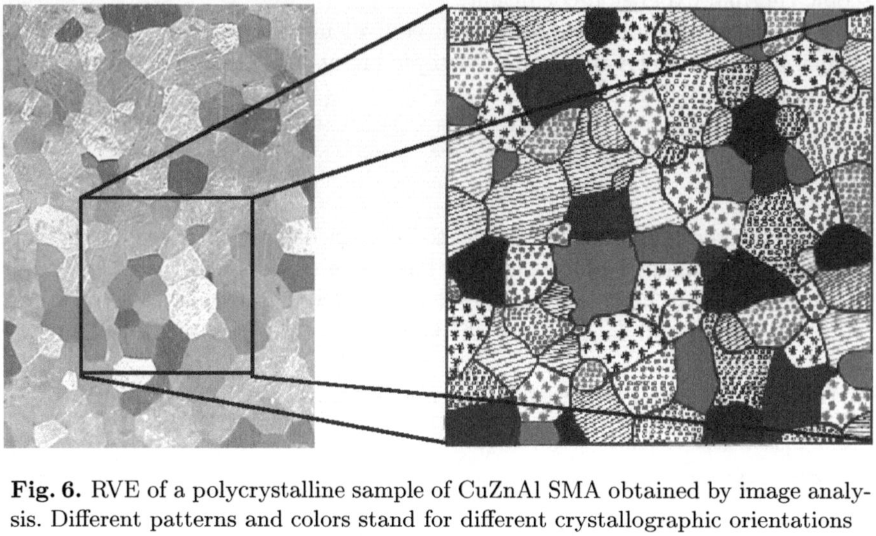

**Fig. 6.** RVE of a polycrystalline sample of CuZnAl SMA obtained by image analysis. Different patterns and colors stand for different crystallographic orientations

It is established, (Suquet, 1984) that for microscopic constitutive equations containing internal variables, "the homogenized law does not reduce to a single equation" on the macroscopic domain. "The knowledge of the macroscopic law requires as data the (mesoscopic) state variables". An alternative and pragmatic attitude is to link the macroscopic thermomechanical variables

together, by prescribing thermomechanical loading and numerically solving the problems on the RVE. The RVE is considered as a virtual sample, and the finite element code as a virtual thermomechanical testing device.
The set of variables used at the mesoscopic level is $(T, \varepsilon, x_k)$, and additional variables such as the stress tensor $\sigma$ or the heat influx vector $q$ can be deduced from the initial set owing to the transition diagram or the heat equation.
On the macroscopic level, we consider the classic and natural macroscopic strain and stress tensors $E$ and $\Sigma$ defined as average values of $\varepsilon$ and $\sigma$:

$$E = < \varepsilon > \quad and \quad \Sigma = < \sigma > \tag{14}$$

In a first time we assume a quasi-homogeneous mesoscopic temperature field, the value of which is identified to the macroscopic temperature. Note that the quasi-homogeneity of the temperature field does not imply the homogeneity of its gradient.At this point $E^{pc}$ , part of macroscopic strain due to phase change, is given by:

$$E^{pc} = E - {A^{\mathrm{hom}}}^{-1} \Sigma - \alpha^{\mathrm{hom}} \theta \tag{15}$$

where $A^{\mathrm{hom}}$ is the macroscopic tensor of elasticity, and $\alpha^{\mathrm{hom}}$ the macroscopic thermal expansion coefficient.

It remains to define $X$ : macroscopic equivalence to the volume proportion of phase change. Deriving $\overline{L}$ , macroscopic latent heat, from $\overline{\rho}\overline{L} = \langle \rho L \rangle$ , we can propose as a definition of $X$ :

$$\overline{\rho}\overline{L}X = \langle \rho L x \rangle \tag{16}$$

This macroscopic proportion has to be regarded as an energy indicator of the advancement of phase change. The value of $X$ is 0 if the polycrystal is completely austenitic, and reaches 1 if completely martensitic. However a value between 0 and 1 can be reached under several microscopic configurations.

Then, it is of some interest to express $E^{pc}$ as:

$$E^{pc} = X \beta^{\mathrm{hom}} \tag{17}$$

where $\beta^{\mathrm{hom}}$ represents the maximum potential strains associated to phase change.
A convenient set of macroscopic variables being established, we finally present an example of the results of finite element simulations on the RVE (Balandraud, 2000).

We numerically performed a pseudoelasticity test consisting of a load-unload path at a room temperature upper than $T_A$ and the different results

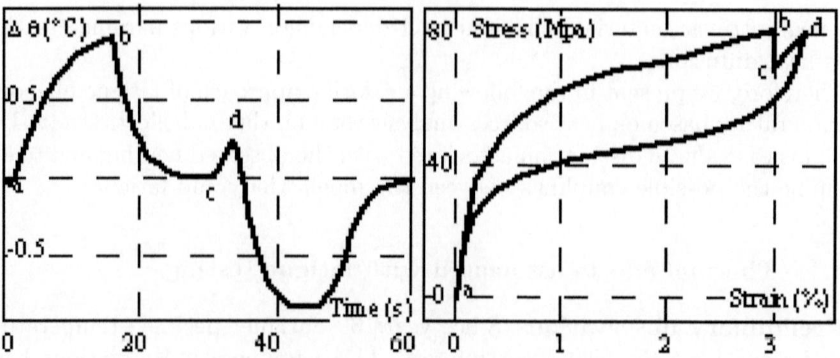


**Fig. 7.** Results of the pseudoelastic numerical test

are plotted in figure 7. The imposed strain is increased from point $a$ to point $d$, and maintained at a certain level between points $b$ and $c$. In the particular stage $bc$, we can observe the role played by thermomechanical couplings. The material keeps on transforming ($X$ increases), the stress relaxes while the temperature returns to the imposed value. Concerning the entire test, the evolution of the different variables is consistent with experimental results obtained on polycrystalline SMA. This macroscopic behavior is of course different from the monocrystalline one and corresponds to the particular ordering of grains chosen in the RVE.
The possibilities of this approach have been widely developed in (Balandraud, 2000). For instance, low-temperature tests allowed the determination of the value of the macroscopic recoverable strain due to variant reorientation at zero stress. This numerical tool was of great use when investigating multiaxial behaviour of polycrystalline SMA and when studying the influence of various dissipation potentials.

## 4 Thermal and dissipative effects associated with fatigue of metallic alloys.

Statistical considerations on time-consuming and expensive tests are needed to evaluate the fatigue resistance of materials and structures. Increasing pressure for shorter design assessment time in industry leads researchers to develop new short-time methods of fatigue limit determination.
In this aim, variations of specimen temperature during fatigue tests have been studied (Bousseau, 1998, Luong, 1998, La Rosa, 2000, Liaw, 2000, Galtier, 2001). These observations incited researchers to propose a new methodology for the determination of fatigue limit. This methodology is based on the analysis of thermal effects associated with fatigue testing. However, some of its application are still problematic. Indeed, the observed heating does not

reveal intrinsic material behaviour as it depends on various parameters such as heat diffusion.
Therefore, we present in the following a calorific approach of fatigue phenomenon rather based on heat sources analysis than on thermal effects study. The aim is to evaluate dissipation associated with the observed heating and to examine the possible couplings between thermoelasticity and fatigue.

### 4.1 Thermal effects associated with fatigue testing

**Preliminary observations** Since years 30, fatigue specimen temperature variations have received great interest. These temperature variations have been measured either by thermocouples or, more recently, using infrared thermography. All the authors (Bousseau, 1998, La Rosa, 2000, Liaw, 2000) agreed on the following observations :

-the greater the applied stress and the frequency, the higher the temperature increase during fatigue test.

-for given stress and frequency, the average specimen temperature increases during the first part of the test (first thousands cycles), then remains constant (steady-state conditions) until an abrupt increase immediately prior to failure.

This suggests that, concerning temperature, the few first thousands cycles are relevant and leads to consider this steady-state temperature as a representative parameter to characterize fatigue.

**Short-time measurements of fatigue limit using thermal analysis** Some authors derived a new method for the determination of the fatigue limit from these preliminary observations. This methodology is based on the evolution of the steady-state temperature as a function of the applied stress amplitude (see (La Rosa, 2000, Luong 1998) for detailed procedure). A specimen is successively loaded at different stress levels for about $10^4$ cycles. For each loading level, the steady-state temperature $\Delta\overline{T}$ (supposed to be reached after $10^4$ cycles) is measured. The development of $\Delta\overline{T}$ the steady-state temperature, plotted as a function of the stress level, exhibits a well-marked break. The stress amplitude corresponding to this break appears to be close to the fatigue limit as evaluated with usual methods.
This new method offers considerable interest compared to usual techniques: it allows to obtain reliable results using a very limited number of specimens in a very short time. In addition, it is easily applicable to structures. Consequently, this methodology, which validity has been confirmed by different applications, has been adopted in industrial development (Bérard, 1998). However, some applications, concerning aluminium alloy for instance, are still critical. Indeed, in that case, the steady-state temperature does not show a clear evolution with respect to stress levels. Hence, it is impossible to establish an obvious relation between temperature variations and fatigue limit.

Anyway, even if in most cases, temperature seems to be a relevant parameter to characterize irreversibilities that occur during fatigue testing, it does not express the intrinsic behaviour of the material. Actually, temperature depends on the specimen geometry, on the external heat exchange conditions and on the homogeneity of the studied phenomenon. In order to take into account these various structural effects, we propose to analyse the fatigue mechanisms considering heat sources instead of temperature.

## 4.2 Dissipation and couplings

The set of state variables to be used in the above-mentioned thermomechanical framework is $(T, \varepsilon, f)$, where $f$ is a scalar parameter standing for the state of fatigue in the material.
The local heat conduction equation is then :

$$\rho C\dot{\theta} - k\Delta\theta = d_1 + \rho T \psi,_{T\varepsilon} : \dot{\varepsilon} + \rho T \psi,_{Tf} : \dot{f} \tag{18}$$

Supposing the mechanical irreversibilities only stem from fatigue effects (i.e. no irreversible stress), the intrinsic dissipation can be written:

$$d_1 = -\rho\psi,_{ff} \tag{19}$$

Assuming that the slight temperature variations ($\theta << T_0$) does not affect the state of fatigue but, on the contrary, that the state of fatigue can modify the thermoelastic response of the material, we can write:

$$\psi,_{Tf} = 0 \tag{20}$$

$$\psi,_{T\varepsilon} = b(f) \tag{21}$$

In the case of a one-dimension thermoelastic behaviour including fatigue state variable, the strain becomes:

$$\varepsilon = \frac{\sigma}{E(f)} + \alpha(f)\theta \tag{22}$$

where $E$ is the Young modulus, and $\alpha$ the thermal expansion coefficient. Equations (20) and (22) lead to:

$$(E\alpha),_f = 0 \tag{23}$$

that is to say $b(f)$ is constant. So, when $E$ decreases, $\alpha$ increases, and the strain amplitude gets greater as the fatigue process progresses. Thermoelastic sources amplitude will consequently increase:

$$\Delta S_{the} = T_0 E \alpha \Delta \dot{\varepsilon} \tag{24}$$

Then, it is possible to estimate the thermoelastic sources amplitude for several loading cycles, supposing the evolution of $f$ is slow, that is the variations of $f$ can be neglected. Steady-state temperature reached after a load step of some thousands cycles will be associated with average dissipation.

### 4.3 Experimental

Fatigue tests were performed on a servohydraulic machine with a 25 kN load cell. These tests were conducted at 10 Hz with a load ratio R=0.1. Successive load steps of $10^4$ cycles were carried out on a same specimen. Surface temperature images of the tested sample were provided using an infrared CCD camera (short waves, NETD˜25mK). High-speed data acquisition capabilities are available at 50 Hz with a full-frame (256*256 pixels) and at 180 Hz with a narrow window (128*128 pixels). An infrared image processing has been developed in order to deduce the distribution of heat sources from the surface temperature measurements by the mean of equation (18). This data processing is based on local least-square fitting (Wattrisse, 2000). Hypothesis needed for its application as well as validation checking are also presented in (Chrysochoos, 2000).
Specimens were flat (1 mm thick): this allows to consider the surface temperature as the average thick-wise temperature. All specimens were coated with a black spay paint to maximise their emissivity. In addition, the paint coating reduces surface reflections and ensures uniformity of the specimen response.
A stainless steel (304L) and an aluminium alloy (AU4G) were studied. For these two materials, thermophysical characteristics needed to determine heat sources are given in table 1.

| | $\rho$ (kg.m$^{-3}$) | C (J.kg$^{-1}$.K$^{-1}$) | $\tau_{th}(s)$ | $\tau_{eq}(s)$ | k (W.m$^{-1}$.K$^{-1}$) |
|---|---|---|---|---|---|
| 304L | 7900 | 480 | 327 | 45 | 60 |
| AU4G | 2800 | 930 | 225 | 18 | 115 |

**Table 1 :** *Thermophysical characteristics of the materials*

$\tau_{th}$ is a time constant related to the heat losses and characterises the lateral heat exchanges between the sample and the surrounding air. If the heat sources are assumed to be uniform, the laplacian operator can be reduced to a linear term. An equivalent time constant $\tau_{eq}$ is then defined :

$$\frac{1}{\tau_{eq}} = \frac{1}{\tau_{th}} + \frac{k}{\rho C}\left(\omega_0^2 + \Omega_0^2\right) \tag{25}$$

where $\omega_0$ et $\Omega_0$ are the first eigen-pulsations of the laplacian operator (Chrysochoos, 2000) In that case, steady-state temperatures are related to the average intrinsic dissipation:

$$\Delta\bar{T} \approx \frac{\tau_{eq}}{\rho C}\bar{d}_1\left(f\right) \tag{26}$$

Besides, temperature variations amplitude for a load cycle is:

$$\Delta T \approx -\frac{\alpha\left(f\right)T_0}{\rho C}\frac{4\pi f_r}{\sqrt{\tau_{eq}^{-2}+4\pi^2 f_r^2}}\sigma_0 \tag{27}$$

with $f_r$ standing for the loading frequency. These temperature variations over one cycle are due to thermoelastic coupling.

## 4.4 First experimental results

During a load step (figure 8), the specimen shows on the one hand an average temperature increase related to energy dissipation, and, on the other hand, sinusoidal fluctuations around this average value. If all other couplings are neglected as supposed in §4.2, these fluctuations are induced by thermoelastic effects.

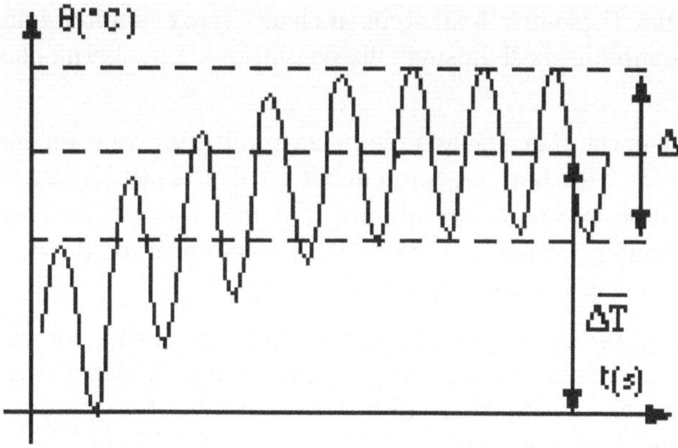


**Fig. 8.** Experimental evolution of temperature during a fatigue cyclic loading

Figure 9 presents results obtained for the whole load steps in the case of two stainless steel specimens. Steady-state temperature evolution is fully

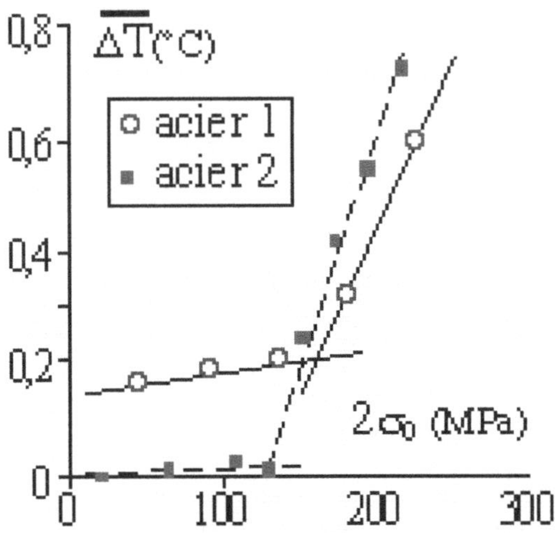


**Fig. 9.** Evolution of $\Delta\bar{T}$ for the A304L steel

consistent with literature results: a well-marked break in the curves is observed.

From a calorific point of view, (Figure 10), thermoelastic sources amplitude evolution is linear for the first load steps, as expected in classical thermoelasticity (constant Young modulus and constant thermal expansion coefficient). After these first load steps, a clear change is noted and might be attributed to coupling mechanisms between fatigue and thermoelasticity (cf. equation 27).

This change in the thermoelastic sources amplitude evolution occurs at the same stress level as for temperature curves break. At last, it is worth noting that average dissipations $\bar{d}_1$ are about $10^3$ times lower than thermoelastic sources for a loading frequency $f_r = 10Hz$ . These average dissipations have been estimated using equation 26.

The results obtained on two aluminium alloy specimens are shown in figure 11. The lack of clear trend when plotting average dissipation (i.e. steady-state temperature, see equation 26) versus stress level is confirmed for this type of material. However, thermoelastic sources development is much more regular and exhibits a marked break for one of the specimens, as it was observed for stainless steel.

It must be noted that the second specimen has been tested at the last stress level for more than $10^7$ cycles without failing. This, combined with the strictly linear development of sources amplitude, suggests the fatigue limit has not been reached in that case.

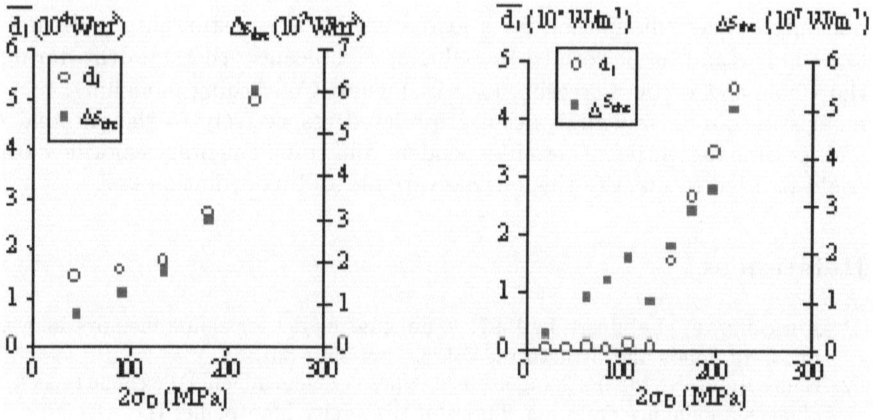


**Figure 10**: *Average dissipation and thermoelastic sources amplitude for A304L steel*

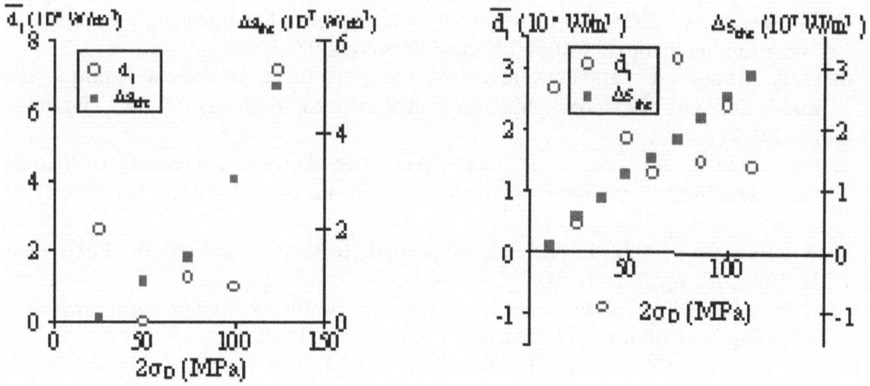


**Figure 11**: *Average dissipation and thermoelastic sources amplitude for aluminium alloy*

## 4.5 Prospects

The scattering of these first results provided by a few tested specimens must be considered from the statistical point of view associated with fatigue testing. It would be premature to point out any conclusion. Nevertheless, the analysis of fatigue effects *via* thermoelastic sources amplitudes appears to be attractive. For instance, in the case of aluminium alloys, the data are more consistent when reported in terms of sources amplitudes than in terms of temperature.

It would be of high interest to develop the same kind of approach considering not only average dissipation for a load step but also instantaneous dissipation. This could be performed by reducing significantly thermal drift during the whole test or by increasing load frequency. Concerning modelling, thermoelastic sources and dissipation are related respectively to the variable $f$ and its time derivative. These independent and complementary calorific data could be used to identify this fatigue variable and its evolution law.

## References

1. Auricchio, F., Lubliner, J., 1997; A uniaxial model for shape memory alloys, Int. J. of Solids and Structures, Vol. 33, pp. 3601-3618.
2. Balandraud, X., 2000; Changement de phase et changements d'é chelle dans les alliages à mémoire de forme, Thesis of University Montpellier II.
3. Ball, J.M., James, R., 1987; Fine phase mixture as minimizers of energy, Arch. Ration. Analysis, 100, pp. 15-52
4. Bérard, J.Y., Rathery, S., Béranger, A.S., 1998; Dé termination de la limite d'endurance des matériaux par thermographie infrarouge. Application sur un bras de suspension. , Matériaux et Techniques, **1-2**, 55-57.
5. Bousseau, M., 1998; Application de la thermographie infrarouge au suivi de l'endommagement de fatigue, PhotoMécanique 98, 199-206.
6. Chrysochoos, A., 1995; Analyse du comportement thermomé canique des matériaux par thermographie infrarouge, Photomécanique 95, Ed. Eyrolles, pp.203-211.
7. Chrysochoos, A., Louche, H., 2000; An infrared image processing to analyse calorific effects accompanying strain localisation, Int. J. of Eng. Sci., 38, 1759-1788.
8. Frémond, M., 1993; Matériaux à mémoire de forme, C. R. Acad. Sc. Paris, série II, Vol. 316, pp.1031-1036.
9. Galtier A., 2001; Influence de la microstructure des aciers sur leurs propriétés mécaniques, Colloque MECAMAT - Aussois 2001, 55-60.
10. Germain, P., 1973; Cours de Mécanique des Milieux Continus, Masson et Cie Ed., p. 417
11. Germain,P., Nguyen Q.S., Suquet, P., 1983; Continuum Thermodynamics, J. of Appl. Mech., Transactions of the ASME, 50, pp. 1010-1020.
12. Halphen, B., Nguyen Q.S., 1975; Sur les Matériaux Standards Généralisés, Journal de Mécanique, Vol. 14, 1.
13. Huon, V., Cousin, B., Maisonneuve, O., 2001; Study of thermal and kinematic phenomena associated with quasi-static deformation and damage process of some concretes, this volume.
14. La Rosa, G., Risitano, A., 2000; Thermographic methodology for rapid determination of the fatigue limit of materials and mechanical components, International Journal of Fatigue, **22**, 1, 65-73.
15. Lexcellent, C., Licht, C., 1991; Some remarks on the modelling of the thermomechanical behaviour of shape memory alloys, J. de Physique, colloque C4, vol. 1, pp. 35-39.
16. Liaw, P. K., Wang, H., Jiang, L., Yang, B., Huang, J.Y., Kuo, R.C., Huang, J.C., 2000; Thermographic detection of fatigue damage of pressure vessel steels at 1000 Hz and 20Hz, Scripta Materialia, **42**, 4, 389-395.

17. Luong, M. P., 1998; Fatigue limit evaluation of metals using an infrared thermographic technique, Mechanics of Materials, **28**, 155-163
18. Muracciole, J.M., Wattrisse, B., Chrysochoos, A., 2001; Local experimental analyses of the thermomechanical heterogeneous behaviour of thermoplastics, this volume.
19. Patoor, E., Eberhardt, A., Berveiller, M., 1987; Potentiel pseudoélastique et plasticité de transformation martensitique dans les mono et polycristaux métalliques. Acta Metall., Vol. 35, pp. 2779-2789.
20. Peyroux, R., Chrysochoos, A., Licht, C., Löbel, M., 1998; Thermomechanical couplings and pseudoelasticity of shape memory alloys, Int. J. of Engng. Sci., Vol. 36, n° 4, pp. 489-509.

# Bioelectrical Impedance Analysis: a Matter of Homogenization of Composites with Imperfect Interfaces

Paolo Bisegna, Franco Maceri

Dipartimento di Ingegneria Civile
Università di Roma "Tor Vergata"
00133 Roma, Italia

## Abstract

Several in vivo methods are currently available for assessment of human body composition. A comprehensive and updated review may be found in [1]. The bioelectrical impedance analysis is probably, at present, the most frequently used method, due mainly to the relatively inexpensive cost of the basic instrument, its ease of operation, and its portability. The bioelectrical impedance measurements are performed using four electrodes: usually two are attached at the wrist and two at the ankle. A weak alternating current is passed through the distal electrodes, while the voltage drop across the body is measured using the proximal electrodes. The body's impedance $Z$ is given by the ratio between the voltage drop and the applied current. Impedance measurements are made over a wide range of frequencies. A fundamental issue is that $Z$ depends on the frequency $f$ of the applied current, and has both resistive and reactive components: $Z = R + jX$. The reactive component $X$ is assumed to be related to cells, which act like capacitors. The experimental curve $f \mapsto Z(f)$ in the complex plane is referred to as impedance locus. It can supply estimates of the extracellular, intracellular and total body water space, and hence it has a considerable clinical significance (e.g., for renal or obese patients), provided it is interpreted on the basis of a clear and rational model [2].

A number of electrical circuits have been used to model the impedance locus [3]. However, it seems obvious that a rational interpretation of the impedance locus must be based on the homogenization theory, because a typical cell dimension is ten micrometers, while the measured impedance is an overall property of the whole body. Following this idea, an ideal tissue, made of a periodic array of unit elements $V$, each containing a cell $V_1$, separated from the extracellular space $V_2$ by the plasma membrane $S$, is considered. The material inside $V_i$ has a conductivity tensor $\boldsymbol{\sigma}_i$ $(i = 1..2)$. The plasma membrane has a capacitance $C$ and a conductance $G$ per unit surface. The unknown electric potential $\phi(\mathbf{x}, t)$, which depends on position $\mathbf{x}$ and time $t$, is represented as the sum of a linear part $-\overline{\mathbf{E}}(t) \cdot \mathbf{x}$, where $\overline{\mathbf{E}}(t)$ is a given electric field, and a $V$-periodic part $\phi^p(\mathbf{x},t)$, which is assumed to be $H^1(V_i)$ ($i = 1..2$), but may be discontinuous across $S$. The governing equations are:

$$\operatorname{div}(\boldsymbol{\sigma}_i \nabla\phi) = 0 \qquad \text{in } V_i, \quad i = 1,2 \tag{1}$$

$$[(\boldsymbol{\sigma}_i \nabla\phi)\cdot\mathbf{n}] = 0 \qquad \text{on } S \tag{2}$$

$$(\boldsymbol{\sigma}_1 \nabla\phi)\cdot\mathbf{n} = G[\phi] + C\frac{d}{dt}[\phi] \qquad \text{on } S \tag{3}$$

where $\mathbf{n}$ is the unit normal vector to $S$, oriented toward $V_2$, and $[\cdot]$ denotes the jump of the enclosed quantity across $S$.

It is pointed out that problem (1)-(3), in the particular case $C = 0$, is well known [4,5], because it models heat conduction in a mixture of two conductors in the presence of a contact resistance between phases. A related problem was considered in [6,7], modeling the elastic behaviour of composites under the assumption that normal and tangential components of interface displacement jumps are proportional to the respective interface traction components. However, the particular case $C = 0$ leads to an elliptic problem, whereas in the present situation ($C \neq 0$), the problem is parabolic, with a concentrated capacity. To the authors' knowledge, homogenization of a parabolic problem with a concentrated capacity has received little attention in the literature.

The problem is approached by a Fourier transform with respect to time $t$. Of course, the unknown of the transformed problem, that is the Fourier transform $\hat{\phi}(\mathbf{x},f)$ of $\phi^p(\mathbf{x},t)$, is a complex quantity. As a consequence, it is convenient to assume as unknowns both the real and the imaginary part of $\hat{\phi} = \varphi + j\chi$. New variational principles involving $\varphi$ and $\chi$ are derived, and are recognized to be saddle-point principles. In order to build up minimum principles, partial Fenchel conjugates of these saddle-point principles are performed. The minimum principles thus obtained are used to bound, for each value of the frequency variable $f$, the homogenized impedance. Voigt-Reuss-type bounds, Hashin-Shtrikman-type bounds, and bounds based on finite-element computations are established .

## References


1. Ellis, K. J. (2000). Human body composition: in vivo methods, *Physiological Reviews* **80**(2) 649–680.
2. DeLorenzo, A., Andreoli, A., Matthie, J., and Withers, P. (1997). Predicting body cell mass with bioimpedance by using theoretical methods: a technological review, *J. Appl. Physiol.* **82** 1542–1558.
3. Cole, K., and Cole, R. H. (1941). Dispersion and absorption in dielectrics. I. Alternating current characteristic, *J. Chem. Physiol.* **9** 341–351.
4. Torquato, S., and Rintoul, M. D. (1995). Effect of the interface on the properties of composite media, *Physical Review Letters* **75**(22) 4067–4070.
5. Lipton, R., and Vernescu, B. (1996). Composites with imperfect interfaces, *Proc. R. Soc. Lond. A* **452** 329–358.

6. Lene, F., and Leguillon, D. (1982). Homogenized constitutive law for a partially cohesive composite material, *Int. J. Solids Structures* **18**(5) 443–458.
7. Hashin, Z. (1992) Extremum principles for elastic heterogenous media with imperfect interfaces and their application to bounding of effective moduli, *J. Mech. Phys. Solids* **40**(4) 767–781.

# Passive Vibration Damping of Piezoactuated Structures by Using Enhanced Shunt Electric Circuits

Giovanni Caruso

Istituto per le Tecnologie della Costruzione
Consiglio Nazionale delle Ricerche
00137 Roma, Italia

**Abstract.** This paper deals with passive vibration damping of piezoactuated elastic structures. An enhanced resonant shunt electric circuit, which is a generalization of the shunt circuits proposed in the literature until now, is analyzed from a theoretical point of view. A modal model is used to evaluate the performances of this shunt in damping structural vibrations; in particular, a numerical optimization of the electrical components is performed according to the pole placement technique. It is shown that the new circuit presented here is able to partially improve the performances of the shunt circuits previously proposed in the literature.

## 1 Introduction

It is well established that piezoelectric sensors and actuators are very suitable for vibration damping and noise suppression of light elastic structures [14]. Many papers in the literature are devoted to the study of active damping (e.g. [9,7,10,2]) and passive damping (e.g. [12,11,15,16,18,17]) of piezoactuated structures. The topic of this paper concerns with the latter methodology, built up by connecting the electrodes of a piezoelectric element, bonded on a vibrating structure, to a shunt electric circuit, as shown in figure 1. The shunt circuit used in a passive damping arrangement can be purely resistive or resonant; in the former case the passive control adds a small damping on a wide frequency band whereas in the latter case the resonant circuit can be properly tuned on a targeted structural eigenmode in order to obtain a very effective control in a narrow frequency band, centred on the selected structural eigenfrequency. An attempt in optimally damp several structural eigenfrequencies by using a single piezoelectric actuator and a multi-resonant shunt circuit is performed in [13] but it presents some difficulties in tuning the different electrical components due to their mutual influence. Different types of resonant shunt circuits have been proposed in the literature; the simple $RL$ series shunt circuit [12], the $RL$ parallel shunt circuit [18], and the shunt circuit composed by the parallel of a $RL$ series branch with a capacitor $C$ [15]. Those circuits have been analyzed in a unified manner in [6], and it turned out that the $RL$ series shunt circuit is the most effective in damping structural vibrations. In practical applications some problems arise concerning the choice of the inductance $L$ when a $RL$ shunt circuit is employed, because the tuning value of the inductance can attain several hundreds of Henry when passive control of low frequency structural eigenmodes is required. This problem can be overcome by using a synthetic inductor [8], which is able to furnish the required inductance. A negative feature of the synthetic inductor is that an high inductance value implies an high value of the inductor electric resistance, which can be higher than the optimal value of resistance required by the control law. Another problem concerns the use of the shunt circuit in a hybrid control; in this case the active control enforces high values of electric potential across the synthetic inductance, which causes a malfunctioning of this component and consequently a non effectiveness of the damping system [1]. To avoid this defect it can be useful to modify the shunt circuit in order to reduce the tuning value of the inductance, as attempted in [15], where the addition of a capacitor $C$ in parallel with the $RL$ series branch reduces the tuning value of $L$ by a factor $1+\alpha$, where $\alpha$ is the ratio between the external capacity $C$ and the inherent piezoelectric capacity $C_p$. The negative effect of this solution is that the reduction of the tuning inductance implies also a reduction of the attained damping, as analytically and experimentally shown in [6]. In this paper a new shunt circuit is proposed and analyzed, in order to look for a compromise between the sake of reducing the tuning inductance $L$ and the sake of having an effective damping system. The considered shunt circuit, as shown in figure 2, contains the parallel of a $R_2L$ branch and a $R_1C$ branch, in series with a resistor $R_3$. The capacitor $C$ is useful to reduce the tuning value of $L$, as in [15], whereas the resistors $R_1$, $R_2$ and $R_3$ must be chosen in order to optimize the energy dissipation and consequently maximize the damping effect on the controlled structure. A modal model of the controlled structure, containing only the structural eigenmode to be controlled, is used for the analysis of the passive control as in [6]. The optimization of the electrical components is based on the pole placement technique [12]; accordingly, the electrical parameters are chosen in order to move as to the left hand side of the complex plane as possible the system poles. Due to the complexity of the shunt circuit, the optimization is performed by numerical simulation, as shown in the next sections. The analysis shows that the proposed shunt circuit is able to partially improve the previously proposed solutions, as it turns out from the comparison with the results relevant to the $RL$-$C$ shunt circuit.

## 2 The model

In this section a modal model is considered, able to describe the passive vibration damping of a linearly elastic piezoactuated structure, as shown in figure 1. As mentioned in the introduction, an oscillating shunt circuit acts on a targeted eigenmode of the controlled structure. Near a resonance, the mechanical behaviour of a vibrating structure can be well described by

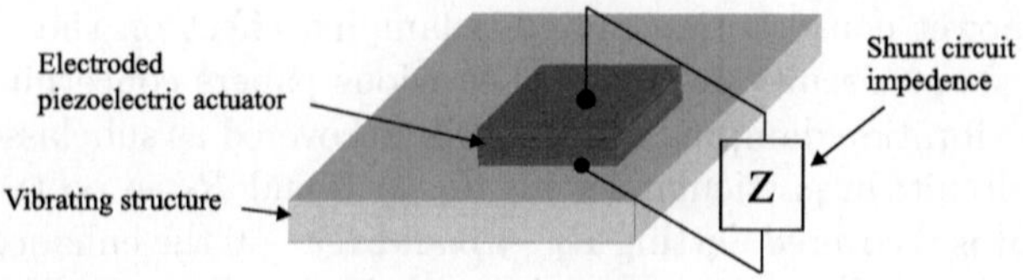


**Fig. 1.** Passively controlled piezoactuated structure

taking into account only the modal displacement relevant to the targeted structural eigenmode. Accordingly the coupled behaviour of a passively controlled piezoactuated structure is described by the following set of ordinary differential equations:

$$\begin{aligned} & m\ddot{y} + d\dot{y} + k_{mm}y + k_{me}v = 0 \\ & k_{me}y - C_p v = -q \\ & \mathcal{F}(v) + \mathcal{G}(q) = 0 \end{aligned} \tag{1}$$

where $y$ is the modal displacement, $v$ is the difference of electric potential between the piezoelectric electrodes, $q$ is the electric charge on the electrodes, $k_{mm}$ is the modal stiffness, $k_{me}$ is the modal piezoelectric coupling stiffness, $d$ is the modal mechanical damping coefficient, $C_p$ is the piezoelectric inherent capacity, $\mathcal{F}$ and $\mathcal{G}$ are linear differential operators with respect to the time $t$ and, finally, a dot means derivation with respect to time. The first equation in (1) is the balance law of the modal forces acting on the structure, the second equation is the electric charge balance law on the piezoelectric electrodes and the last one is the Kirchhoff equation relevant to the oscillating electric shunt circuit, which relates the electric charge $q$ to the voltage $v$. It can be noted that $k_{me}v$ and $k_{me}y$ are the coupling terms due to the converse and direct piezoelectric effect, respectively; in particular the former is the modal force on the structure and the latter is the electric charge induced on the electrodes due to the coupling between strain and electric field, which characterizes piezoelectric materials. A numerical method able to estimate the modal parameters in system (1) is given in [3], employing a finite element formulation and a modal reduction. The coupled system is schematically reported in figure 2, where the mechanical oscillator in the left side represents the controlled structural eigenmode whereas, on the right side, the electric shunt circuit is drawn; the piezoelectric actuator is electrically equivalent to the parallel of the capacitor $C_p$ and the electric charge generator $-k_{me}y$ (see equation $(1)_2$). The capacitor $C$ is used to reduce the tuning value of the inductance $L$ [15]. The resistances $R_1$, $R_2$ and $R_3$ must be chosen as to optimize the electric current flows in the different circuit branches in order to maximize the power consumption and the damping effect on the vibrating structure. The shunt circuits proposed in previous papers concerning piezoelectric passive vibration damping can be easily recovered as subclasses of the present shunt circuit; in particular letting $R_3 \to 0$ and $R_1 \to \infty$ the simple $RL$ series circuit is recovered, letting $R_1 \to 0$ and $R_3 \to 0$ the enhanced shunt circuit analyzed in [15,6] is obtained and finally, letting $R_3 \to 0$, $R_2 \to 0$ and $C \to \infty$ the present circuit reduces to the $RL$ parallel shunt circuit proposed in [18].

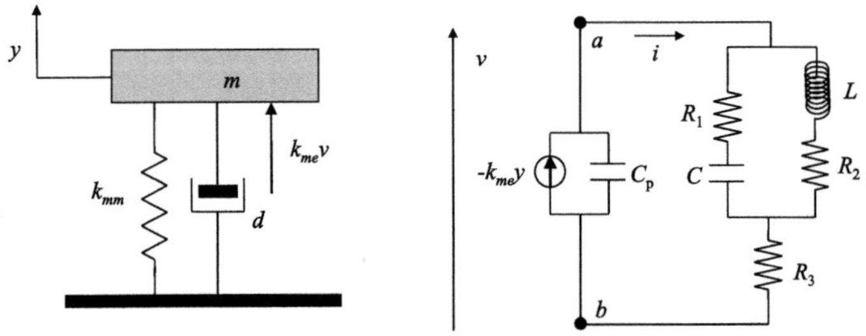


**Fig. 2.** Electromechanical coupled structure

In order to analyze the considered electric shunt circuit, system (1) is rewritten in the Laplace domain; assuming vanishing initial conditions, the following equations are obtained:

$$\begin{aligned} &(ms^2 + ds + k_{mm})\overline{y} + k_{me}\overline{v} = 0 \\ &k_{me}\overline{y} - C_p\overline{v} = -\overline{q} \\ &Z(s)s\overline{q} + \overline{v} = 0 \end{aligned} \tag{2}$$

where $s$ is the Laplace variable, the overlined unknowns are referred to the Laplace domain and $Z(s)$ is the shunt circuit impedance, given by:

$$\frac{L(R_1 + R_3)Cs^2 + [L + R_1R_2C + R_3(R_1 + R_2)C]s + (R_2 + R_3)}{LCs^2 + (R_1 + R_2)Cs + 1} \tag{3}$$

From equation $(2)_2$ the following expression of $\overline{v}$ can be written:

$$\overline{v} = \frac{\overline{q}}{C_p} + \frac{k_{me}}{C_p}\overline{y} \tag{4}$$

Substituting equation (4) in system (2) the following algebraic system is obtained:

$$\begin{aligned} &\left(ms^2 + ds + k_{mm} + \frac{k_{me}^2}{C_p}\right)\overline{y} + \frac{k_{me}}{C_p}\overline{q} = 0 \\ &\left(sZ(s) + \frac{1}{C_p}\right)\overline{q} + \frac{k_{me}}{C_p}\overline{y} = 0 \end{aligned} \tag{5}$$

In order to perform an optimization of the electrical components of the shunt circuit it is useful to consider a dimensionless version of the governing system (5). Assuming that $t = \tilde{t}T$, $\overline{y} = \tilde{y}Y$, $\overline{q} = \tilde{q}Q$, where a capital letter is used to distinguish dimensionless quantities and a tilde is used for the dimensional scales, and remembering that the Laplace variable $s$ has the physical dimension of the inverse of time, system (5) reads as:

$$\begin{aligned} &(S^2 + \nu S + 1 + \kappa^2)Y + \kappa Q = 0 \\ &[\omega_m C_p S Z(\omega_m S) + 1]Q + \kappa Y = 0 \end{aligned} \tag{6}$$

where the following positions are understood:

$$\begin{aligned} &\tilde{t} = \frac{1}{\omega_m} \qquad \frac{\tilde{y}}{\tilde{q}} = \sqrt{\frac{1}{k_{mm} C_p}} \\ &\nu = \frac{d}{\sqrt{m k_{mm}}} \qquad \kappa = \frac{k_{me}}{\sqrt{k_{mm} C_p}} \end{aligned} \tag{7}$$

In equations (7) $\omega_m = \sqrt{k_{mm}/m}$ is the mechanical eigenfrequency at shorted electrodes ($v$=0 in equation $(1)_1$) and has been chosen for the definition of the time scale $\tilde{t}$, $\nu > 0$ is the dimensionless mechanical damping coefficient and, finally, $\kappa > 0$ is the modal piezoelectric coupling coefficient. The dimensionless system (6) holds for any choice of the external shunt circuit; in order to adapt this general system to the present case it is convenient to give a dimensionless version of the term $\omega_m C_p S Z(\omega_m S)$, therefore the following positions relevant to the electric components $L$, $C$, $R_1$, $R_2$ and $R_3$ are made:

$$\begin{aligned} &\frac{\omega_e}{\omega_m} = \delta_1 \qquad \frac{C}{C_p} = \delta_2 \\ &R_1 \omega_m C_p = \delta_3 \qquad R_2 \omega_m C_p = \delta_4 \qquad R_3 \omega_m C_p = \delta_5 \end{aligned} \tag{8}$$

where $\omega_e = \sqrt{1/(LC)}$ and $\delta_i$, $i = 1..5$ are dimensionless quantities. In terms of the previous positions $\omega_m C_p S Z(\omega_m S)$ can be expressed as:

$$\frac{(\delta_3 + \delta_5)S^3 + [(1/\delta_2) + \delta_1^2 \delta_2 \delta_3 \delta_4 + \delta_1^2 \delta_2 (\delta_3 + \delta_4)\delta_5] S^2 + [\delta_1^2(\delta_4 + \delta_5)]S}{S^2 + [\delta_1^2 \delta_2 (\delta_3 + \delta_4)]S + \delta_1^2} \tag{9}$$

Accordingly, the following system of two algebraic equations can be written:

$$\begin{aligned} &(S^2 + \nu S + 1 + \kappa^2)Y + \kappa Q = 0 \\ &(\delta_3 + \delta_5)S^3 + [(1/\delta_2) + \delta_1^2 \delta_2 \delta_3 \delta_4 + \delta_1^2 \delta_2 (\delta_3 + \delta_4)\delta_5] S^2 + [\delta_1^2(\delta_4 + \delta_5)]S + \\ &+ S^2 + [\delta_1^2 \delta_2 (\delta_3 + \delta_4)]S + \delta_1^2)Q + [S^2 + (\delta_1^2 \delta_2 (\delta_3 + \delta_4))S + \delta_1^2]\kappa Y = 0 \end{aligned} \tag{10}$$

System (10) contains seven dimensionless parameters; $\kappa$ and $\nu$ depend only on the mechanical structure and are given, whereas $\delta_i$, $i = 1..5$, are univocally related to the shunt circuit parameters $L$, $C$, $R_1$, $R_2$ and $R_3$ through the relations (8) and must be optimally chosen in order to maximize the damping effect due to piezoelectric passive control.

## 3 Shunt circuit optimization

In this section the optimization of the electrical parameters relevant to the proposed shunt circuit is performed, according to the pole placement technique [12,5,6]. More specifically, these parameters are chosen in order to maximize the exponential time decay rate of the solution of the governing system (10), which can be written in the time domain as:

$$\sum_{i=1}^{5} C_i(T)\exp(\Lambda_i T) \tag{11}$$

In equation (11) $\Lambda_i$ are the complex roots of the characteristic fifth degree polynomial relevant to system (10), whereas $C_i$ are polynomials in $T$ depending on the initial conditions. Accordingly, the optimization of the shunt circuit components is performed by maximizing the following expression:

$$\Lambda = \min_i\{|\Re(\Lambda_i)|\} \tag{12}$$

where $\Lambda$ denotes the exponential time decay of the solution. The evaluation of $\Lambda$ is now numerically performed by assuming $\kappa = 0.15$ and $\nu = 0$, and the obtained results are compared with those reported in [6] and relevant to the case of $RL$ series, $RL$ parallel and $RL$-$C$ parallel shunt circuits. In particular the effect of the components $R_1$ and $R_5$, which are not present in any of the previously considered shunt circuits, are investigated separately, in order to evaluate their influence in the passive control. For the sake of comparison with the results in [6] relevant to the $RL$-$C$ parallel circuit, the value of the external capacitance $C$ is fixed equal to the piezoelectric inherent capacity $C_p$, implying $\delta_2 = 1$. In figure 3 the results are plotted relevant to the case

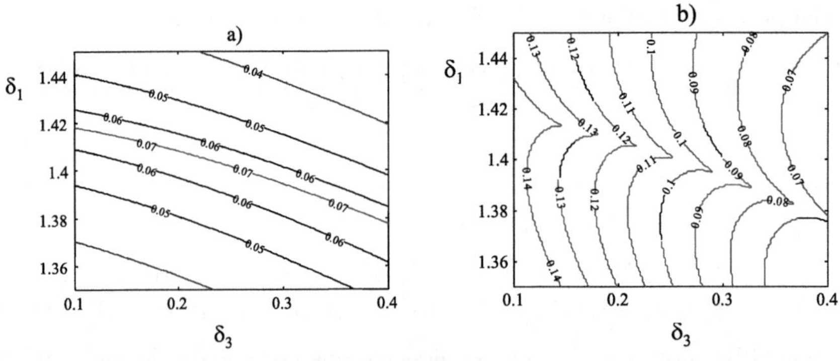


**Fig. 3.** Exponential time decay rate coefficient $\Lambda$ (a) and optimal value of $\delta_4$ (b) versus $\delta_1$ and $\delta_3$; $\delta_2 = 1$, $\delta_5 = 0$, $\kappa = 0.15$, $\nu = 0$

in which the resistance $R_3$ is kept fixed to zero (implying $\delta_5$=0) whereas $R_1$

(and therefore $\delta_3$) is variable. In particular in figure 3-a the isolevel curves of the optimal exponential time decay $\Lambda$ are plotted against $\delta_3$, $R_1$ and $\delta_1$; in figure 3-b the corresponding optimal value of $\delta_4$ is reported, proportional to the value of the resistance $R_2$. From the numerical simulation it turns out that the exponential time decay $\Lambda$ increases with the increasing of the resistance $\delta_3$, according to the following linear law $\Lambda = 0.071 + 0.01(\delta_3 - 0.1)$ whereas the optimal value of $\delta_1$ slightly decreases and therefore the tuning value of the inductance $L$ increases. The limit case $\delta_3 = 0$ corresponds to the case of the *RL-C* shunt circuit, in which the tuning value of the inductance $L$ is almost reduced by a factor $1 + \alpha$ and $\Lambda$ is almost reduced by a factor $\sqrt{1+\alpha}$ with respect to the case of the *RL* series shunt circuit [6]. This latter case is recovered by taking the limit $\delta_3 \to \infty$ of the present solution.

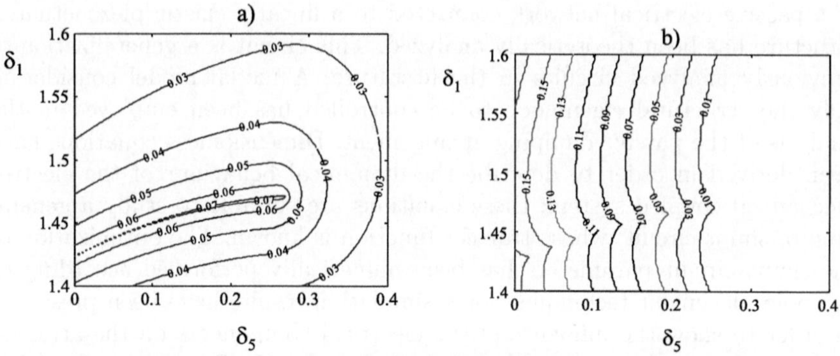


**Fig. 4.** Exponential time decay rate coefficient $\Lambda$ (a) and optimal value of $\delta_4$ (b) versus $\delta_1$ and $\delta_5$; $\delta_2 = 1$, $\delta_3 = 0$, $\kappa = 0.15$, $\nu = 0$

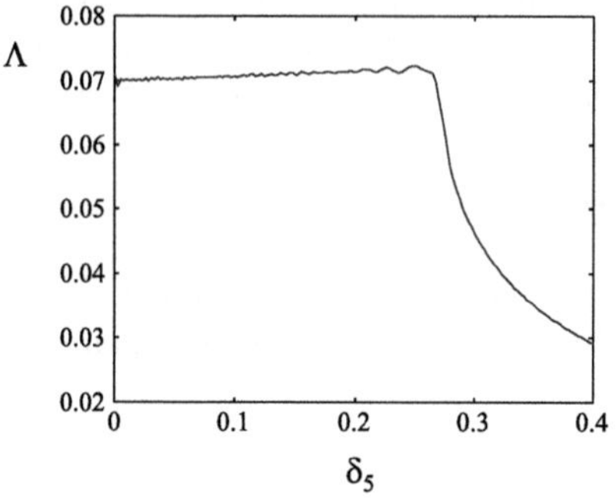


**Fig. 5.** Optimal exponential time decay rate coefficient $\Lambda$ versus $\delta_5$; $\delta_2 = 1$, $\delta_3 = 0$, $\kappa = 0.15$, $\nu = 0$

In figures 4-5 the effect of the resistance $R_3$ is investigated, keeping $R_1 = 0$. It turns out that increasing $\delta_5$, the optimal value of $\delta_4$ decreases (figure 4-b) and the optimal exponential decay $\Lambda$ slightly increases till a limit value of $\delta_5$ (figure 5) for which $\delta_4 = 0$ (figure 4-b); further increasing the value of $\delta_5$ the attained damping $\Lambda$ rapidly decreases since the value of $\delta_4$ can't be lowered below zero. Moreover the optimal value of $\delta_1$ increases with the increasing of $\delta_5$ (figure 4-a); the presence of the resistance $R_3$ is then proved to be useful to further reduce the tuning value of $L$ below the value obtained with the simple $RL$-$C$ shunt circuit, corresponding to $\delta_5 = 0$, without reducing the value of $\Lambda$.

## 4 Conclusions

In this paper a new oscillating shunt circuit for vibration damping by means of a passive electrical network connected to a linearly elastic piezoactuated structure has been theoretically analyzed. This circuit is a generalization of previously analyzed circuits in the literature. A modal model considering only the structural eigenmode to be controlled has been employed in the analysis of the passive damping arrangement. Dimensionless equations have been derived in order to describe the dynamical behaviour of the electro-mechanical coupled system; these equations are suitable to study a general kind of shunt circuit, whose transfer function is known. The optimization of the shunt circuit parameters has been numerically performed according to the pole placement technique. Some simulation results have been presented in order to show the influence of the electrical components on the attained damping and on the tuning value of the inductance $L$. The results show that the present circuit is able to reduce the tuning value of $L$ more than the reduction attainable with the $RL$-$C$ parallel shunt circuit previously proposed in the literature, without decreasing the attained damping and can be suitable in the implementation of a hybrid control using piezoelectric actuators. Future work could be devoted to the analysis of multi-resonant shunt circuits, for damping several eigenmodes at the same time by using a single piezoelectric actuator, and to the use of methods based on power dissipation, in order to perform an easier optimization of the shunt circuit electrical components.

## Acknowledgments

The author wishes to thank Prof. Franco Maceri for the discussions about the topic here studied and for his valuable comments on this paper.

## References

1. Agnes, G. S. (1995) Development of a modal model for simultaneous active and passive piezoelectric vibration suppression. J. Intell. Mater. Syst. Struct. **6**, 482–487

2. Blanguernon, A., Lené, F. and Bernadou, M. (1999) Active control of a beam using a piezoceramic element. Smart Mater. Struct. **8**, 116–124
3. Bisegna, P. and Caruso, G. (2000) Mindlin-type finite elements for piezoelectric sandwich plates. J. Intell. Mater. Syst. Struct. **11**, 14–25
4. Bisegna, P., Caruso, G. et al. (2000) Semi-active control of a thin piezoactuated structure. SPIE Proc. **3989**, 300–311
5. Caruso, G. (2000) Laminati piezoelettrici, modellazione, algoritmi di calcolo ed ottimizzazione della risposta dinamica. Doctoral Thesis, Dep. of Civil Engineering, University of Rome "Tor Vergata".
6. Caruso, G. (2000) Smorzmento passivo di vibrazioni mediante materiali piezoelettrici e circuiti elettrici risonanti. Proc. XXIX National Congress AIAS, Lucca, Italy, 307–316
7. Chandrashekhara, K., Shyser, C. P. and Agarwal, S. (1998) Dynamic modeling and neural control of composite shells using piezoelectric devices. J. Intell. Mater. Syst. Struct. **9**, 29–43
8. Edberg, D. L., Bicos, A. S. and Fechter, J. S. (1991) On piezoelectric energy conversion for electronic passive damping enhancement. Paper GBA-1, Proceedings of Damping, San Diego, CA
9. Ghiringhelli, G. L. (1997) Structural vibrations and noise reduction by piezoelectric active control. J. Struct. Control **4**, 41–68
10. Lam, K. Y. and Ng, T. Y. (1999) Active control of composite plates with integrated piezoelectric sensor and actuators under various dynamic loading conditions. Smart Mater. Struct. **8**, 223–237
11. Lesieutre, G. A. (1998) Vibration damping and control using shunted piezoelectric materials. Shock vibr. digest **30**, 181–190
12. Hagood, N. W. and von Flotow, A. (1991) Damping of structural vibrations with piezoelectric materials and passive electrical networks. J. Sound Vib. **146**, 243–268
13. Hollkamp, J. J. (1994) Multimodal passive vibration suppression with piezoelectric materials and resonant shunts. J. Intell. Mater. Syst. Struct. **5**, 49–57
14. Inman, D. J. (1997) Vibration suppression trough smart damping. Proc. of the Fifth International Congress on Sound and Vibration, Adelaide, South Australia
15. Park, C. H., Kabeya, K. and Inman, D. J. (1998) Enhanced piezoelectric shunt design. Adaptive Structurs and smart material systems. Trans. ASME AD **57**, 149–155
16. Park, C. H. and Inman, D. J. (1999) A uniform model for series R-L and parallel R-L shunt circuits and power consumption. SPIE Proc. **3668**, 797–804
17. Ryu, Y. H., Kim, J. and Cheong, C. C. (2000) Piezoelectric shunting estimation using electrical impedance models. SPIE Proc. **3989**, 39–48
18. Wu, S. Y. (1996) Piezoelectric shunts with a parallel R-L circuit for structural damping and vibration control. SPIE Proc. **2720**, 259–269

# Flexure Waves in Electroelastic Plates

Giovanni Lancioni, Giuseppe Tomassetti

Dipartimento di Ingegneria Civile
Università di Roma "Tor Vergata"
00133 Roma, Italia

**Abstract.** The near-cutoff propagation of free waves of flexure in a transversely isotropic, linearly electroelastic plate is studied, in two cases: for the simplest kinematics, when both the mechanical displacement and the electric potential are taken linear in the thickness variable; and for the enriched, third-order kinematics. The dispersion curves are four in the second case, only two in the first. By a comparison with the first four dispersion curves obtained by solving the corresponding three-dimensional problem, it is shown that perhaps the most definite advantage of adopting an enriched kinematics is a better approximation of the low-cutoff curves.

## 1 Introduction

The purpose of plate theories is to construct simple and yet accurate two-dimensional models of a plate-like three-dimensional body. When dealing with free-wave propagation, as we here do, the accuracy of a given plate theory can be checked by comparing the dispersion curves the theory furnishes with the curves obtained by solving a corresponding three-dimensional problem.

For plate theories deduced from a three-dimensional parent theory a representation for the possible kinematics is chosen *a priori*. Thus, the resulting plate, whose motion class is intrinsically less rich, is generally "stiffer" than the three-dimensional body it models; in particular, waves of the same wavelength propagate at higher frequencies. The most frequently used remedies to attenuate this discrepancy are:

- To introduce correction factors in the constitutive equations.

This was done by Mindlin in [7] and [8], where two and, respectively, three factors were used to correct the cutoff frequencies of the thickness-stretch and thickness-shear modes. The number of correction terms grows with the order in the thickness coordinate of a plate theory (see, *e.g.* [15]); in [4], two factors are used to adjust the behaviour of the flexure dispersion branch.

- To use a more descriptive kinematics.

Typically, this is done by a representation based on high-order polynomials in the thickness variable. For the elastostatic problem, it has been shown [1] that a two-dimensional polynomial solution converges in the energy norm to the three-dimensional solution when the number of terms increases. In [14], a comparison of the frequency spectra from elastic plate theories of different orders shows that a third-order theory is more accurate than the first-order one with correction factors.

- To use suitable special functions, not powers, of the thickness coordinate in the representations of the fields of interest.

In [3,6],the dependence of the in-plane displacement on the thickness coordinate is assumed to be the sum of a cosine series and a linear term; this allows for fairly accurate predictions of the cutoff frequencies. In [10], the dependence on the thickness coordinate is accounted for by means of the complete set of functions appearing in the exact solution of the three-dimensional plate problem at cutoff.

In our present study we adopt the second of the above remedies, and compare the results that can be achieved by the first-order representation for the displacement and the electric potential stipulated in [9] and [11] with the corresponding results when a third-order representation is used.

We employ the linear theory of electroelastic plates developed in [9], where the plate's material is assumed to be *transversely isotropic* and *coherently oriented.* Within that theory, two uncoupled evolution problems govern the "membrane" and "flexure" regimes. "Membrane" waves have been studied in [11]; we here focus on "flexure" waves, as described within both a first- and a third-order kinematics. The first-order theory is dealt with in Sections 2 and 3; precisely, in Section 2 we list the general equations for flexure vibrations and, in Section 3, we determine the conditions for propagation of flexure waves. The enriched, third-order theory is briefly considered in Section 4, where we restrict attention to free-wave propagation; in particular, we compute and plot the dispersion curves for a representative electroelastic ceramic, the lead titanate-zirconate P1-88. Finally, in Section 5, we compare the dispersion branches of our first- and third-order plate theories – two branches in the case of the first-order theory, four in the other case – with the first four dispersion branches given by the three-dimensional theory. As expected, increasing the order of the plate theory yields a better match between the first two branches, while the cutoff frequencies are modestly affected; both the third and the fourth cutoff frequencies and branch shapes differ much more from the corresponding three-dimensional modes.

## 2 Free flexure vibrations of electroelastic plates

Let the region $\Omega$ be *plate-like*, *i.e.*, let $\Omega$ be a right cylinder of axis $\mathbf{z}$ and typical point $p = x + \zeta\mathbf{z}$, with $(x, \zeta) \in \mathcal{P} \times (-\varepsilon, +\varepsilon)$, and with the cross section $\mathcal{P}$ a flat domain with smooth boundary $\partial\mathcal{P}$. For $\{o; \mathbf{c}_1, \mathbf{c}_2, \mathbf{z}\}$ an orthonormal frame, the position vector of $x \in \mathcal{P}$ with respect to the frame origin $o$ is $x_\alpha \mathbf{c}_\alpha$

($\alpha = 1, 2$), while the transverse fiber through $x$ is the set of points $p \in \Omega$ such that $p - x = \zeta\,\mathbf{z}$, $\zeta \in (-\varepsilon, +\varepsilon)$. Assume that: (**i**) the material is homogeneous and has *transversely isotropic* response with respect to the axis $\mathbf{z}$ [*vid.* [9]]; (**ii**) the displacement and potential fields have the representations

$$\mathbf{u}(x,\zeta,t) = \mathbf{a}(x,t) + w(x,t)\mathbf{z} + \zeta\left(\mathbf{b}(x,t) + u(x,t)\mathbf{z}\right), \ \mathbf{a}\cdot\mathbf{z} = \mathbf{b}\cdot\mathbf{z} = 0, \quad (1)$$

$$\phi(x,\zeta,t) = \phi_0(x,t) + \zeta\phi_1(x,t), \quad (2)$$

It is convenient to split the representations (1) and (2) into their "membrane" and "flexure" parts, namely,

$$\begin{array}{ll} \mathbf{u}_m(x,\zeta,t) = \mathbf{a}(x,t) + \zeta\, u(x,t)\,\mathbf{z}, & \phi_m(x,\zeta,t) = \zeta\phi_1(x,t), \\ \mathbf{u}_f(x,\zeta,t) = w(x,t)\,\mathbf{z} + \zeta\,\mathbf{b}(x,t), & \phi_f(x,\zeta,t) = \phi_0(x,t), \end{array} \quad (3)$$

where $\mathbf{a}\cdot\mathbf{z} = \mathbf{b}\cdot\mathbf{z} = 0$. The membrane displacement $\mathbf{u}_m$ is mirror symmetric with respect to the middle plane $\mathcal{P}$ while the flexure displacement $\mathbf{u}_f$ is mirror antisymmetric.

If we neglect noninertial body loads and body electric charges over $\Omega \times \Re$ and furthermore, if we assume null contact loads and surface charges over $\partial\Omega \times \Re$ (we are interested in free vibration), then the virtual work principle states that

$$\int_\Omega (\mathbf{S}\cdot\nabla\mathbf{v} + \mathbf{d}\cdot\nabla\psi) + \int_\Omega \varrho\ddot{\mathbf{u}}\cdot\mathbf{v} = 0 \quad (4)$$

for all test fields $(\mathbf{v}, \psi)$. Here $\mathbf{S}$ is the Cauchy stress, $\mathbf{d}$ the electric displacement and $\varrho > 0$ is the mass density.

Under the assumptions (**i**) and (**ii**), the membrane and flexure descriptors $(\mathbf{u}_m, \phi_m)$ and $(\mathbf{u}_f, \phi_f)$ turn out to be energetically orthogonal [9]; hence, the dynamical behavior of a linearly, electroelastic plate is described by two independent systems of equations which are deduced by (4) after integration over the plate thickness (see [9] and [2]). The two-dimensional flexure problem follows:

$$\begin{array}{l} \eta\,\mathrm{Div}\,(\nabla w + \mathbf{b}) + \delta_1\Delta\phi_0 = \rho\ddot{w}, \\ \frac{1}{3}\varepsilon^2\left(\mu\Delta\mathbf{b} + (\lambda+\mu)\nabla\left(\mathrm{Div}\ \mathbf{b}\right)\right) - \eta\left(\nabla w + \mathbf{b}\right) - \delta_1\nabla\phi_0 = \frac{1}{3}\varepsilon^2\rho\ddot{\mathbf{b}}, \\ \delta_1\mathrm{Div}\,(\nabla w + \mathbf{b}) - \gamma_1\Delta\phi_0 = 0. \end{array} \quad (5)$$

where

$$\lambda + 2\mu = \mathbb{C}_{1111},\ \lambda = \mathbb{C}_{1122},\ \tau_1 = \mathbb{C}_{3333},\ \tau_2 = \mathbb{C}_{1133},\ \tau_3 = \mathbb{C}_{2323};$$

$$\gamma_1 = \mathbf{C}_{11},\ \gamma_2 = \mathbf{C}_{33};\ \delta_1 = \mathbb{C}_{131},\ \delta_2 = \mathbb{C}_{311},\ \delta_3 = \mathbb{C}_{333}\,. \quad (6)$$

We are going to study solutions of system (5) having the form of plane progressive waves.

## 3 Flexure waves

We decompose $\mathbf{b}(x,t)$ into its irrotational and solenoidal parts as follows:

$$\mathbf{b}(x,t) = \nabla b(x,t) + \mathbf{Z}\nabla\tilde{b}(x,t), \tag{7}$$

where $\mathbf{Z}$ is such that $\mathbf{Z}\mathbf{v} = \mathbf{z} \times \mathbf{v}$ for all $\mathbf{v} \in \mathcal{V}$, and $b(x,t)$ and $\tilde{b}(x,t)$ are scalar fields.

Applying the Div and Curl operators to equation $(5)_2$, and using the decomposition (7), system (5) splits into the following *bending* and *twist* problems:

$$\begin{aligned} &\eta\,\Delta\,(w+b) + \delta_1\Delta\phi_0 - \rho\ddot{w} = 0, \\ &\frac{1}{3}\varepsilon^2\,(\lambda+2\mu)\,\Delta\Delta b - \rho\ddot{w} - \frac{1}{3}\varepsilon^2\rho\Delta\ddot{b} = 0, \\ &\delta_1\Delta\,(w+b) - \gamma_1\Delta\phi_0 = 0, \end{aligned} \tag{8}$$

in the unknowns $w, b, \phi_0$ and

$$\varepsilon^2\mu\Delta\Delta\tilde{b} - 3\eta\Delta\tilde{b} - \varepsilon^2\varrho\Delta\ddot{\tilde{b}} = 0,$$

which determines $\tilde{b}$.

We look for solutions $(w, b, \phi_0)$ of (8) having the form

$$(w, b, \phi_0) = (W, B, \Phi_0)\, g(x,t)\,, \tag{9}$$

where $W$, $B$, and $\Phi_0$, are complex amplitudes and the *wave shape* is

$$g\,(x,t) = \exp\left(2\pi i\,(\mathbf{k}\cdot(x-o) - ft)\right), \tag{10}$$

with $\mathbf{k} = k\mathbf{v}$, where $k$ is a complex number, $f$ is a positive, real number and $\mathbf{v}$ is a unit vector such that $\mathbf{v}\cdot\mathbf{z} = \mathbf{0}$. We call these motions *flexure waves.*

Inserting (9)-(10) into (8) we obtain an equation of the form

$$\mathbf{N}(K,F)\,\mathbf{d} = \mathbf{0}\,, \tag{11}$$

where the vector $\mathbf{d}$ stands for the triplet $(W, B, \Phi_0)$, and $\mathbf{N}(K,F)$ is a 3 by 3 matrix whose entries are functions of the dimensionless real quantities $K = {L_\varepsilon}^2\,k^2$ and $F = {T_\varepsilon}^2\,f^2$, with $L_\varepsilon = 2\pi\varepsilon$ and $T_\varepsilon = 2\pi\varepsilon\sqrt{\rho/\mu}$ the length and time scale we select for the problem. Equation (11) admits nontrivial solutions when

$$\det\mathbf{N}\,(K,F) = 0\,. \tag{12}$$

For each fixed positive value of $F$, the dispersion relation (12) admits the following real roots:

$$K^{\pm}(F) = (\mu(\varphi+\gamma_1\psi)F \pm 3\sqrt{d(F)})/(2\varphi\psi)\;, \tag{13}$$

with $\varphi = (\delta_1^2 + \gamma_1\eta)$, $\psi = \lambda + 2\mu$ and $d(F) = (\varphi - \gamma_1\psi)^2\mu^2F^2 + 12\varphi^2\psi\mu F > 0$. The dispersion relation has two solution branches (while the original three-dimensional problem has infinitely many of them).

Note that the in-plane component of the displacement field is parallel to $\mathbf{v}$, and thus the related mechanical waves are longitudinal. The wave-number $k$ is either real or imaginary according to whether $K$ is positive or negative. The first case corresponds to a *progressive* wave (a motion with a net energy flux in the direction $\mathbf{v}$), the second, to a localized *standing* wave. Accordingly, we call *cutoff* frequencies the values $F^{\pm}$ such that $K^{\pm}(F) = 0$.

It is worth pointing out some qualitative features of the solution for $K = 0$ (*i.e.*, at infinite wavelength): for $F = 0$, we have $\mathbf{b} \cdot \mathbf{v} = 0$, $w \neq 0$ and $\mathbf{e} = \mathbf{0}$; for $F = 3(\eta + \delta_1^2/\gamma_1)/\mu$ we have $\mathbf{b} \cdot \mathbf{v} \neq 0$, $w = 0$ and $\mathbf{e} = -(\delta_1/\gamma_1)\mathbf{b}$. Thus, the limit motion at $F = 0$ is an "up and down" rigid oscillation and there is no electric field, while the one at $F = 3(\eta + \delta_1^2/\gamma_1)/\mu$ is an in-plane shear oscillation of the "deck of cards" type and it is accompanied by a longitudinal electric field. In addition, both motions are isochoric. Inspired by these limit situations, we call the mechanical waves related to $K^+(F)$ and $K^-(F)$ *shear waves* and *in-plane shear waves*, respectively. We remark that shear waves propagate at all frequencies, while in-plane shear waves are standing below the cut-off frequency $f_x = L_\varepsilon^{-1}\sqrt{(3\varphi/\rho\gamma_1)}$, progressive for all frequencies above cut-off.

## 4 Enriched electroelastic plate theory

We have already remarked in the Introduction that structural theories obtained by imposing kinematical restrictions are "stiffer" than their parent three-dimensional theory, and hence lead to an overestimation of the vibration frequencies. To improve the approximation achievable by means of a first-order theory we here allow for a wider set of motions.

The plate-like body we consider continues to be homogeneous and transversely isotropic with respect to $\mathbf{z}$. However, the representations we take for the displacement and the potential in the "flexure" problem are now as follows:

$$\begin{aligned} \mathbf{u}_f(x,\zeta,t) &= w(x,\zeta,t)\mathbf{z} + \mathbf{b}(x,\zeta,t), \quad \mathbf{b}(x,\zeta,t)\cdot\mathbf{z} = 0, \\ \phi_f(x,\zeta,t) &= \phi_0(x,t) + \zeta^2\phi_2(x,t), \end{aligned} \tag{14}$$

with

$$\mathbf{b}(x,\zeta,t) = \zeta\mathbf{b}_1(x,t) + \zeta^3\mathbf{b}_3(x,t), \quad w(x,\zeta,t) = w_0(x,t) + \zeta^2 w_2(x,t) \tag{15}$$

(see [1] for a justification of (14)-(15)).

Again, the virtual work equation (4) is used to deduce two uncoupled evolution problems. The Helmoltz decomposition of $\mathbf{b}_1$ and $\mathbf{b}_3$ splits the original problem into a system of four partial differential equations (*bending* problem) and a system of two partial differential equations (*twist* problem).

For the bending problem, we select harmonic-wave solutions and arrive to an algebraic system; the corresponding dispersion relations give four dispersion curves. Figure 1 shows the dispersion curves computed for a plate in lead titanate-zirconate (P1-88), for both the simple and the enriched kinematics (resp., dashed and solid lines). Note that the kinematic enrichment, beside adding two new dispersion branches, produces a welcomed frequency shift: for a given $K$, the third-order theory leads to a smaller value of $F$, and thus to a smaller frequency $f$.

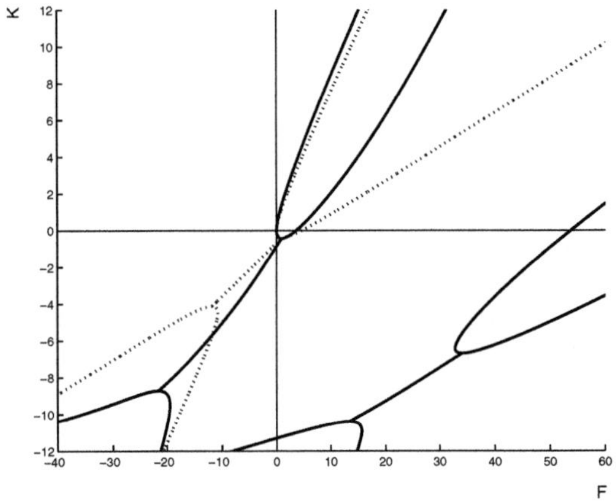


**Fig. 1.** Dispersion curves for bending waves: dashed line ≡ 1st-order theory; solid line ≡ 3d-order theory.

The cutoff frequencies turn out to be:

$$F_1 = 0,$$
$$F_{2,3} = (45\,\gamma_1\,\eta + 10\,{\delta_1}^2 \mp \sqrt{1605\,\gamma_1^2\,\eta^2 + 480\,\gamma_1\,\eta\,{\delta_1}^2 + 100\,{\delta_1}^4})/(2\gamma_1\,\mu),$$
$$F_4 = 15\,({\delta_3}^2 + \tau_1\,\gamma_2)/(\mu\,\gamma_2),$$

At cutoff ($K = 0$) we have:

- For $F = F_1$, $\mathbf{b}\cdot\mathbf{v} = 0$, $w = w_0 \neq 0$, $\mathbf{e}\cdot\mathbf{v} = \mathbf{e}\cdot\mathbf{z} = 0$. The motion is an "up and down" rigid oscillation.
- For $F = F_{2,3}$, $\mathbf{b}\cdot\mathbf{v} = (\zeta\mathbf{b}_1 + \zeta^3\,\mathbf{b}_3)\cdot\mathbf{v} \neq 0$, $w = 0$, $\mathbf{e}\cdot\mathbf{v} \neq 0$. We call the motion an in-plane shear oscillation of the "deck-of-cards" type.
- For $F = F_4$ , $\mathbf{b}\cdot\mathbf{v} = 0$, $w = w_0 + \zeta^2 w_2 \neq 0$, $\mathbf{e}\cdot\mathbf{v} = 0$, $\mathbf{e}\cdot\mathbf{z} = 2\zeta\phi_0 \neq 0$. The motion is an "up and down" voluminal oscillation; the volume change is given by the quadratic term.

We remark that the in-plane shear motions are associated to longitudinal electric fields, while a transverse electric field is associated to "up and down" voluminal oscillation.

## 5 Comparison with the three-dimensional theory

In this section we compare the dispersion curves predicted by our two "flexure" plate theories with those given by a three-dimensional linear electroelastic theory. We follow the path suggested by Tiersten [13] to deduce the dispersion relations for the same *transversely isotropic* electroelastic plate (see also [12] and [5]). We begin with the mechanical and electrical field equations

$$\mathrm{Div}\,(\mathbb{C}\nabla\mathbf{u} + \mathbb{c}^t\nabla\varphi) = \varrho\ddot{\mathbf{u}}, \quad \mathrm{Div}\,(\mathbb{c}\nabla\mathbf{u} - \mathbf{C}\nabla\varphi) = 0, \tag{16}$$

with Neumann boundary conditions on the bases $\zeta = \pm\varepsilon$:

$$(\mathbb{C}\nabla\mathbf{u} + \mathbb{c}^t\nabla\varphi)\mathbf{n} = 0, \quad (\mathbb{c}\nabla\mathbf{u} - \mathbf{C}\nabla\varphi)\cdot\mathbf{n} = 0. \tag{17}$$

We consider solutions $(\mathbf{u}(x,\zeta,t), \phi(x,\zeta,t))$ of (16) and (17) having the form:

$$\mathbf{u} = (w(\zeta)\mathbf{z} + v(\zeta)\mathbf{c}_1(\zeta))g\,(x,t)\,, \quad \phi = \phi(\zeta)g\,(x,t)\,, \quad \mathbf{c}_1\cdot\mathbf{z} = 0, \tag{18}$$

where $g\,(x,t)$ is given in (10) and the propagation direction $\mathbf{v}$ coincides with $\mathbf{c}_1$. Since the chosen displacement lies in the plane formed by $\mathbf{c}_1$ and $\mathbf{z}$ and it is independent from the $x_2$ coordinate, it defines a planar displacement wave. When the representation (18) is substituted into the field equations (16) it turns out that solutions must have the form:

$$\begin{aligned} w(\zeta) &= \textstyle\sum_{i=1}^3 \left(A_i^f\cos(p_i\zeta) + A_i^m\sin(p_i\zeta)\right)W_i, \\ v(\zeta) &= \textstyle\sum_{i=1}^3 \left(A_i^f\sin(p_i\zeta) + A_i^m\cos(p_i\zeta)\right)V_i, \\ \phi(\zeta) &= \textstyle\sum_{i=1}^3 \left(A_i^f\cos(p_i\zeta) + A_i^m\sin(p_i\zeta)\right)\Psi_i, \end{aligned} \tag{19}$$

where $(p_i, (W_i, V_i, \Psi_i)^T; i = 1, ..., 3)$ are the solutions of a suitable eigenvalue problem derived from (16). The coefficients $A_i^f$ and $A_i^m$ are obtained by imposing the boundary conditions (17). Nontrivial solutions for $A_i^f$ or $A_i^m$ exist, provided the frequency $f$ and the wave number $k$ satisfy a dispersion equation. In Fig. 2 we plot the first four flexure dispersion curves (dash-dot line) for a lead titanate-zirconate P1-88 plate. We compare them with the ones from the first- and third-order plate theories (dashed and solid line respectively).

As the order increases, a plate theory would seem to lead to a better and better matching of more and more three-dimensional dispersion curves. In our case, the third-order theory brings in two new approximate dispersion curves and improves the description of the first two modes; the new branches are however quite different from the corresponding three-dimensional modes, both by cutoff frequency and by shape.

## Acknowledgments

We thank Prof. Paolo Podio-Guidugli for his valuable comments and suggestions. We also acknowledge the help of Dr. Vincenzo Nicotra for the calculation of the dispersion curves.

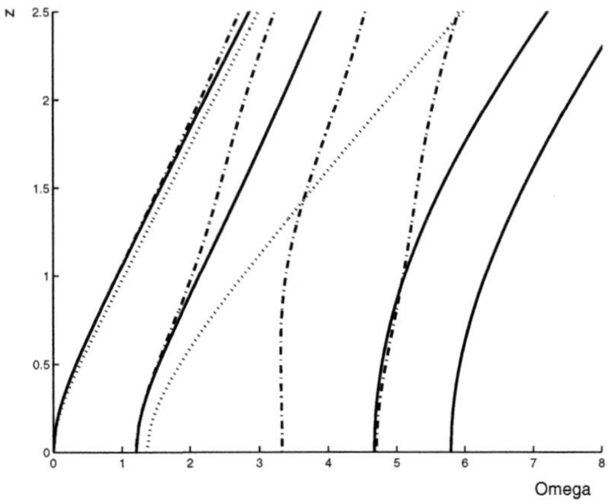


**Fig. 2.** Dispersion curves for flexure waves: dashed line ≡ 1st-order theory; solid line ≡ 3d-order theory; dash-dot line≡ 3D theory (here $z = 4\varepsilon k$ and $\Omega = 4\varepsilon\sqrt{(\varrho/\mu)}f$).

# References


1. S. Carillo, P. Podio-Guidugli, and G. Vergara Caffarelli, "Energy estimates in hierarchical plate theories", *J. Math. Anal. Appl.* 248, 216-232 (2000).
2. A. DiCarlo, P. Podio-Guidugli, and W.O. Williams, "Shells with thickness distension", to appear in *Int. J. Solids Struct.*
3. P. C. Y. Lee, and W. S. Lin, "Piezoelectrically forced vibrations of rectangular SC-cut quartz plate", *J. Appl. Phys.*, 83 (12), 7822-7833, (1998).
4. P. C. Y. Lee, S. Syngellakis, and J. P. Hou, "A two-dimensional theory for high-frequency vibrations of piezoelectric cristal plates with or without electrodes", *J. Appl. Phys.*, 61 (4), 1249-1262, (1987).
5. P. C. Y. Lee, and J-D Yu, "Governing equation for a piezoelectric plate with graded properties across the thickness", *IEEE Trensactions on Ultrasonics, ferroelectrics, and Frequency Control*, 45 (1), 236-250, (1998).
6. P. C. Y. Lee, J. D. Yu, and W. S. Lin, "A new two-dimensional theory for vibrations of piezoelectric crystal plates with electroded faces", *J. Appl. Phys*, 83 (3), 1213-1223, (1998).
7. R. D. Mindlin, "High frequency vibrations of piezoelectric crystal plates", Mindlin. *Int. J. Solids Structures*, 8 ,895-906, (1972).
8. R. D. Mindlin, "Frequencies of piezoelectrically forced vibrations of electroded, doubly rotated, quartz plates", *Int. J. Solids Structures*, 20 (2),141-157 , (1984).
9. V. Nicotra and P. Podio-Guidugli, "Piezoelectric plates with changing thickness", *J. Struct. Control*, 5 (2), 73-86 (1998).
10. R. C. Peach, "A normal mode expansion for piezoelectric plates and certain of its applications", *IEEE Trensactions on Ultrasonics, ferroelectrics, and Frequency Control*, 35 (5), (1988).
11. P. Podio-Guidugli and G. Tomassetti, "Thickness waves in electroelastic plates", to appear in *Wave Motion.*
12. S. Syngellakis and P. C. Y. Lee, "Piezoelectric wave dispersion curves for infinite anisotropic plates", *J. Appl. Phys.*, 73 (11), 7155-7161, (1993).

13. H.F. Tiersten, *Linear Piezoelectric Plate Vibrations*, Plenum Press, New York (1969).
14. Y-K Yong, Z. Zhang, and J. Hou, "On the accuracy of plate theories for the prediction of unwanted modes near the fundamental thickness shear mode", *IEEE Intrenational Frequency Control Symposium*, 755-760, 1995.
15. J. D. Yu, "Second-order analysis of free vibrations of piezoelectric ceramic actuators", *IEEE International Frequency Control Symposium*, 695-702, (1998).

# On the Plating of Reinforced Concrete Beams with Composite Laminates

Luigi Ascione, Luciano Feo, Fernando Fraternali

Dipartimento di Ingegneria Civile
Università di Salerno
84084 Fisciano, Italia

**Abstract.** This paper deals with the formulation of a one-dimensional mechanical model for studying the statical behaviour of a RC beam strengthened with Fibre-Reinforced Polymers (FRP) plates. Some numerical results obtained via finite element method (FEM) are also given and show the efficiency of the proposed model in evaluating the stresses distributions at the interface between the concrete core and the composite laminate.

## 1 Introduction

In the last few years the use of Fiber Reinforced Polymers (FRP) to strengthen reinforced concrete beams is becoming more and more frequent in the field of Civil Engineering. Nowadays, all the drawbacks related to Hermite's traditional technique, which uses steel plates externally bonded to concrete structures, can be easily overcome by applying composite laminates at the tension side of beams. They are characterized by high strength-to-weight ratio, low maintenance cost and corrosion resistance.

A static problem of relevant technical interest is represented by the transmission of shear and normal stresses at the plate-core interface. In fact, these stresses can produce the brittle failure of the strengthened beam because of a sudden and premature debonding of the composite laminate or cracking of the concrete cover along the level of internal steel reinforcements.
The problem has been studied by many authors both from a theoretical and from a numerical and an experimental point of view. Relevant features of this problem are: the shear and normal stresses concentrations at the ends of the composite overlay (cut-off cross-sections) [1-4]; the non-uniform distributions of such stresses along the reinforced boundaries [5-7].
This work presents a one-dimensional mechanical model of such strengthened beam able to analyze all the above mentioned aspects under general hypotheses of loading. The paper also presents some theoretical predictions of the interlaminar stresses distributions along the beam axis and the strengthened boundaries.

The authors are going to compare them with the experimental results they are carrying out at the Testing Laboratory of the Department of Civil Engineering of the University of Salerno.

# 2 Kinematical assumptions

Consider a straight reinforced concrete beam strengthened with an external FRP laminate. It is useful to schematize the natural configuration B of the strengthened beam as an assembly of two one-dimensional components corresponding to the concrete core $B^{(1)}$ and to the composite overlay $B^{(2)}$, respectively (Fig.1).

Further let {O, x, y, z} be a right-handed orthonormal reference frame and assume that the origin O belongs to one of the beam bases. The unit vectors of x, y and z axes will be denoted by i, j and k respectively and the displacement components of a generic point P of the strengthened beam by u, v and w.

It is useful to introduce a local reference frame relative to a generic point of the laminate cross-section middle line $\rho$ (Fig.1). Let {n, t, k} be the unit vectors of the local reference axes: n and t are respectively normal and tangent to the middle line $\rho$. The vector basis {n, t, k} is supposed to be right-handed as the basis {i, j, k}.

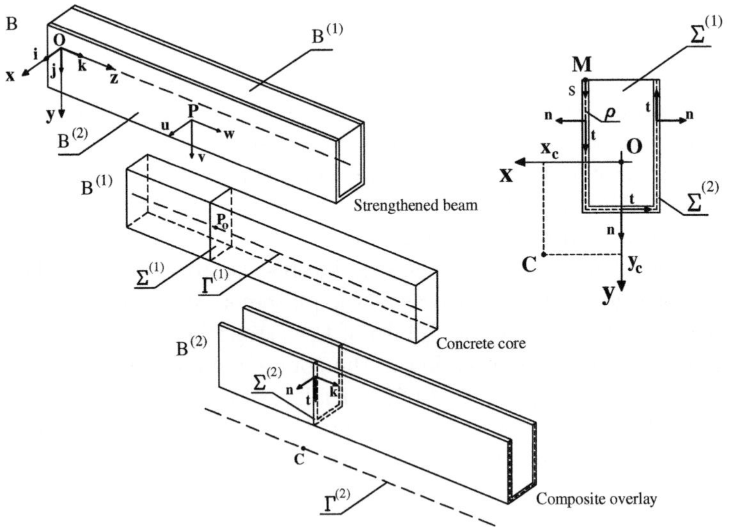


**Fig. 1.** Natural configuration of a RC beam strengthened with a composite overlay.

## 2.1 Kinematical model of the concrete core ($B^{(1)}$)

The kinematical model of the concrete core $B^{(1)}$ is driven from a suitable power expansion of the displacement field components with respect to the coordinates x and y starting from a suitable point $P_0(x_0, y_0)$ of its cross-section $\Sigma^{(1)}$ [5].

That is:

$$u^{(1)}(x,y,z) = \\ = u_0^{(1)}(z) - \theta^{(1)}(z)(y-y_0) + \sum_{i,j} u_{ij}^{(1)}(z)(x-x_0)^i(y-y_0)^j \ , \tag{1a}$$

$$v^{(1)}(x,y,z) = \\ = v_0^{(1)}(z) - \theta^{(1)}(z)(x-x_0) + \sum_{m,n} \nu_{mn}^{(1)}(z)(x-x_0)^m(y-y_0)^n \ , \tag{1b}$$

$$w^{(1)}(x,y,z) = \\ = w_0^{(1)}(z) - \phi^{(1)}(z)(y-y_0) - \psi^{(1)}(z)(x-x_0) + \\ + \sum_{h,k} w_{ij}^{(1)}(z)(x-x_0)^h(y-y_0)^k \ , \tag{1c}$$

where: $u_0^{(1)}$, $v_0^{(1)}$ and $w_0^{(1)}$ are the displacement components of $P_0$; $\varphi^{(1)}(z)$, $\psi^{(1)}(z)$ and $\theta^{(1)}(z)$ are the rotation vector components of $\Sigma^{(1)}$ , and the quantities $u_{ij}^{(1)}$, $v_{mn}^{(1)}$ and $w_{hk}^{(1)}$ are terms of higher order in the power expansion.

The terms in $u^{(1)}$ and $v^{(1)}$ , related to $u_{ij}^{(1)}$ and $v_{mn}^{(1)}$ , model an in-plane deformation of the cross-section $\Sigma^{(1)}$ ; the terms in $w^{(1)}$ , related to $w_{hk}^{(1)}$ , model, instead, an out-of-plane warping of $\Sigma^{(1)}$ .

## 2.2 Kinematical model of the FRP laminate ($B^{(2)}$)

The kinematics of the FRP laminate $B^{(2)}$ is based upon the following assumptions: classically, a generic cross-section $\Sigma^{(2)}$ exhibits a rigid rotation about a suitable point C ($x_C, y_C$ ) of its own plane (Fig.1) and a superimposed out-of-plane warping [8]. Shearing deformations $\gamma_{tz}$ of the middle surface are also allowed: they are related to the weak shear stiffness opposed by the composite overlay because of the low values exhibited by the elastic tangential modulus of the plastic matrix.

Consequently, the following displacement field is assumed:

$$u^{(2)}(x,y,z) = u_C^{(2)}(z) - \theta^{(2)}(z)(y-y_C) \ , \tag{2a}$$

$$v^{(2)}(x,y,z) = v_C^{(2)}(z) - \theta^{(2)}(z)(x-x_C) \ , \tag{2b}$$

$$w^{(2)}(s,z) = w_C^{(2)}(z) - \beta^{(2)}(z)x(s) + \alpha^{(2)}(z)y(s) + \\ -\dot{\theta}^{(2)}(z)\omega^{(2)}(s) + \sum_p \gamma_p^{(2)}(z)\omega_p^{(2)}(s) \ , \tag{2c}$$

where: $u_C^{(2)}(z)$, $v_C^{(2)}(z)$ and $w_C^{(2)}(z)$ are the displacement components of C; $\alpha^{(2)}(z)$ and $\beta^{(2)}(z)$ are the cross-section flexural rotations; $\dot{\theta}^{(2)}$ is the derivative of the twisting rotation $\theta^{(2)}$ with respect to the $z$ coordinate and $\omega^{(2)}$ is the sectorial area of the classical Vlasov's theory [9]; $\gamma_p^{(2)}(z)$ are further generalized displacement components and $\omega_p^{(2)}$ are geometrical quantities defined as :

$$\omega_p^{(2)} = \int_M^Q f_p^*(s)ds \ , \tag{3}$$

$f_p^*(s)$ being suitable shape functions of the curvilinear coordinate s. In Eq. (3) $M$ is the origin of the curvilinear coordinate s and $Q$ is a generic point of $\rho$. The term $\dot{\theta}^{(2)}(z)\omega^{(2)}(s)$ corresponds to a warping of the cross-section without shear deformations of the middle surface, while the term $\gamma_p^{(2)}(z)\omega_p^{(2)}(s)$ corresponds to a warping of the cross-section in presence of shear deformations of the middle surface, depending on $s$ and $z$. The above kinematical model coincides with Vlasov's one if functions $\gamma_{xz}^{(2)} = \dot{u}_C^{(2)} - \beta^{(2)}$, $\gamma_{yz}^{(2)} = \dot{v}_C^{(2)} + \alpha^{(2)}$ and $\gamma_p^{(2)}$ are equal to zero. It is interesting to remark that $\gamma_{xz}^{(2)}$ and $\gamma_{yz}^{(2)}$ correspond to sliding deformation modes of a generic beam element of length dz in the plane $(x, z)$ and $(y, z)$ respectively. The further terms $\gamma_p^{(2)}$ allows us to simulate a more realistic behaviour of the FRP overlay with regard to the effects of shear deformability.

### 2.3 Interface

The mechanical model here presented assumes that the composite overlay $B^{(2)}$ is bonded to the concrete core $B^{(1)}$ by continuous distributions of bilateral elastic springs, arranged along the three axes of the cartesian reference system. The spring reactions give an approximation of the interlaminar stresses.

A relevant problem in studying composite plating is the concentration of shear stresses and peeling forces near the anchorage zone (cut-off cross-sections) [1,2,4]. It depends on the actual diffusion of such interactions from the concrete core to the strengthened beam, which is a tridimensional phenomenon.

Nevertheless, within a one-dimensional model, the transmission of the interface bond-stresses can be simulated [4] by assuming that the FRP plates present a "virtual" increment of thickness from 0 to the actual value $t_p$ over a suitable length $l_d$ which can be related to the characteristic length exhibited by an elastic beam axially loaded and constrained by a continuous distribution of elastic springs along its own axis:

$$l_d = \chi\sqrt{\frac{E_p}{G_g}t_p t_g} \ . \tag{4}$$

In Eq.(4) $E_p$ denote the plate Young's longitudinal modulus, $G_g$ is the shear modulus of the adhesive, $t_p$ and $t_g$ are the thicknesses of the plate and of the adhesive respectively, and $\chi$ is an appropriate real number. This kind of modeling has been already used by the authors in [4] and provides good agreement with the experimental results presented by Swamy in [1].

## 3 Numerical results and conclusions

In this section we present some numerical results concerning a rectangular RC beam strengthened both in flexural and in shear with externally bonded Carbon FRP plates. The beam is simply supported and subjected to two concentrates forces (F), symmetrically placed from the mid-span, as shown in Fig.2.

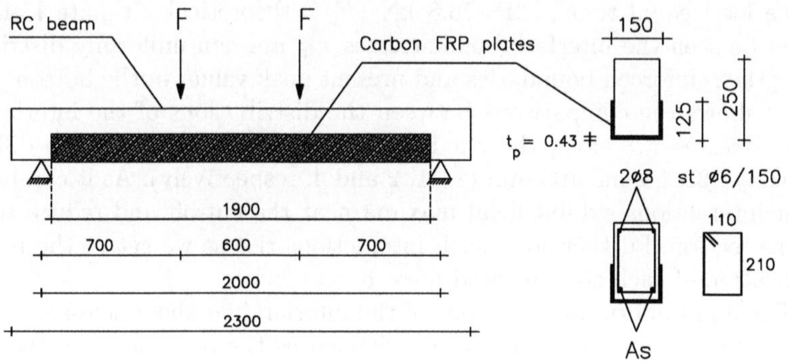


**Fig. 2.** Geometry and loading configuration of a RC beam strengthened with carbon FRP plates (Dimensions in mm - not to scale).

The numerical analysis has been performed by using the finite element model presented earlier by the authors in [8]. The materials constituting the FRP plates have been supposed to be linearly elastic and orthotropic. Within the finite element programme developed by the authors, the behaviours of the concrete core and of the internal steel rebars have been modeled both as linearly elastic or elasto-plastic. The possibility of a different behaviour of the concrete in tension and in compression has been also taken into account. The materials elastic properties used in the numerical analysis are summarized in Table 1. The finite element model uses Hermitian cubic interpolation shape functions to approximate all the kinematical unknowns. The corresponding discrete problem, which is non-linear because of the constitutive hypotheses, has been solved by the Newton-Raphson method.

**Table 1.** Material properties (Fig.1)

| | | | |
|---|---|---|---|
| Concrete | $E_c$= 28460 N/mm$^2$ | $G_c$ = 14230 N/mm$^2$ | $\nu_c$= 0 |
| Steel | $E_s$= 210000 N/mm$^2$ | $G_s$ = 105000 N/mm$^2$ | $\nu_s$= 0 |
| | $E_t$= 215480 N/mm$^2$ | $G_{tn}$ = 42430 N/mm$^2$ | $\nu_{nt}$= 0.016 |
| FRP Plates | $E_n$= 86250 N/mm$^2$ | $G_{tk}$ = 86190 N/mm$^2$ | $\nu_{kt}$= 0.25 |
| | $E_k$= 215480 N/mm$^2$ | $G_{nk}$ = 42430 N/mm$^2$ | $\nu_{kn}$= 0.016 |
| Adhesive | $E_g$= 400 N/mm$^2$ | $G_g$ = 160 N/mm$^2$ | $\nu_g$= 0.25 |

Fig. 3 shows the distributions of shear and normal stresses along the cross-section boundary at a distance from the supports of 56mm. They are relative to a value of the diffusion length $l_d$ equal to 25 mm ($\chi = 1$) and to a value of the load equal to $F_u$=2F=76.8 kN ($F_u$ =theoretical ultimate load). As it can be seen the interface bond-stresses $\tau_{nz}$ are non-uniformly distributed along the reinforced boundaries and present peak values in the bottom plate. Fig. 4 shows the comparisons between the distributions of the interlaminar shear stresses $\tau_{nz}$ along the z-axis relative to three values of the diffusion length $l_d$ : 25, 50 and 100 mm ($\chi$=1, 2 and 4, respectively). As it can be seen shear interactions exhibit local maxima near the cut-off and reduce to zero in this section. Furthermore, such interactions rise as we get to the point of application of each concentrated force F.

The non uniform distributions of the interlaminar shear stresses $\tau_{nz}$ suggest us to introduce a coefficient $\alpha$ , defined as the ratio between the maximum value of the shear stresses $\tau_{nz}$ and the average one $\bar{\tau}_{avg}(\alpha = \bar{\tau}_{max}/\bar{\tau}_{avg})$ calculated by the Jourawsky's formula. This coefficient could be relevant for technical purposes.

It is interesting to express the coefficient $\alpha$ (interlaminar shear stress multiplier) as product of two further coefficients:

$$\alpha = \alpha_1 \alpha_2 \ , \tag{5}$$

where:

- $\alpha_1$ is a partial interlaminar shear stress multiplier due to the shear and normal stresses concentrations at the ends of the composite overlay (cut-off cross-sections),

- $\alpha_2$ is a partial interlaminar shear stress multiplier due to the non-uniform distributions of such stresses along the reinforced boundaries.

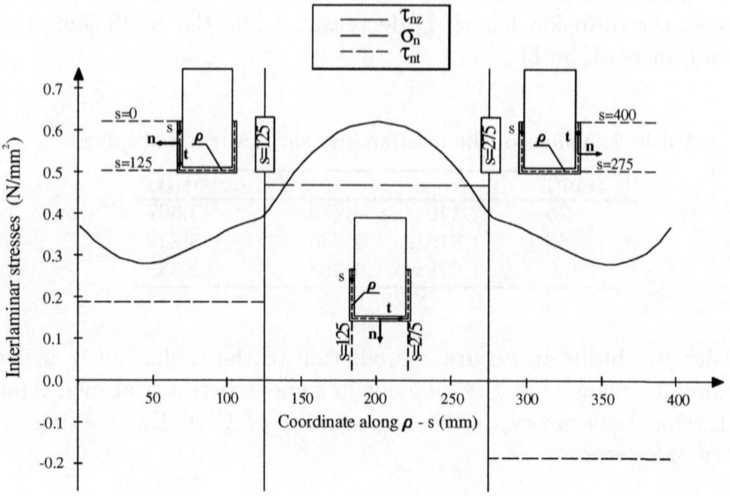


**Fig. 3.** Interlaminar stresses along $\rho$ (z=56 mm - $l_d$=25 mm - F=38.4 kN).

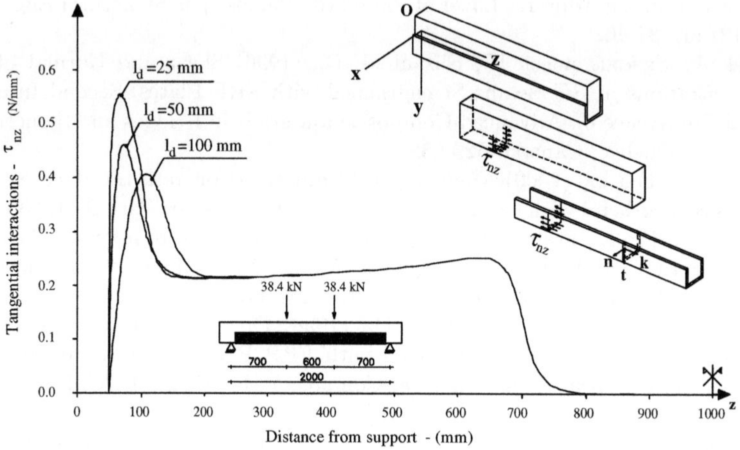


**Fig. 4.** Tangential interactions $\tau_{nz}$ along the beam axis (s=200 mm - F=38.4 kN).

Table 2 summarizes the values of the above coefficients corresponding to the three different diffusion lengths $l_d$. As it can be seen the coefficient $\alpha$ increases as the diffusion length $l_d$ decreases, while the coefficient $\alpha_2$ does not depend, instead, on $l_d$.

**Table 2.** Values of the interlaminar shear stress multipliers

| $l_d$ (mm) | $\alpha_1$ | $\alpha_1$ | $\alpha = \alpha_1\alpha_2$ |
|---|---|---|---|
| 25 | 2.140 | 1.779 | 3.807 |
| 50 | 1.819 | 1.786 | 3.249 |
| 100 | 1.628 | 1.803 | 2.935 |

In order to obtain an accurate prediction of the coefficient $\chi$ in Eq.(4) the authors are engaged at the moment in some experimental investigations at the Testing Laboratory of the Department of Civil Engineering of the University of Salerno.

## References


1. Jones R., Swamy R. N., Charif A., (1988) Plate Separation and Anchorage of Reinforced Concrete Beams Strengthened by Epoxy-Bonded Steel Plates, The Structural Engineer, 66 (5/1): 85-94.
2. Roberts T. M., (1989) Approximate Analysis of Shear and Normal Stress Concentrations in the Adhesive Layer of Plated RC Beams, The Structural Engineer, 67 (12/20): 37-46.
3. Malek A., Saadatmanesh H., Ehsani M. R., (1996) Shear and Normal Stress Concentrations in RC Beams Strengthened with FRP Plates, Second International Conference on Advanced Composite Materials in Bridges and Structures, Montréal, Québec, Canada: 629-638.
4. Ascione L., Feo L., (2000) Modeling of Composite/Concrete Interface of R/C Beams Strengthend with Composite Laminates, Composites Part B 31: 535-540.
5. Ascione L., Feo L., Fraternali F., (1997) The Wrapping of Reinforced Concrete Beams with FRP Plates: A Mechanical Model, Proceedings of International Congress "Advancing with Composites '97": 155-170.
6. Feo L., D'Agostino G., Tartaglione D., (1998) On the Statical Behaviour of Reinforced Concrete Beams Wrapped with FRP Plates: an Experimental Investigation, European Conference On Composite Materials – ECCM-8, Napoli, Volume 2, pp. 189-196.
7. Ascione L., Feo L., Fraternali F., (1998) Stress Analysis of Reinforced Concrete Beams Wrapped with FRP Plates, European Conference On Composite Materials – ECCM/8, 2:197-204.
8. Ascione L., Feo L., Mancusi G., (2000) On the Statical Behaviour of Fibre-Reinforced Thin-Walled Beams, Composites Part B 31: 643-654.
9. Vlasov V.Z., (1961) Thin-Walled Elastic Beams, Pergamon Press, New York.

# Modelling of Fibre Reinforced Concrete

Antonio Grimaldi[1], Raimondo Luciano[2]

[1] Dipartimento di Ingegneria Civile
Università di Roma "Tor Vergata"
00133 Roma, Italia
[2] DiMSAT
Università di Cassino
03043 Cassino, Italia

**Abstract.** In this paper, the non linear tensile behaviour of fibre reinforced concrete is analyzed. In order to determine the possible failure micromechanisms and the tensile strength of the material, a non linear analysis is developed. Specifically, the possible slippage between fibre and matrix and the different concrete behaviour, under tensile and compressive load, are considered. Finally, by using the finite element method, numerical results are obtained for several fibre volume fractions and comparisons with analytical results are carried out.

## 1 Introduction

The possible applications of concrete reinforced by short fibres and the study of its mechanical behavior is an important issue in the field of the Civil Engineering. The most difficult problem is to obtain the constitutive equations of the concrete reinforced with fibres in the linear and nonlinear phase. This material can be considered as a composite made by a matrix of concrete with a brittle behavior and by steel short fibres randomly distributed in the matrix with a volume fraction between 1% and 2%. The aim of such a reinforcement is to improve the mechanical behavior of the material when tensile forces are acting. Infact, a limit tensile strength with ductile behavior is ensured to the material. The tensile uniaxial behavior of the composite is usually schematized by the constitutive relationship represented in figure 1. This behavior has been observed in many experiments [1].
If the material is loaded by an uniaxial strain, it is possible to observe the following characteristic phases of the behavior of the material:

a) Initial linear elastic phase. It is characterized by an elastic modulus and by the peak strength of the concrete $\sigma_{mt}$.

b) When the limit strain $\varepsilon_c$ of the matrix is attained, the composite changes configuration for values of the strain greater than the limit one. Furthermore, cracks develop in the matrix and the resulting stress is sustained by the composite material made of the cracked concrete and the fibres.

c) The value of the stress $\sigma_r$ at damaged conditions, represents the minimum value of stress for strains greater than $\varepsilon_c$. As a matter of fact, increasing the applied strain, several damage phenomena can occur: slipping between fibre and matrix or plastic compressive strains in the matrix. The failure condition due to one of the two previous damage phenomena is attained for values of the strain quite greater than $\varepsilon_c$ but with ultimate stress close to $\sigma_r$. This experimental observation justifies the use of the constitutive behavior shown schematically in figure 1 [1].
For this reason, the value of the stress $\sigma_r$, represents the maximum stress corresponding to a sufficiently ductile material behavior. Because of such a ductile behavior the performance of structural elements in f.r.c. improves considerably.
As matter of fact, though the residual stress $\sigma_r$ , is less than the traction strength of the matrix, in the case of structures, for example reinforced concrete beams, the presence of the fibers can increase the bending strength of the structural element. It is worth noting that the limit bending moment of the beam grows by increasing the fiber volume fraction and, in several cases, it can became greater than the limit bending moment of beams with unreinforced concrete.

In summary, the use of fibre reinforcement increases the strength and the ductility of concrete structures and, in order to model such a behavior, it is very important to evaluate the residual stress $\sigma_r$.

Aim of this work is the evaluation of $\sigma_r$ by using micromechanics [2-9]. Initially this task is reached by modelling the concrete as a no tension material [10] and the fiber as linear elastic. The no-tension material is characterized by elastic strain in the compressive stress directions and by cracks in the quasi-zero stress directions. Then, the global strength of material reinforced by elastic fibres can be obtained by using homogenization methods and the strain and complementary energy principles, i.e. by using techniques similar to the ones used in [3-6]. Further, by assuming for the matrix a limit value for the compressive stress and a rigid plastic behavior at the interface between fibres and matrix, it is possible to evaluate the limit strength of the composite by using the formulations of limit analysis for no tension materials [10].

In the case of periodic arrangement of the fibres, the analysis can be developed by using the finite element method [9]. In this work a two-dimensional (2-D) model characterized by rectangular fibres is presented.

## 2 Constitutive law of the constituents

**Matrix:** An elastoplastic constitutive model characterized by a very low tensile strength and a finite compressive strength schematises the behaviour of the concrete. Specifically, for general stress states, a Rankine-type failure

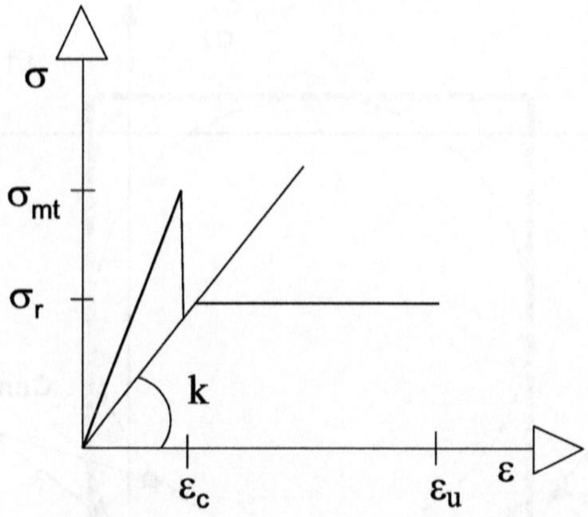


**Fig. 1.** Experimental behaviour of fiber reinforced concrete.

surface has been considered.
The material experiences the failure when the principal stresses reach the tensile or compressive yield value. In this way, the different behaviour of the concrete in tensile or compressive stress can be modelled.
The elastic domain of the concrete is represented by a cube in the space of the principal stresses ($\sigma_1$,$\sigma_2$,$\sigma_3$), moved with respect to the origin. In the case considered here, the stress state is 2D, then the stress $\sigma_3$ is zero, and, if the elastic domain is expressed in the plane of the principal stresses $\sigma_1$ and $\sigma_2$, the following relations can represent it:

$$f_1 = \sigma_1 - \sigma_t = 0; f_2 = \sigma_2 - \sigma_t = 0; f_3 = -\sigma_1 + \sigma_c = 0; f_4 = -\sigma_2 + \sigma_c = 0 \tag{2}$$

where $\sigma_t$ and $\sigma_c$ are the tensile and compressive limit strength, respectively. Further, in the computations equation (2) has been replaced by a continuous and differentiable surface.
Specifically, the failure surface has the following expression in the plane of the principal stresses:

$$\left(\frac{\sigma_1 + w}{a}\right)^p + \left(\frac{\sigma_2 + w}{a}\right)^p = 1 \tag{3}$$

with $a, w > 0$ and p a non-negative even integer (see figure 2).

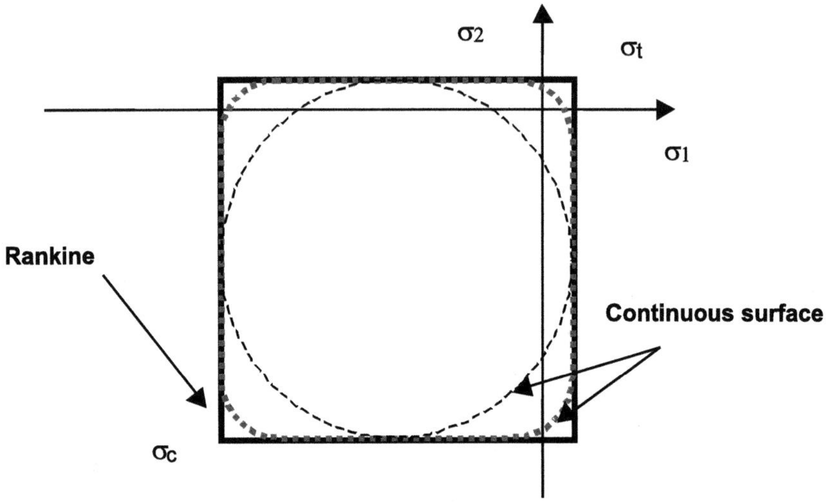


**Fig. 2.** Limit surface.

The parameter $p$ determinates the shape of the limit surface so that it represents a circle with centre the origin and radius a when $p = 2$. Increasing p, the shape of the surface tends to become a square with side $2a$.

The parameter $p$, allows approximating the rectangular domain of Rankine. The other parameters are:

$$w = -\frac{\sigma_T + \sigma_C}{2}; \qquad a = \frac{\sigma_T - \sigma_C}{2} \tag{4}$$

which represent the abscissa and the ordinate of the centre and the radius of the circle obtained for p=2, respectively.

In the following p is chosen equal to 6.

**Fibre:** The steel fibres are considered linear elastic. Such a hypothesis is consistent with the model since the stress in the fibres does not reach the yield limit.

**Fibre-matrix contact:** The behaviour of the possible slip between concrete and steel is assumed elastoplastic with limit slip-stress $\tau_o$.

## 3 Geometrical model

The model considered in this paper is a concrete reinforced by short fibres parallel to the applied load. The fibres have a unique geometry and they are spread periodically in the matrix.

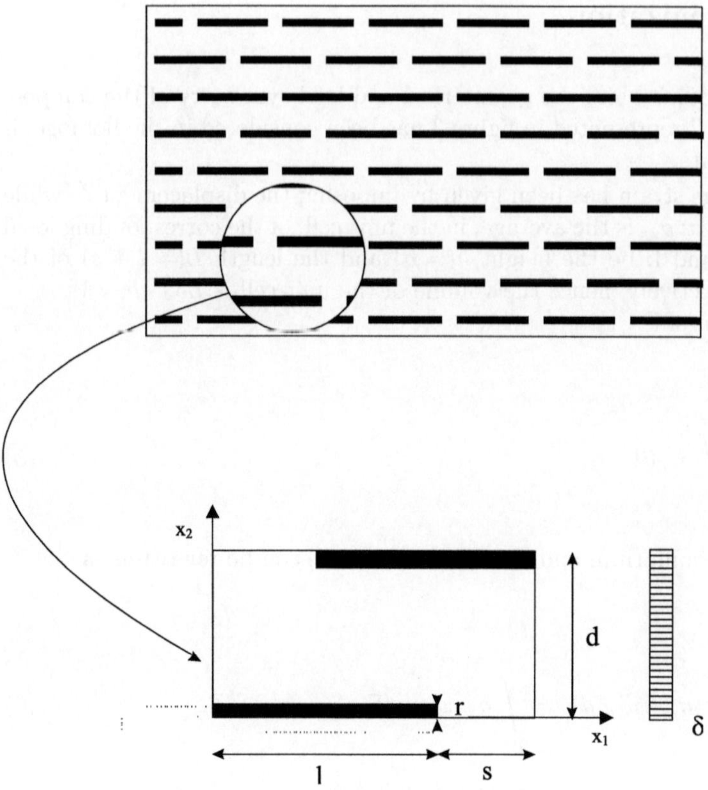


**Fig. 3.** Geometrical and load scheme of the unit cell.

Therefore their length, the distance and the overlapping between two of them define the pattern of the fibres.

The periodicity assumption allows to consider the behaviour of the unit cell instead of considering the entire composite.
Further, it is assumed that the model is plane with thickness t and the matrix is reinforced by rectangular fibres.
In figure 3, the unit cell considered in the analysis is shown. It is composed by two steel fibres with sizes 2r and 2l in a concrete matrix. The distance between two fibres is d and their overlapping is s.
The unit cell is constrained vertically along the horizontal sides and horizontally along the vertical sides. Further, along one vertical side, a displacement $\delta$ in the direction of the fibres is enforced.
In the analysis, the behaviour of the unit cell is studied until failure of the model for several volume fractions of the fibres (2% to 14%) and by taking fixed some geometrical parameters (l and s).

# 4 Homogenization

Since the periodicity and the geometrical and load symmetry of the composite, the unit cell represented in figure 3 has been considered in the homogenisation procedure.

The average strain has been given by imposing the displacement $\delta$, while the overall stress $\widehat{\sigma}_{ij}$ is the average, in the unit cell, of the corresponding local stress. Let h and L be the height $h(=d)$ and the length $L(=l+s)$ of the unit cell, respectively, hence the volume of the unit cell is $L \cdot t \cdot h = V$. The expression of $\widehat{\sigma}_{ij}$ is:

$$\widehat{\sigma}_{ij} = \frac{1}{V} \int\limits_{V} \sigma_{ij} \, dV \tag{5}$$

that, for the equilibrium and the Gauss theorem, can be rewritten as

$$V\widehat{\sigma}_{ij} = \int\limits_{V} \left(\sigma_{ik} x_j\right)_{,k} dV = \int\limits_{\partial V} \sigma_{ik} \, x_j n_k dS \tag{6}$$

where $n_k$ are the components of the outwards normal of the unit cell and $x_j$ are the coordinates in the assumed reference framework. The main component of the average stress is $\widehat{\sigma}_{11}$, which can be evaluated by using (6) as:

$$V\widehat{\sigma}_{11} = L \cdot t \cdot \int\limits_{h} \sigma_{11} \, n_1 dx_2 \tag{7}$$

Further, by denoting $\sigma_{11} \, n_1 = s_1$, the total reaction on the face of normal 1 is $R_1 = t \cdot \int_h s_1 \, dx_2$. Therefore, we obtain $\widehat{\sigma}_{11}$ as

$$\widehat{\sigma}_{11} = R_1 / \left(t \cdot h\right) \tag{8}$$

In summary, the overall material response can be evaluated by computing the total reactions on the sides of the unit cell at each load step by solving the corresponding elasto-plastic problem.

## 5 A finite element analysis

In this section, the homogenisation problem will be approached via finite elements.
The unit cell has been discretised by a mesh of four node isoparametric elements.
Further, since the non-linearity of the problem, an iterative procedure has been implemented for solving the problem at each load step. Specifically, the Return Mapping scheme [11] has been adopted, by using the failure surface presented in the previous section.

The contact element between the steel and the concrete has been realised by two springs directed in two orthogonal directions. The computation of the plastic slip at each load step is developed by *Return Mapping* for one-dimensional elements [11]. The numerical algorithms have been implemented in FORTRAN and the Finite Element Code FEAP has been adopted (see [12]).

## 6 Limit analysis

The results obtained by the finite elements formulation can be compared with the ones provided by the limit analysis developed in [8]. At this aim, the expressions of the main theoretical results are presented in the following [8, 9].
Specifically, only two failure mechanisms have been considered.
The first one is the slip mechanism characterised by plastic compressive deformations in the central zone. By using such a micromechanism, the following upper bound on the failure load of the unit cell is obtained:

$$\overline{\sigma}_{11}^{+} = \frac{\sigma_c\, s}{2\, d}. \tag{9}$$

On the other hand, by considering a micromechanism characterised by slippage between fibres and matrix, the upper bound becomes

$$\overline{\sigma}_{11}^{+} = \frac{\tau_0\, s}{d}. \tag{10}$$

Where $\tau_0$ is the yield strength at the interface. The value of $\tau_0$ is usually less than $\sigma_c/2$ and, therefore, the second expression provides the upper bound for the limit average stress of the homogenised material.
The static theorem of limit analysis provides the lower bound on the strength of the model which can be expressed as:

$$\overline{\sigma}_{11} = \frac{\tau_0 \; s \; sin(\beta)}{d \; sin(\widehat{\beta})} \cdot \left(1 - \frac{\tan(\beta/2)}{\tan(\beta)}\right) \tag{11}$$

where $\tan(\beta) = s/d$ while $\sin(\widehat{\beta}) = 2\tau_0/\sigma_c$ . Such a situation is valid only when $\beta < \widehat{\beta}$ which represents the failure of the concrete.
In the case $\beta > \widehat{\beta}$ , i.e. failure for slippage of the fibres, the following relation holds:

$$\overline{\sigma}_{11} = \frac{\tau_0 \; s}{d} \cdot \left(1 - \frac{\tan\left(\widehat{\beta}/2\right)}{\tan(\beta)}\right) . \tag{12}$$

From the previous equation it can be remarked that $\overline{\sigma}_{11}$ depends on the ratio between the overlapping fibre-matrix and the distance between two fibres.

## 7 Numerical applications

In this sections several numerical applications and comparisons will be presented.
The mechanical parameters of the constituents of the model are reported in tab. 1, where the force is measured in kg and the length in centimetres. The strength of the concrete used in the computations is $\sigma_c = 300kg/cm^2$.
The limit tensile strength of the concrete has been chosen as small as possible in relation to the numerical model adopted.

These assumptions are equivalent to study the model of fibre-reinforced concrete by considering the matrix unable to support tensile stresses.
The stiffness of the contact element (spring) is assumed of the same order of magnitude than the one of the concrete.
Several comparisons and models have been studied.
On the other hand, several values of the strength $\tau_o$ have been considered.
In order to limit the number of the cases to study, the length and the thickness of the model and the overlapping between the fibres have been taken fixed: l=2, r=.1 and s=1.
The distance between the fibres is the only varying parameter.
Several volume fractions of the fibres have been considered in the applications, hence several values of $\beta$ have been examined.
Further, two different values of $\tau_0$ ($\tau_0 = 30kg/cm^2$, $\tau_0 = 120kg/cm^2$) have been considered in order to examine the failure of the model due to the compression of the matrix or the slippage of the fibres.

| Materials | Elastic Modulus $[Kg/cm^2]$ | $\nu$ | $\sigma_t[Kg/cm^2]$ | $\sigma_v[Kg/cm^2]$ | $\tau_0[Kg/cm^2]$ |
|---|---|---|---|---|---|
| Stell | 2.100.000 | 0.3 | $\infty$ | $\infty$ | / |
| Concrete | 2.100.000 | 0.1 | 8div 14 | 300 | / |
| Contact | 300.000 | 0.1 | $\infty$ | $\infty$ | 30div 120 |

**Table 1**

It is worth noting that, for $\tau_0$=30, $\widehat{\beta}$ is equal to 11.50°, while for $\tau_0$=120, $\widehat{\beta}$ is equal to 53.13°.

In figure 4 the relation $\widehat{\sigma}_{11}$ (average stress in the direction 1) - $\varepsilon_{11}$ is presented for $\widehat{\beta}$ = 11.50° and $\widehat{\beta}$ =53.13° while a volume fraction of fibre equal to 11 (i.e. for $\beta$ = 45° has been considered.
The behaviour of the material obtained via FEM does not reflect completely the experimental behaviour in figure 1, because the analysis has been limited to the phase $b$ (cracked medium) and the phase $c$ (failure) of the model.
In figure 4, three points of the stress state of the overall composite have been reported (A, B, and C). They respectively detect:
(A) the elastic state
(B) the cracked state
(C) the failure.
In order to determine if the failure is reached for compression or for slippage, the contact tractions between concrete and steel have been obtained.

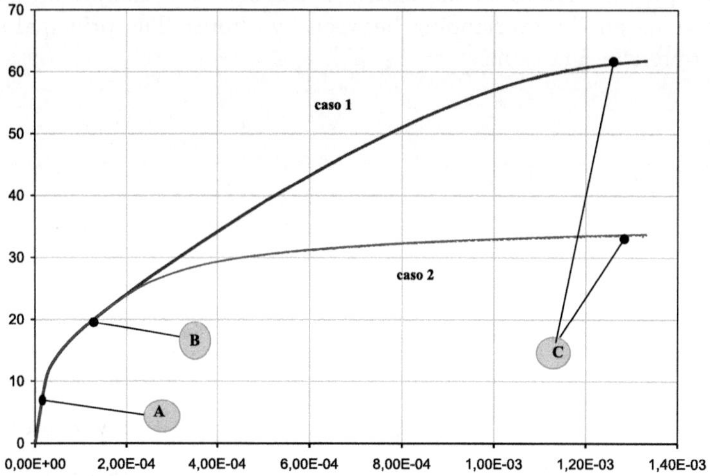


**Fig. 4.** Homogenised relation : $(\overline{\sigma}_{11} - \varepsilon_{11})$for $\widehat{\beta}$= 11.50° (case 1) and $\widehat{\beta}$=53.13° (case 2).

Since the two cases considered ( $\widehat{\beta} = 11.50°$ and $\widehat{\beta}$ =53.13) present the same geometrical scheme and have different limit slippage between fibre and matrix (the values of $\sigma_t$ and $\sigma_c$ are the same for both models), the behaviour of the two models coincide until the limit tangential stresses, at the interface, are reached. Then the behaviour follows two different paths.
In the first case ($\widehat{\beta} = 11.50°$) the stresses can increase until the compression failure of the matrix is reached.
In the second case, the stresses cannot increase since the slippage mechanism is already activated.
In summary, the following conclusions can be stated:
*Point A*
The stress pattern in the two cases is the same. In this phase, the concrete supports tensile stresses. Further, the distribution of tensile stresses localises at the end of the fibre.
*Point B*
In both models, the limit tensile strength of the concrete is reached. In the whole matrix the principal tensile stresses $\sigma_1$ are maximum, while the compression stresses $\sigma_2$ increase slowly only in an inclined zone of the unit cell (compressed strut).
*Point C*
Case $\beta$=53°
The compression failure of the matrix is reached. The compressed strut develops along all the overlapping between two fibres. The principal stress is the compression limit one.

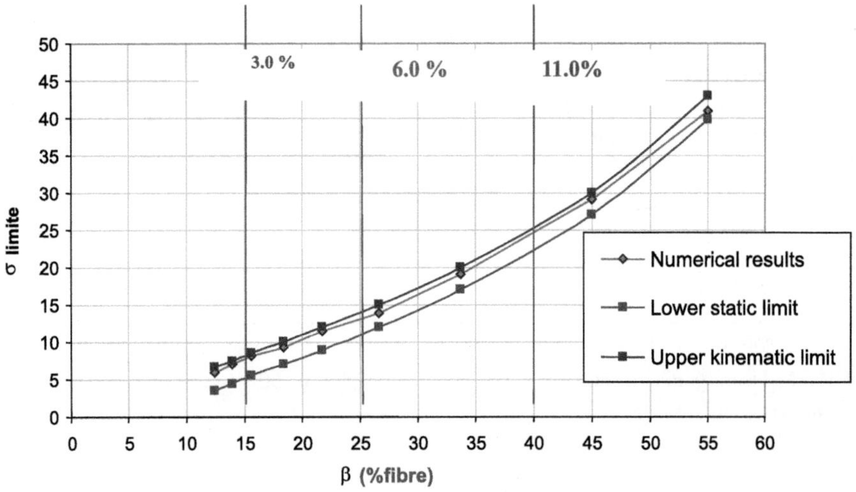


**Fig. 5.** Numerical comparison for $\widehat{\beta}$=11.50°

Case $\beta$=11°

The stresses values are lower than the ones of the case $\beta = 53°$. Further, the compression stress in the strut is lower than the compression strength $\sigma_c$, while on the interface between fibres and matrix the shear stress reaches the limit value $\tau_0$.

## 8 Comparisons

The failure mean stress values of the model corresponding to the analytical and numerical results have been compared.

In figures 5 and 6, the numerical FEM results, and, the lower and upper bounds, are given as functions of the parameter $\beta$. The corresponding values of the volume fraction are also given for three values of $\beta$. It is worth noting that the numerical results are bounded by the lower static and upper kinematic limits, and that, for small values of $\widehat{\beta}$ , the limit analysis bounds are almost coincident.

## References


1. Lim,T.Y., Paramasivam, P., and Lee, S. L., *Analytical model for tensile behavior of steel-fiber concrete.* ACI Material Journal July/August, (1987) 286-298.
2. Tvergaard V., *Model studies of fibre breakage and debonding in a metal reinforced by short fibres.* J. Mech. Phys Solids 41 (1993) 1309-1326.


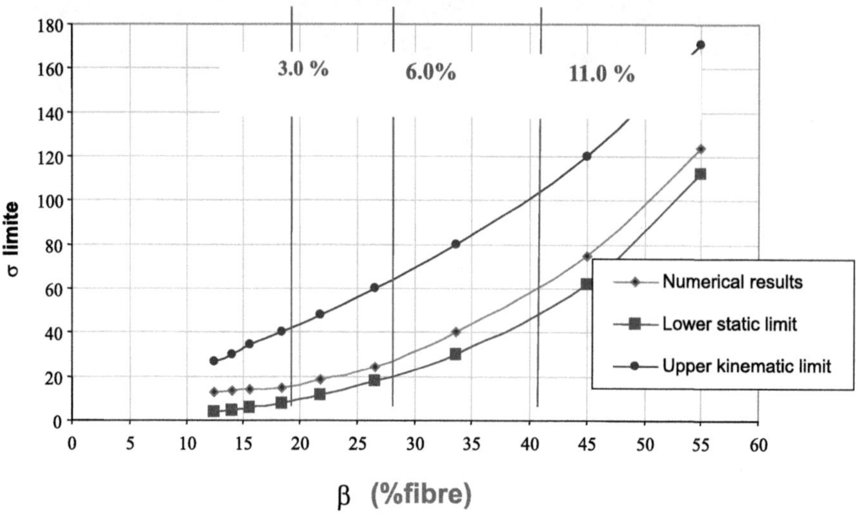


**Fig. 6.** Numerical comparison for $\widehat{\beta}$=53.13°

3. Talbot D. and Willis J. R., *Variational principles for inhomogeneous non linear media.* IMA J.Appl.Math 35 (1985) 39-54.
4. Willis J. R., *On methods for bounding the overall properties of non linear composites.* J. Mech. Phys. Solids 39 (1991) 73–86.
5. Ponte Castaneda P., *New variational principles in plasticity and their application to composite materials.* J. Mech. Phys. Solids 44 (1992) 827-862.
6. Suquet P., *Overall potentials and flow stresses of ideally plastic or power law materials.* J. Mech. Phys. Solids 41 (1993) 981-1002.
7. De Buhan P. and Taliercio A., *A homogenization approach to the yield strenght of composites.* European J.Mechanics: A/Solids 10 (1991) 129-154.
8. Grimaldi A., Luciano R., (1995) *Modellazione micromeccanica e valutazione della resistenza a rottura di calcestruzzi rinforzati con fibre.* XII Congresso AIMETA, Napoli, Ottobre 1995.
9. Grimaldi A., Luciano R., *Tensile stiffness and strength of fibre-reinforced concrete.* J. Mech. Phys. Solids 48 (2000) 1987–2008.
10. Giaquinta M., Giusti E. , *Researches on the equilibrium of masonry structures.* Arch. Rat. Mech. Annal. 88, (1985) 359-392.
11. Crisfield M.A., Non linear finite element analysis of solids and structures, Vol 1-2, (1991).
12. FEAP, Copyright prof. R.L.Taylor, Departement of Civil Eng. Univ. of California Berkeley,1996.

# Influence of the Steel Properties on the Ductility of R.C. Structures

Antonio Grimaldi,[1] Zila Rinaldi[2]

[1] Dipartimento di Ingegneria Civile
Università di Roma "Tor Vergata"
00133 Roma, Italia
[2] DiMSAT
Università di Cassino
03043 Cassino, Italia

**Abstract.** Aim of this paper is the evaluation of the steel properties influence on the behaviour of reinforced concrete structures, with particular reference to the ductility characteristics. A simplified beam model, subjected to bending moment and axial force, is developed and the relation between average strength and mean deformation is obtained. Furthermore the phenomenon of strain localization in the steel is pointed out.

## 1 Introduction

In these last years the problem of the influence of the steel properties on the global behaviour of r.c. structures has arisen a great interest both in the scientific and industrial fields. Recent studies and researches, reported in the quoted references, witness the need to state the minimum requirement on the ductility characteristics of the reinforcement bars. This aspect is strictly connected with the spreading in the whole Europe of new kinds of weldable steels, made with the Tempcore process, characterised by high strengths (grade 500), but affected by a low ratio between the ultimate ($f_{us}$) and yield ($f_{ys}$) strengths (hardening ratio), and by a reduced ultimate strain ($\varepsilon_{us}$). The use of these steels can increase the local strength, but can reduce the local and global ductility of the structure.

This production has been accepted by the more recent European codes [EC2 1993, EC8 1994, ENV 10080 1995] in which a steel degree named B500B has been introduced, with yield strength equal to 500 N/mm$^2$. Nevertheless the same codes require the use of high ductility steels corresponding to a minimum hardening ratio of 1.2 and a minimum ultimate strain $\varepsilon_{us} = 9\%$, for structures in seismic zone.

In this paper the steel properties influence on the ultimate behaviour of beam elements and simple framed structures is pointed out. The strength and ductility characteristics of reinforced concrete members are evaluated in a simplified way, taking account of the cracking phenomenon and the materials behaviour. In particular the possibility of steel strain localization near the crack, at the rebar yielding is considered and highlighted. In order to evaluate the rotation capacity of r.c. beams many models, even much sophisticated, are available in literature and, for the great number of parameters involved in the problem, quite often a numerical solution is used. In this paper a model that allows to obtain an approximate, but simple evaluation of the mean curvature and of the plastic rotation of r.c. elements is proposed.

## 2 Beam model behaviour

The evaluation of the ductility capacity has been firstly developed with reference to a beam element, with a length ($l$) equal to the crack distance, subjected to tensile load. A similar analysis has been performed for a beam model under bending and axial forces. The mean strain (axial strain or mean curvature) applied at the element is the loading parameter and the structural behaviour is analysed for subsequent steps, up to the failure.

After the initial elastic behaviour (Fig. 3 - Phase 1), cracking in the concrete and slip occur with a stress and strain redistribution (Fig. 3 - Phase 2). By increasing the applied strain, the steel yield is reached in the cracked middle section firstly, then the plastic deformations spread along the element (Fig 3 - Phase 3) up to the ultimate strain in the steel or in the concrete. Simple material constitutive relations are used in the analysis, as given in Fig. 1 in order to clearly point out the parameters that influence the behaviour of the model.

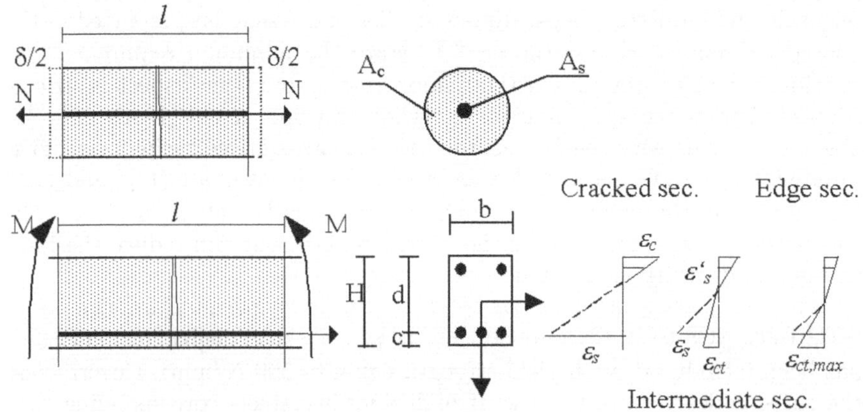


**Fig. 1.** Beam model

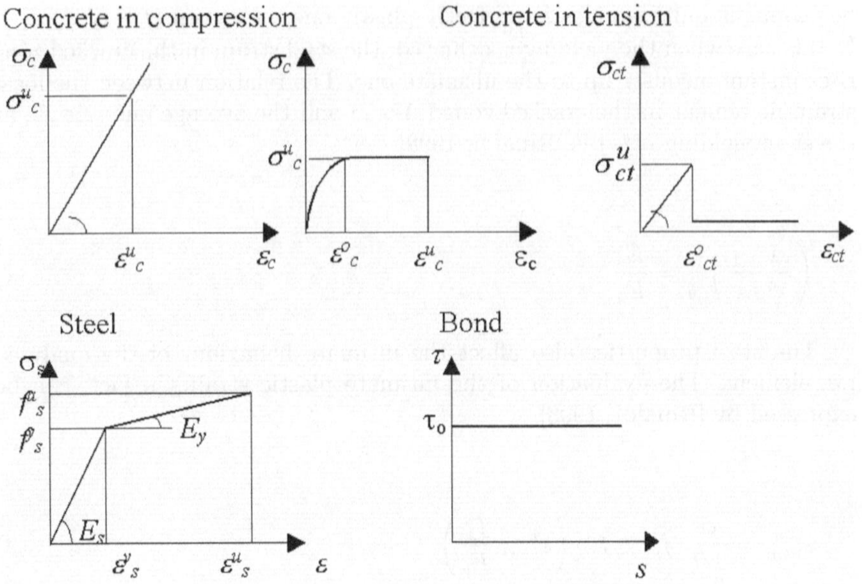


**Fig. 2.** Constitutive relations of the materials

The tensile behaviour of the element (Fig 1) is analysed by increasing the applied mean deformation $\varepsilon_m = \delta/l$, and by taking account of the slip between the concrete and the steel. This model has been already studied both theoretically and experimentally [Marti et al. 1998]. At the first crack, in the middle section, the tensile strength in concrete vanishes and a slip between the two materials occurs. In this phase the model assumes a constant value of the bond stress along the bar, equal to $\tau_o$ (Fig. 2). The equilibrium condition gives the linear stress distribution in the steel and in the concrete.

After the stabilization of the crack formation, the bond strength and the crack distance are related to the tensile concrete ultimate strength ($\sigma^u{}_{ct}$ in Fig. 2), that is reached in the edge section of the element. This condition corresponds to the hypothesis of maximum cracks spacing. As well known a minimum cracks distance can be evaluated according to the fracture energy criteria [Bazant et al., 1983]. It is worth noting that in the model equation the product $\tau l$ appears as a parameter depending only on the ultimate tensile concrete strength.

The model allows to highlight the localization of the steel deformation near the cracks, during the yielding phase. In fact, when in a cracked section the steel reaches the yield strength, a further load increment leads to an increase of the steel deformation as higher than the average one, as lower is the hardening ratio. Near the cracked zones localization of steel deformation can occur and then the ultimate strain and the failure of the bar can be quickly achieved. This phenomenon is particularly remarkable when the steel behaviour is quite an elastic-perfectly plastic one, i.e. when $(f^u{}_s/f^y{}_s) \approx 1$. In this case when the yielding is achieved, the steel strain in the cracked zone rises instantaneously up to the ultimate one. The relation between the local strain increment in the cracked zone ($\Delta\varepsilon_{s,1}$) and the average one ($\Delta\varepsilon_m$), at the steel yielding ($\varepsilon^y{}_s$) is [Rinaldi, 1998]:

$$\left(\frac{\Delta\varepsilon_{s,1}}{\Delta\varepsilon_m}\right)_{\varepsilon_s^y} = \frac{E_s}{E_y} \tag{1}$$

The steel properties also affect the ultimate behaviour of the analysed r.c. element. The evaluation of the ultimate plastic strain, in fact, can be expressed by [Rinaldi, 1998]:

$$\varepsilon_{m,p}^u = \frac{\phi}{2\tau l}\left(f_s^u - f_s^y\right)\left(\varepsilon_s^u - \frac{f_s^u}{E_s}\right) \tag{2}$$

being ∅ the bar diameter. The equation points out the influence of the steel characteristics on the plastic ultimate deformation and furthermore the localization phenomenon is confirmed. In particular, due to the lack of plasticization spread along the element, the plastic deformation vanishes when the steel behaviour is elastic perfectly plastic.

A similar kind of analysis has been then applied to a beam member loaded by constant bending and axial actions. A beam element with rectangular cross-section and a length $l$ equal to the cracks distance is considered, aiming at evaluating the stress and strain distributions and the local and mean curvatures. The classical hypothesis of plane sections is adopted only for the middle cracked section, where the tensile concrete stress vanishes; the assumption of perfect bond between steel and concrete is removed, and the slip between the two materials is taken into account. In all the cross-sections of the element the concrete strain pattern is assumed as a bilinear one, (Fig.1), as also proposed in [Gambarova et al., 1998], and as shown by experimental works, [Giuriani et al., 1979] and numerical analyses [Plauk et al., 1981, Ngo et al., 1967]. Also in this case the cracks are supposed to be diffused at constant spacing. The ultimate concrete strength has been assumed at the edge of the element. The constitutive relations for the materials are given in Fig. 2.

The formulation of a simple relationship between the bending moment and the mean curvature is achieved by following the behaviour phases related to an increase of the mean curvature up to the failure, defined as the reaching of the limit strain in the steel or in the concrete. At the concrete cracking a slip along the whole element occurs and according to the bond rigid-plastic behaviour the bond stress distribution is constant and the steel one is linear

from the middle to the edge section of the element. The neutral axis ($x_c$) and concrete stress patterns, instead, are not linear with $z$ along the element. The first step of the procedure is the solution of the equilibrium equations in the cracked section, where the tensile concrete stress is assumed to vanish. The parameter $\tau l$ can be derived in a simple way by the equilibrium condition of the rebar, at the first crack formation. The value of this parameter is assumed to be constant, during the load process. This hypothesis allows to solve the equilibrium equations of the edge section. The analysis is then repeated in all the intermediate sections and the stress and strain distributions in steel and concrete are obtained. After the yielding, the steel stress is always linear along the rebar, while the strain pattern is bilinear along the element according to the steel constitutive model.
Of particular interest is the analysis of the intermediate section where the rebar strain is equal to the yield one. The distance ($m$) between this section and the middle cracked one defines the length of the yielded steel, and is related to the deformation level and to the steel and concrete properties, according to the relation:

$$\frac{m}{l} = \frac{\phi}{4\tau l} E_y \left(\varepsilon_s^1 - \varepsilon_s^y\right) \qquad \frac{m}{l} = \frac{\phi}{4\tau l} \frac{f_s^u - f_s^y}{\varepsilon_s^u - \varepsilon_s^y} \left(\varepsilon_s^1 - \varepsilon_s^y\right) \tag{3}$$

In this expression the parameter $\tau l/\emptyset$ is related to the ultimate concrete tensile strength. The steel ultimate strain affects the plastic zone length only when the collapse occurs with concrete failure. On the contrary, the steel deformation in the cracked section $\varepsilon_{s1}$ is equal to the ultimate steel strain $\varepsilon_{us}$.

The procedure can be summarised as follows:
1) Analysis of the cracked middle section (Known: $x_c$; $\sigma_{ct}$): evaluation of the strain distribution ($\sigma_s$; $x_c$) by solving equilibrium equations;
2) Evaluation of the $\tau$l/$\emptyset$ factor by equilibrium condition along the steel bar;
3) Analysis of the edge cross section (known $\sigma_s$; $\sigma_{ct} = \sigma_{oct}$): - evaluation of the strain distribution ($\sigma_c$; $x_c$) by solving equilibrium equations;
4) Analysis of the yielded cross section (known $\sigma_s = \sigma_{ys}$; $x_c$): - evaluation of strain distribution ($\sigma_c$; $\sigma_{ct}$) by solving equilibrium equations.

The local curvature is defined, according to the European Model Code 90 [CEB-FIP, 1991] as the ratio between the sum of the deformation in the tensile steel and compressed concrete and the effective depth ($d$). The mean curvature has been evaluated assuming the curvature $\rho(z)$ as a linear function from the middle cracked section to the intermediate yielded one and still linear to this section to the edge one.

An example of the moment-mean curvature ($\rho_m$) relationship is plotted in the Fig. 4. The cross section geometry and the materials parameters are given in the Fig. 3; different values of steel hardening ratios but constant ultimate steel stress ad strain are considered.

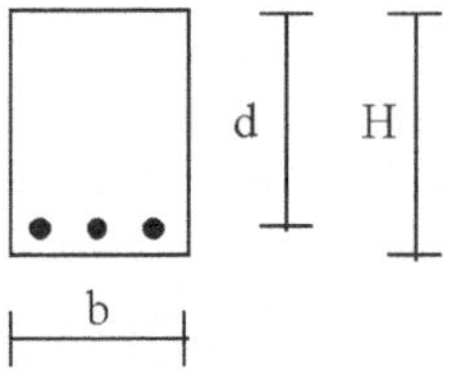


| b [cm] | H [cm] | $\sigma_c^u$ [N/mm²] | $\varepsilon^u_s$ | $\sigma^u_s$ [N/mm²] | $E_s$ [N/mm²] |
|---|---|---|---|---|---|
| 30 | 60 | 35 | 0.008 | 550 | 210000 |

**Fig. 3.** Geometrical and materials characteristics

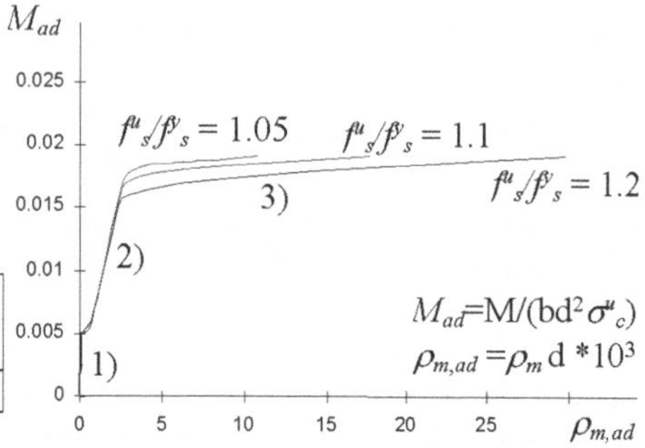


**Fig. 4.** Moment – mean curvature relationship

It is worth noting the relevant influence of the hardening ratio on the element behaviour; the reduction of this parameter from 1.2 to 1.05, leads to small increments of the bending moment, but produces a large ductility reduction.
Finally, we observe that the mean curvature $\rho_m$ can be evaluated with reference to the steel strain according to the approximate relation:

$$\rho_m = \int_o^{l/2} \frac{\varepsilon_s(z)}{d - x_c(z)} dz \tag{4}$$

When the element failure corresponds to the ultimate strain in steel bar, then the mean ultimate curvature can be simply expressed as:

$$\rho_m^u = \frac{1}{d - \widehat{x}} \left\{ \varepsilon_s^y \left( \frac{f_s^u}{f_s^y} - 1 \right) \left[ \frac{\phi E_s}{4\tau l} \left( \varepsilon_s^u - \varepsilon_s^y \right) \left( 1 - \frac{E_y}{E_s} \right) + 1 \right] + \varepsilon_s^y - \frac{\tau l}{\phi E_s} \right\} \tag{5}$$

where ($\widehat{x}$ ) is a mean value of the neutral axis depth inside the element The difference between this simplified formulation and the more exact procedure is practically negligible.

## 3 Structural ductility evaluation

The proposed formulation has been applied first to a single beam element under constant bending moment, then to simple models subjected to bending moment with linear distribution. Finally the frame models in Fig. 11 have been examined. The results concerning the beam element under constant bending moment, given in Fig. 5 and Fig. 6, are discussed. The hardening ratio (varying from 1.05 to 1.20), the tensile steel percentage $\mu$ (from 0.13% to 1%) and the steel amount in compression, expressed as a percentage of the tensile one, are the model parameters. In the next pictures the ratio between the ultimate and the yield mean curvatures (curvature ratio $\rho_{um}/\rho_{ym}$) is plotted versus the steel tensile percentage for two values of the compression steel; the curvature ratio is then represented versus the steel ultimate strain, in the case of rebar failure, for two values of the hardening ratio. The different behaviour of the element is well highlighted in Fig.5. A sharp tip ($\mu_{cr} \approx$ 0.3%) separates the range corresponding to the steel failure ($\mu$ ¡$\mu_{cr}$), to the one related to the concrete collapse ($\mu_{cr}$ ¿$\mu_{cr}$). These results agree with the experimental and theoretical ones [Eligehausen et al., 1987]. The influence of the hardening ratio is relevant particularly in the case of steel failure, when a variation of $f_{us}/f_{ys}$ from 1.05 to 1.2 gives a very large curvature ratio decrease.

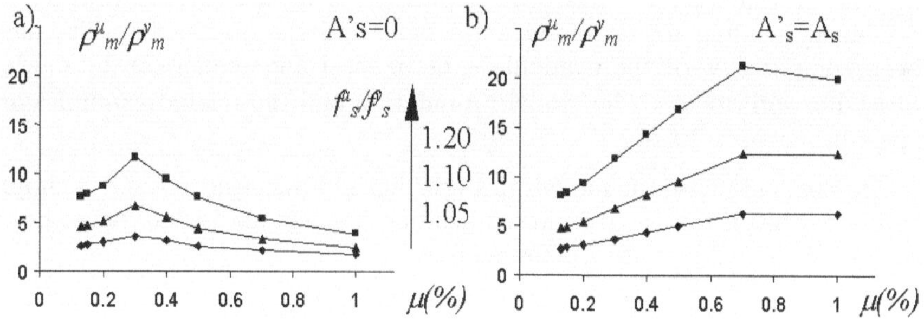


**Fig. 5.** Mean curvature versus steel percentage

When steel bars are also in the compression zone the structural behaviour is modified (Fig 5.b), and the hardening ratio effects are relevant also for tensile steel percentage of about 1%. In Fig. 5 the influence of the ultimate deformation on the curvature ratio is shown.

The relation between $\rho_{um}/\rho_{ym}$ and $\varepsilon_{us}$ is almost linear, and the ultimate steel deformation effects are particularly evident for steels with higher hardening ratio. A ultimate steel strain variation from 2% to 12% gives a ductility increase of about 200% for steel hardening ratio $f_{us}/f_{ys}$=1.05, and of about 400% if $f_{us}/f_{ys}$=1.20.

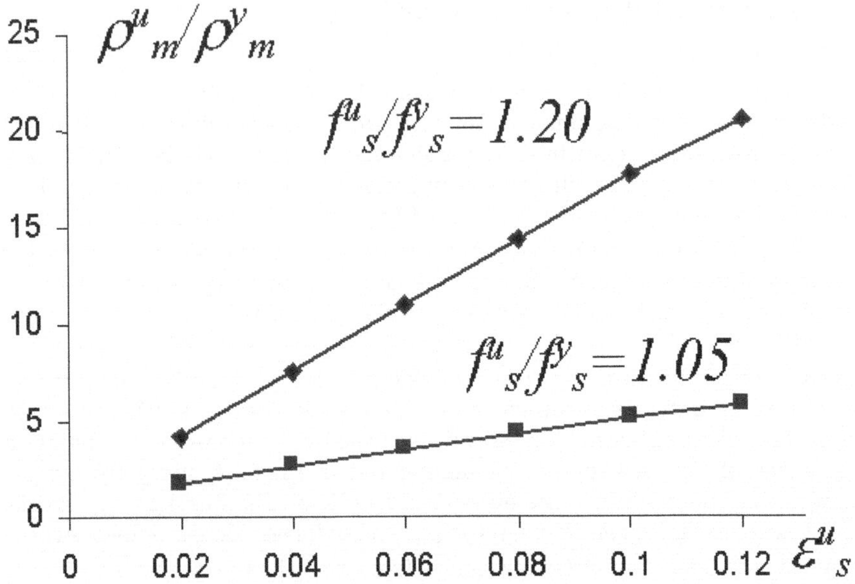


**Fig. 3.** Mean curvature versus ultimate steel strain

The procedure, then, has been applied to a beam element subjected to combined bending moment and axial force and the obtained results have been compared with the available experimental and numerical data, with satisfying agreements. The model formulation and the related examples are given in [Rinaldi, 1998].

In a second phase the models in Fig. 7 have been considered, i.e. a simply supported beam subjected to linear bending moment, and a cantilever beam subjected to bending and compression.

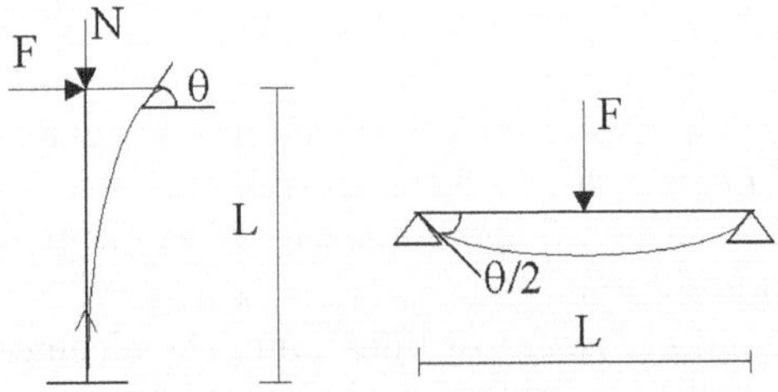


**Fig. 7.** Beams and column models

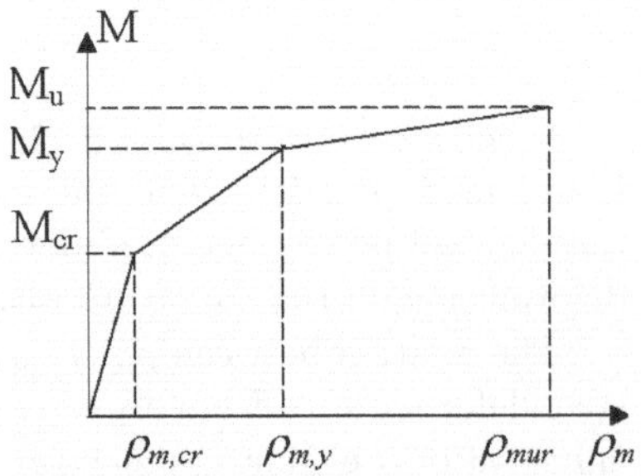


**Fig. 8.** Simplified moment – mean curvature relation

The plastic rotation of the end section ($\theta_{pl}$) has been assumed as a plastic deformation parameter and evaluated according to the bending moment-mean curvature relationship, approximated with a piecewise linear relation defined by the cracking, first yielding and ultimate bending moments, as given in Fig. 8.
Numerical examples and parametric studies have been developed for a simply supported beam with a length of 6.00 m and a rectangular 30x60 cm cross section. The steel hardening ratio and the tensile steel percentage are the parameters, while the ultimate steel strength is constant. The concrete behaviour is given Fig. 2. Some of the obtained results are given in Fig. 9 and Fig. 10, where the plastic rotation versus the steel percentage is plotted, for three hardening ratio and for two values of reinforcement in compression.

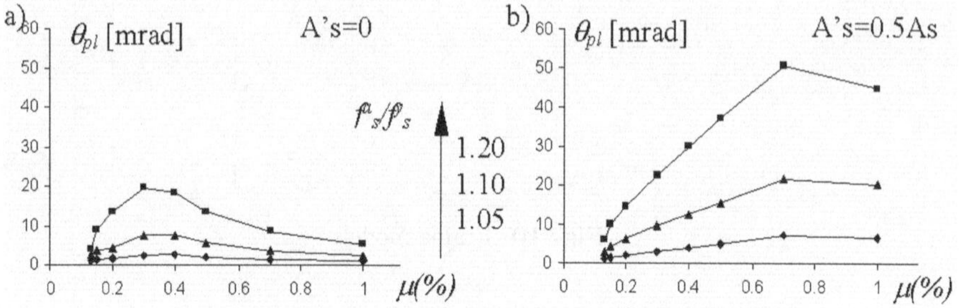


**Fig. 9.** Plastic Rotation versus steel percentage

The obtained patterns are similar to those already discussed for the single beam-element. Once again in the absence of compressed steel bars, the ranges related to the failure of the two materials are well evident. The influence of the steel properties is relevant when the collapse is due to the steel, and the plastic rotation can be reduced of a factor ten when the ratio $f_{us}/f_{ys}$ varies from 1.20 to 1.05. The rotation capacity reduction is not relevant only for steel percentage higher than 1%. When the compressed bars are considered the ductility is mainly related to the hardening parameters and quite independent from the tensile steel percentage. The influence of the ultimate steel strain on the plastic rotation is similar to the single beam element one.
Finally some simple framed schemes have been analysed. The behaviour of these structures is more complex and governed by the plastic hinge number and location, and by the rotation capacity they are able to exploit. The present design criteria aim at obtaining structures able to grant a ductile ultimate behaviour, and to dissipate the energy in a plastic field. The adoption of steel with low values of hardening ratio and ultimate strain could lead to a brittle collapse due to an untimely failure of a single section. In order to study this aspect a first parametric analysis has been performed for the schemes in

Fig. 11. The considered steel (named Feb44 e B500 Tempcore) properties are shown in the same figure. The frames are subjected to a constant uniform vertical load and to horizontal forces, with a linear distribution, increasing up to the collapse of the structure. The model corresponds to the case of framed structures under seismic static forces. The schemes have been analysed with a numerical analysis according to a finite element discretization of the beams and the columns. The previous given beam model has been applied to the single element, assumed to be loaded by constant bending moment and axial force.

| | Feb44 | B500 Tempcore |
|---|---|---|
| $f_s$ [Mpa] | 440 | 525 |
| $f_s$ [Mpa] | 600 | 600 |
| $\varepsilon^u_s$ [%] | 12 | 8 |

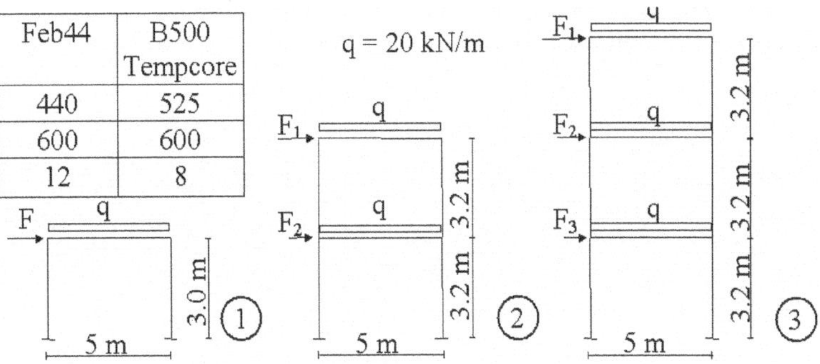


**Fig. 10.** Frame models

The results obtained for the frames characterised by equal beams and columns sections (30 cm x 50 cm) symmetrically reinforced with As = A's = 6 cm2 are summarised in Fig. 11. For each scheme and for the two analysed steel properties the base shear (V) is plotted versus the top displacement $\delta$.

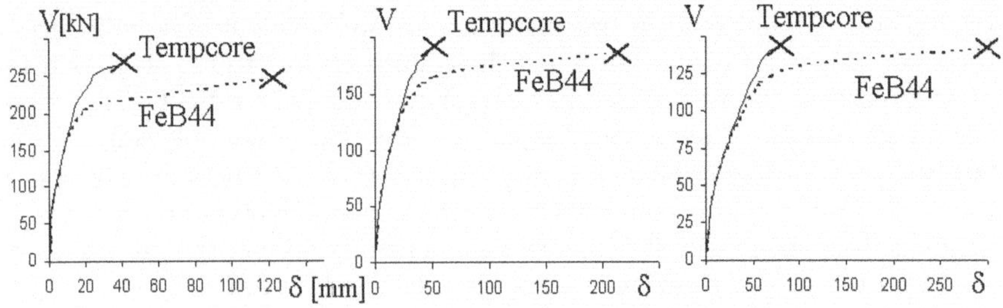


**Fig. 11.** Base shear force versus top displacement

The single-storey frame (Fig. 11) reinforced with Tempcore bars, compared with to the same scheme with FeB44 steel, shows a large ultimate displacement reduction, of about 1/3, but a similar ultimate strength. When the number of storeys increases, the ultimate load for the two frames is almost equal, but the ultimate displacement is reduced of about $\frac{1}{4}$ for the B500 reinforced schemes. In these cases the untimely collapse of one section (at the base of the first column), does not allow the formation of an adequate number of yielded zones with sufficient rotational capacity, and therefore an ultimate failure with a global collapse mechanism is not attained.

The obtained results have been interpreted according to a traditional limit analysis, by introducing plastic hinges in which the yield effects are concentrated. Since the sixties the problem of the plastic hinge length has been studied leading to simplified formulations [Baker, Amarakone 1964, Sawyer 1964, Corley 1966, Mattock 1967, Riva, Cohn 1991, Cosenza et al. 1993, Manfredi, Pecce 1997]. The proposed model allows the evaluation of an equivalent plastic hinge length ($lp$), based on the material properties, and on the bending moment distribution. In particular, when the moment variation is linear along the beam the length ($l_p$) can be expressed as:

$$\frac{l_p}{z} = \frac{\phi}{4\tau l}\left(1 - \frac{M_y}{M_u}\right)(\sigma_u^s - f_s^y) \tag{6}$$

where $z$ is the distance between the sections with zero and maximum bending moment, $f^y{}_s$ the yield steel stress and $\sigma^s{}_u$ the steel stress related to the ultimate curvature (equal to $f^u{}_s$ in the case of steel failure). The $\tau l/\emptyset$ parameter, related to the ultimate tensile stress in the concrete, to the section geometry and to the amount of axial forces, can be easily evaluated by means of simple expressions [Grimaldi et al. 1997, Rinaldi, 1998]. The obtained results have been compared with the formulation developed by [Riva e Cohn, 1994], showing a good agreement.[Como et al., 1999].

The ductility differences in the examined frame schemes, can be explained on the base of the plastic yield length. In the case of frames reinforced with more ductile steels the failure mechanism is characterised by plastic hinge length comparable with the section depth. When using rebars with low values of hardening ratio, a brittle collapse occurs due to the failure of one section. In this case the plastic hinge lengths are smaller, with values equal to 1/4 - 1/5 than the one corresponding to more ductile steels (Feb 44).

## 4 Conclusions

The developed analysis and the numerical studies have pointed out and confirmed the relation between the ultimate behaviour of r.c. structural elements and the steel mechanical properties.The basic model is a beam element loaded by axial force and constant bending moment, with length corresponding to the distance between stabilised cracks.

The model allows to point out that at yielding a relevant strain increment in the steel bar can occur, near the crack, and this strain concentration is highly dependent on the hardening steel modulus. In particular, in the limit case of

elastic-perfectly plastic behaviour of the steel, at yielding, strain localization occurs causing rebar crisis and a global brittle collapse.
The steel properties strongly affect the element behaviour, also increasing the applied deformation up to the final failure of the structure.
The influence of the hardening ratio and the ultimate steel strain on the structural ductility has been emphasised by numerical studies on single elements.
Similar analyses have been developed for simple beams and framed models. The obtained results show mainly the influence of the hardening ratio on the rotation capacity or on the global ductility of the structures, and therefore the necessity, particularly in the case of seismic actions, of requiring suitable minimum values of the steel ductility parameters.

## References

1. Baker A.L.L, Amarakone A. M. L. (1964), "Inelastic Hyperstatic Frames Analysis", Proc. of the International Symposium on the Flexural Mechanics of Reinforced Concrete. ASCE-ACI, Miami, pp.85-142.
2. Bazant Z. P., Oh B. H. (1983) "Spacing of cracks in Reinforced Concrete" Journal of Structural Engineering, ASCE, Vol.109, No. 9.
3. Beeby A.W. (1997), "Ductility in reinforced concrete: why is it needed and how is it achieved?" The Structural Engineering Vol. 75/No 18, pp.311-318.
4. CEB Bulletin d'Information n° 242 (1998), "Ductility of Reinforced Concrete Structures".
5. CEB FIP (1991), "Model Code 1990", Bulletin d'Infornation n° 203
6. Commission of the European Community (1993) "Eurocode 2. Common unified rules for concrete structures".
7. Commission of the European Community (1994) "Eurocode 8. Design provision for earthquake resistance of structures". CEN/TC250/SC8, ENV 1998-1-1/2/3, Lisbon.
8. Como M., Grimaldi A., Rinaldi Z. (1999), "Duttilità e resistenza ultima di telai in c.a.: influenza delle caratteristiche degli acciai.", *Proc. VIII Convegno Nazionale ANIDIS "L'Ingegneria Sismica in Italia"*, Torino, Italy.
9. Corley W.G. (1966), "Rotational Capacity of Reinforced Concrete Beams", *Journal of Structural Division, ASCE*, Vol. 92 ST5, pp.121-146.
10. Eligehausen R., Langer P. (1987), "Rotation capacity of plastic hinges and allowable moment redistribution", CEB Bulletin d'Information No.175.
11. ENV 10080 (1995) "Steel for reinforced of concrete-Weldable ribbed reinforcing steel B500 – Technical delivery conditions for bars coils and welded fabric". UNI.
12. Franchi A. Riva P. Ronca P. (1997), "Meccanismi di rottura di armature al piede di pilastri in c.a. soggetti a carichi ciclici", *Atti delle Giornate AICAP 1997*, Roma
13. Gambarova P.G., Iori I., Vallini P. (1998) "Correlazione tra curvatura media e curvatura locale in elementi monodimensionali in conglomerato armato". *Relazione conclusiva gruppo di studio CNR "Rapporto tra curvatura media e locale in elementi in c.a."*.

14. Giuriani E., Ronca P. (1979), "Il metodo di moirè per trasparenza per lo studio di travi inflesse in cemento armato", *Proc. VII Convegno Nazionale A.I.A.S.* pp. 6.55-6.68.
15. Grimaldi A., Rinaldi Z. (1997), "Influenza delle caratteristiche degli acciai sulla duttilità di elementi inflessi in c.a.", *Proc. VII Convegno Nazionale ANIDIS "L'Ingegneria Sismica in Italia"*, Taormina,.
16. Macchi G., Pinto P.E., Sanpaolesi L. (1996), "Ductility requirementes for reinforcement under Eurocodes" *IABSE* No.4.
17. Manfredi G, Pecce M. (1997), "Influenza delle proprietà dell'acciaio sulla duttilità di colonne in c.a." *Proc. VII Convegno Nazionale ANIDIS "L'Ingegneria Sismica in Italia"*, Taormina, Italy.
18. Marti P., Alvarez M., Kaufmann W., Sigrist V. (1998), "Tension chord model for structural concrete" *Structural Engineering International* 4/98; pp. 287-198 IABSE
19. Mattock A.H. (1967), Discussion of "Rotational Capacity of Reinforced Concrete Beams" , by W.G. Corley. *Journal of Structural Division, ASCE*, Vol. 93, ST2, April 1967, pp.399-492.
20. Ngo D., Scordelis A. C. (1967), "Finite element analysis of reinforced concrete beams", *ACI Journal.* Vol. 64(3). Pp. 152-163.
21. Plauk G., Hees G. (1981), "Finite element analysis of reinforced concrete beams with special regard to bond behaviour", *IABSE Colloquium on Advanced Mechanics of Reinforced Concrete.* Delft.
22. Rinaldi Z. (1998), "Duttilità e resistenza di elementi in c.a.: influenza della localizzazione delle deformazione nell'acciaio" PhD Thesis, University of Rome "Tor Vergata".
23. Riva P., Cohn M.Z, (1990), "Engineering Approach to Nonlinear Analysis of Concrete Structures", *Journal of Structural Engineering*, Vol.116, N. 8.
24. Riva P., Cohn M.Z. (1991), "Plastic Rotation Capacity of structural concrete member" *Università degli studi di Brescia – Dipartimento di Ingegneria Civile, Rapporti tecnici.*
25. Sawyer H.A. (1964), "Design of Concrete Frames for Two Failure States", *Proc. of the International Symposium on the Flexural Mechanics of Reinforced Concrete.* ASCE-ACI, Miami, pp.405-431.
26. Grimaldi A., Zila Rinaldi. (2000) "Influence of the steel properties on the ductility of r.c. structures" *Proc. 12 World Conference on Earthquake Engineering*, Auckland, New Zealand, 2000

# Wavelet Analysis of Transient Signals in Civil Engineering

Pierre Argoul, Thien Phu Le

LAMI
Ecole Nationale des Ponts et Chaussées
77455 Marne–la-Vallée, Cedex 2, France

**Abstract.** In this paper, the use of wavelet analysis is examined for the processing of transient signals in civil engineering. These signals generally decay with time; they are frequency modulated and they can be considered asymptotic. After an introduction about the use of wavelet analysis in the processing of transient signals, the procedure for the treatment of asymptotic signals with wavelet transform is shortly presented and the local time and frequency resolutions are given when the mother wavelet is the Cauchy wavelet. The procedure is then applied to two sets of signals obtained from civil engineering experiments. The applications concern the improvement of the impact-echo method and the processing of free oscillations of buildings after non-destructive shock tests.

## 1 Introduction

Since its first definition in the beginning of the 1980s by French researchers, especially Grossmann and Morlet [1], the growth of wavelet research in mathematics has been explosive with significant contributions from numerous authors; and its application in signal and image processing has been successful. Although wavelet analysis is becoming more and more frequently used in engineering applications, it is still far from being systematically used in civil and mechanical engineering, certainly due to the ignorance of the properties of the wavelet transform (WT). Williams and Amaratunga in [2] highlighted the potential benefits of using wavelets for the analysis of engineering data and for finding wavelet solutions of partial differential equations. For example, the two-dimensional wavelet decomposition has been applied to stiffness matrices in order to determine the lowest eigenfrequencies which are generally the most critical for the behavior of a structure during an earthquake. Another use of the WT consists in projecting the governing differential equations on a subspace spanned by a finite number of wavelets. The parameters of the system in the resulting algebraic equations are unknowns; and the inversion of this algebraic problem in order to identify the parameters, is the next step. This procedure is applied to linear time-varying discrete dynamical systems. These systems are usually appropriate for the description of the dynamical behavior of most structural systems that accumulate damage or fatigue under both service loads and environmental excitations [3].

In classical parameter identification techniques, it is generally assumed that there exists some kind of mathematical operator which relates the model parameters to the data. The first step of our research is to look for a "more adapted" representation of the measured data. This "better" representation highlights the maximum amount of information within the data and will make the parameter identification procedure much easier. Our efforts were first focused on the representation of the transient responses of civil engineering structures that can in the case of linear under-damped filters, be modelled by exponentially decaying time signals. For these signals, Tang [6] proposed the use of new harmonic wavelets that provide a high flexibility in the choice of transform bandwith. He emphasized the facility this analysis provides to estimate the decay constant of frequency components for signals within a relatively narrow bandwith. In this paper, the use of the wavelet transform with progressive mother wavelets (the complex-valued Cauchy wavelets) is studied and proposed for processing the time variation of the spectral contents of such signals. The analysis performed is based on the properties of the WT applied to asymptotic signals. Important theoretical and numerical progress has recently been made for the time-frequency analysis of such signals (cf. [4] and [5]).
This article is divided into two parts. The first one is devoted to the presentation of the procedure using wavelet analysis for extracting information within transient signals. In the second part, the procedure is applied to two sets of data. The first set comes from non-destructive impact-echo technique for evaluating thickness or detecting voids within concrete slabs. The second set concerns the accelerometric responses of buildings under non destructive shocks for the assessment of characteristics of their mechanical behavior.

## 2 Time-frequency analysis of transient signals

### 2.1 The wavelet transform with a progressive mother wavelet

We hold the following definition for the wavelet transform (WT) of $u(t)$

$$T_{\psi_n}[u](b,a) = \frac{1}{a}\int_{-\infty}^{+\infty} u(t)\,\overline{\psi}_n\left(\frac{t-b}{a}\right) dt \tag{1}$$

where the mother wavelet $\psi_n$ is the Cauchy wavelet of $n$ order for $n \geq 1$

$$\psi_n(t) = \left(\frac{i}{t+i}\right)^{n+1} = \left[\frac{1}{\sqrt{t^2+1}}\right]^{n+1} e^{i\,(n+1)\mathrm{Arctg}(t)} \tag{2}$$

whose Fourier transform (FT) is : $\widehat{\psi}_n(\omega) = \frac{2\pi\omega^n e^{-\omega}}{n!} H(\omega)$ where $H(\omega)$ is the Heaviside function. Cauchy wavelets are progressive : $\widehat{\psi}_n(\omega) = 0$ for $\omega \leq 0$. The local resolutions of the WT in time $\Delta t$ and in frequency $\Delta\omega$ are stated in terms of Root Mean Square(RMS) value (cf [7])

$$\Delta t = a\Delta t_\psi \qquad \Delta\omega = \frac{\Delta\omega_\psi}{a} \tag{3}$$

where $\Delta t_\psi$ and $\Delta\omega_\psi$ are the RMS duration and the RMS bandwidth of the mother wavelet. For Cauchy wavelets, the time and frequency centres are $t_0 = 0$ and $\omega_0 = n + \frac{1}{2}$. Finally we got

$$\Delta t_\psi = \sqrt{\frac{1}{2n-1}} \qquad \Delta\omega_\psi = \frac{1}{2}\sqrt{2n+1} \tag{4}$$

## 2.2 Wavelet transform of asymptotic signals

For multi-degree of freedom mechanical systems, free oscillations $u(t)$ can be written as a sum of frequency and amplitude modulated components $u_k(t)$

$$u(t) = \sum_k u_k(t) = \sum_k A_k(t)\cos(\alpha_k(t)) \tag{5}$$

It can be shown that under certain conditions, these signals can be considered as asymptotic, which essentially means that the oscillations coming from the phase term are much faster than the variations coming from the amplitude term. Some conditions for free oscillations of a mechanical system to be considered as asymptotic are when the system is slightly damped linear or when the system is weakly non-linear or when the non-linear system has small oscillations around an equilibrium point.
For asymptotic signals, the analytic signal $U(t)$ associated to $u(t)$ can be then approximated in the form

$$U(t) = \sum_k U_k(t) = \sum_k A_k(t)\, e^{i\alpha_k(t)} \tag{6}$$

The main feature of the time-frequency transform of asymptotic signals is that it is concentrated along curves in the time-frequency domain called "ridges" and that the restriction of the wavelet to each ridge called the "skeleton" of the transform contains a maximal information : it is very close to the component of the signal itself (cf [5]). For asymptotic transient signals defined by relation (5), each component $u_k(t)$ within the global signal will produce one ridge, say $a_{rk}(b)$. There are several ways to define a ridge $a_{rk}(b)$. Here, for each value of $b$, we look for the value of $a$ such as $|T_{\psi_n}[u_k(t)](b, a_{rk}(b))| = \max_a |T_{\psi_n}[u_k(t)](b,a)|$. More precisely, we search for the local maxima of the absolute value of the WT and the ridge is defined by

$$a_{rk}(b) = \frac{\omega_{0_n}}{\dot\alpha_k(b)} \tag{7}$$

where $\omega_{0_n}$ is the angular frequency corresponding to the maximum of $\widehat{\psi}_n(\omega)$ ; $\omega_{0_n} = n$ for the Cauchy wavelet. Torrésani [4] also showed that in the case of a progressive wavelet, the value of the wavelet transform of each component $u_k(t)$ of the signal on the associated ridge can be expressed as

$$T_{\psi_n}\left[u_k(t)\right](b, a_{rk}(b)) = \frac{1}{2} A_k(b)\, e^{i\alpha_k(b)} \overline{\widehat{\psi}_n}(\omega_{0_n}) + r(b, a_{rk}(b)) \tag{8}$$

Depending on the properties of the mother wavelet and on the regularity of the component $u_k(t)$, Torrésani gives in [4] some upper limits for the remainder $r(b,a)$. When the instantaneous frequency is constant ($\dot{\alpha}_k(b) = \omega_k$), along the ridge $a_{rk}(b) = \frac{\omega_{0n}}{\omega_k}$ that is constant, the local resolutions $\Delta t_k$ in time and in frequency $\Delta\omega_k$ given in relation (4) become

$$\Delta t_k = \frac{1}{\omega_k}\sqrt{\frac{n^2}{2n-1}} \qquad\qquad \Delta\omega_k = \frac{\omega_k}{2}\sqrt{\frac{2n+1}{n^2}} \tag{9}$$

# 3 Applications

## 3.1 Non destructive evaluation of concrete structures by the impact-echo method

This application initiated by O. Abraham in LCPC Nantes where the experiments are conducted, deals with the non destructive impact-echo technique. This method is based on the analysis in the frequency domain of the transient seismic response of a structure to an impact. It allows to estimate thicknesses and more generally to detect voids and the presence of interfaces between materials with different mechanical impedances. A transient elastic stress wave, generated by a shock on the surface, propagates through the structure as the sum of surface (Rayleigh), compression (P) and shear (S) waves. Within a slab, these waves are first reflected by faults (whose dimensions are in comparison with the wavelength) and by the bottom of the slab. Then, they alternate between the fault, the top and the bottom of the slab. Once the Rayleigh wave has dissipated, the energy of P-wave is predominant compared to that of the S-wave. The impact-echo method characterizes the position of faults existing in the structure by the frequency analysis of the P-wave echoes within the response signal. The medium within the slab is assumed to be homogeneous, isotropic and linear elastic; the thickness $e$ of the slab and the depth $d$ of an eventual void can be then expressed from

$$e = \frac{V_P}{2\,f_e} \qquad\qquad d = \frac{V_P}{2\,f_d} \tag{10}$$

where the resonance frequency $f_e$ (respectively $f_d$) correspond to the travel of the P-wave from the top of the slab to the bottom (respectively the edge of the void) and back again and $V_P$ is the P-wave velocity. Since the mid-1980s, this technique has been applied successfully for the search of faults in concrete. But in some cases, the current method fails or the diagnosis is not straightforward for example when the thickness resonance frequency may be not clearly apparent and it becomes difficult in the spectrum to decide for the thickness resonance frequency. Sansalone et al reduced R-wave contributions to spectrum by clipping technique [8]. Abraham et al [9] proposed to use a moving time window to select the time interval where P-wave dominates. With Abraham, we started to study the use of the wavelet transform for the

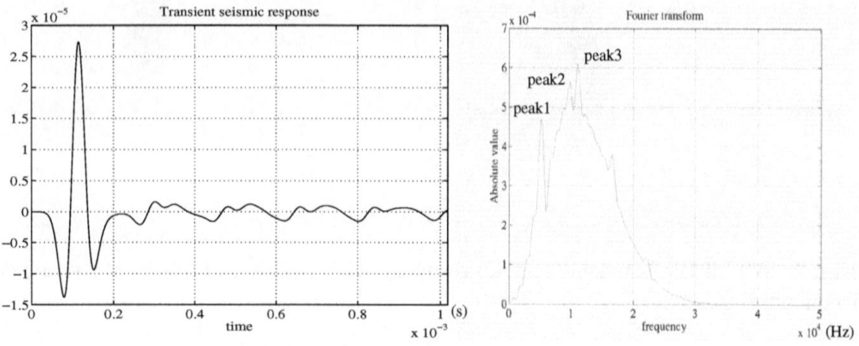


**Fig. 1.** Transient seismic signal and its Fourier transform

processing of such signals and some preliminary results are presented. The processed signal comes from numerical simulations made by Abraham with the finite element software CESAR, the measurement point is located under the impact point. The exact value for the thickness of the concrete slab and the P-wave velocity are $e = 40cm$ and $V_P = 4470m/s$. The lateral dimensions of the slab are assumed to be large relative to the thickness. The seismic signal is plotted in Fig. 1 with the absolute value of its Fourier transform. Three peaks are present in the spectrum with similar amplitudes; with the current approach, it is difficult to choose the 'right" peak. To remedy this problem, the wavelet analysis of this signal is then performed with two different values of $n$ ($n = 30$ and $n = 50$) in order to investigate the effect of local time and frequency resolutions. The ridges are also extracted with the procedure previously described. In Fig. 2, wavelet transforms with local ridges are plotted. According to relation (9), it is obvious that when $n$ is increasing, frequency resolution $\Delta\omega_k$ along a ridge is increasing while time resolution $\Delta t_k$ is decreasing. Moreover for $n$ fixed, relation (9) shows that $\Delta t_k$ is proportional to $\frac{1}{\omega_k}$, so time resolution is "better" for higher frequencies.

The effect of the impact force (energy concentration, starting and ending time), the effect of the R-wave reflection from the lateral edges (similar form to that of impact, starting time after the one of P-wave) and the effect of P-wave echoes (starting time) can be easily seen in Fig. 2.

Thus, for all those reasons, $f = 5270Hz$ (first peak) is chosen to be the thickness resonance frequency. Thickness is then estimated by relation (10) : $\tilde{e} = 42,41(cm)$ with a relative error of $\triangle e = 6.03(\%)$.

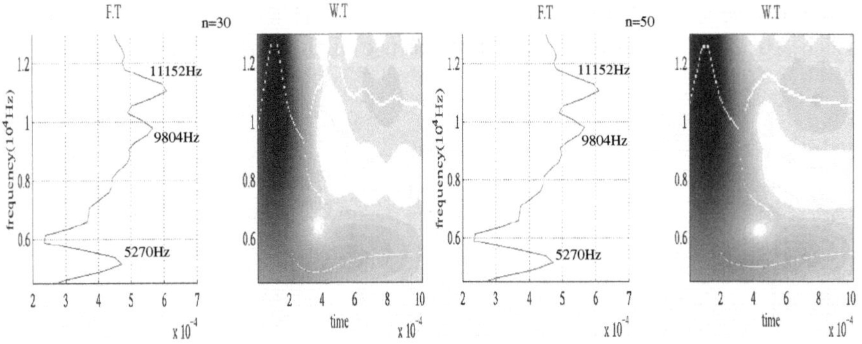


**Fig. 2.** WT of the transient seismic response with position of the local resonance frequencies

### 3.2 Assessment of the mechanical characteristics of buildings submitted to non destructive shocks

This research is made in collaboration with C. Boutin and S. Hans of the “Laboratoire des Géomateriaux” in Lyon (France) where the experiments are conducted. The data comes from a research program on the seismic vulnerability of buildings. The aim of this program is to improve the estimation of the seismic risk in urban environment; it is based on non destructive tests applied to modern buildings built during the period 1955-1975. The responses of several buildings under shock tests have been processed by the continuous wavelet transform. Cauchy wavelet of order $n = 100$ is normally used for numerical computation. Higher order $n = 1000$ has been chosen for buildings exhibing very close eigen frequencies (cf [10]). The building under study is a seven storied one built in 1970 and constituted of concrete shear walls and reinforced concrete floors. It is 30 metres long, 14 metres wide and 22 metres high. Non destructive shocks were applied by a mechanical shovel on the façade of the building. Accelerometers installed at floors 1, 3, 5 and 7 of the building allow the collection of free oscillations of the building after the shock. The level of the measured accelerations is about $10^{-2}g$ involving that the behavior of the structure stays in the quasi-elastic domain. Fourier and wavelet analysis of these signals are performed and the results are plotted in Fig. 3 for the longitudinal response of the seventh floor. From the absolute value of the WT of accelerometric responses, the ridge and the corresponding amplitude and phase term for the two first modes are extracted. For each

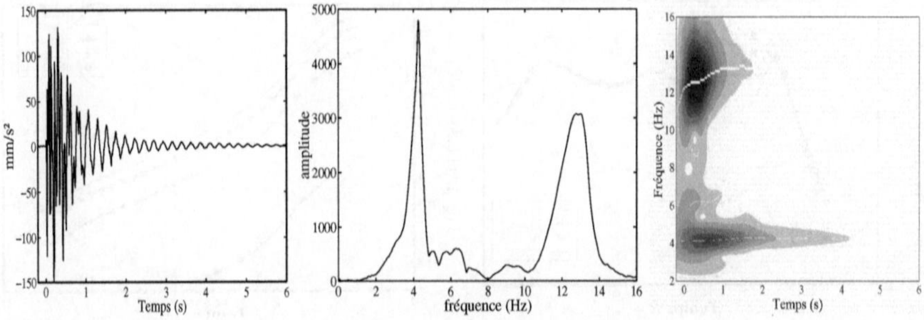


**Fig. 3.** Measured and processed signal successively with Fourier and wavelet analysis

ridge $a_{rkj}(b)$, the ratio $r_{kj}(b)$ of the extracted amplitude of the response at the $j$th floor to that of the response at a reference floor is computed as well as the difference $d_{kj}(b)$ between the phase term of the response at the $j$th floor and that of the response at a reference floor (here the floor 7 is the reference floor). The time evolution of both these parameters for the first mode are plotted in Fig. 4. It is noticed that the shape of the $k$th modal vector can be estimated from $r_{kj}(b)$ obtained at each instrumented floor (cf [10]).
The processing reveals a slight increase of the eigenfrequency just after the shock (when $t \in [1,2]$ sec ). The variation of the square of the eigenfrequency versus roof displacement amplitude exhibited a quasi-linear decrease (cf. [10]). This decrease (also observed by Trifunac in [11] with harmonic tests) has been interpreted as a softening of global rigidity of the structure and soil foundation with the increase of the excitation amplitude. This weakly non-linear behavior is usual for geo-materials such as concrete or sand. Moreover, when the behavior is assumed to be linear with the Basile assumption for the viscous damping, the two parameters $r_{kj}(b)$ and $d_{kj}(b)$ previously defined are constant versus time. We propose the deviation of these parameters from the "linear" behavior as an indicator of the presence of non-linear effects. For the first mode, we notice for lower floor that the time evolution of $d_{11}(b)$ is no longer constant when $t \in [1,2]$ sec, pointing out that the origin of the non-linear effect can be attributed to the non-linear behaviour of the soil-structure interaction.
In conclusion, the use of time-frequency analysis has several advantages for the processing of transient signals in civil engineering such as simplicity of implementation, quality of the obtained information, and allows a better comprehension of the studied phenomenon.

## References

1. Grossmann A., Morlet J. (1984) Decomposition of Hardy functions into square integrable wavelets of constant shape. SIAM J. Math Anal. **15**, 723–736

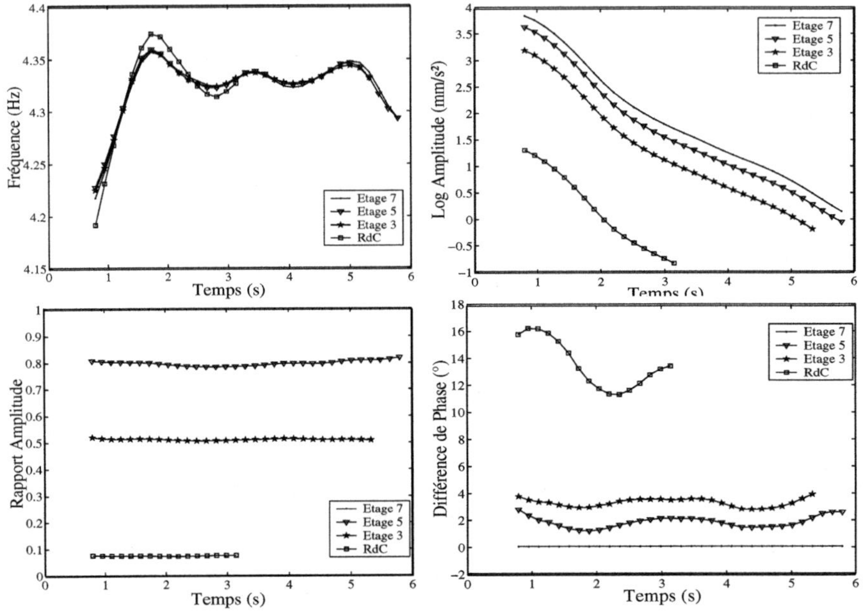


**Fig. 4.** Results of the extraction procedure (ridge, amplitude and phase term associated to the first longitudinal mode).


2. Williams, J. R., Amaratunga, K. (1994) Introduction to wavelets in engineering. Int. J. Num. Methods. in Eng. **37**, 2365–2388
3. Ghanem, R., Romeo, F. (2000) A wavelet-based approach for the identification of linear time-varying dynamical systems. J. Sound and Vib. **234**(4), 555–576
4. Torrésani B. (1995) Analyse Continue Par Ondelettes. InterEditions & CNRS Editions, Paris
5. Carmona R., Hwang W-L., Torrésani B. (1998) Practical Time-Frequency Analysis. Academic Press, San Diego, London
6. Tang S.K. (2000) On the time-frequency analysis of signals that decay exponentially with time. J. Sound and Vib. **324**(2), 241–258
7. Gram-Hansen K., Dorize C. (1992) On the choice of parameters for time-frequency analysis. Proc. "Wavelets and Applications", Marseille, 86–92
8. Sansalone M and Streett W.B. (1997) Impact-echo: Nondestructive Evaluation of Concrete and Masonry. Bullbrier Press, USA
9. Abraham O., Leonard C., Cote and Piwakowski B. (2000) Time frequency analysis of impact-echo signals : numerical modeling and experimental validation. ACI Materials J. **97**(6), 645–657
10. Argoul P., Hans S., Conti F., Boutin C. (2000) Time-frequency analysis of free oscillations of mechanical structures. Application to the identification of the mechanical behavior of buildings under shocks. Proc. COST F3 Conf. "System Identification and Structural Health Monitoring", Madrid, Spain, 283–292
11. Trifunac M.D. (1972) Comparisons between ambient and force vibrations experiments. Earthquake Engin. and Struct. Dyn. **1**, 133-150

# Dynamic Identification of Physical Parameters with Subspace State Models

Bernard Crosnier, François Le Rohellec

Laboratoire de Mécanique et Génie Civil,
cc 048, Université Montpellier II,
F-34095 Montpellier Cedex 5, France

## 1 Introduction

The ability to track the dynamic and physical behaviour of a class of large and flexible structures – typically discrete spatial structures or adaptable structures for which inertia or stiffness may vary significantly - is of great importance. Likewise, the development of new structural concepts and original applications such as tensegrity systems [1] made up of both cables and bars, shell's folding/unfolding and deployable covers involves creating skillfull tools to accurately identify the structure even after marked alterations.
In contrast to traditional approaches, we decided to build the model directly on the basis of the experimental results – the selected control and measurement points – without any preliminary theoretical model. The recent and relatively unknown subspace-state methods, which identify a discrete space state model (being composed of the following quadruple of matrices: $A$, $B$, $C$ and $D$) of the studied system, has proven [2] very efficient for solving such problems.
Besides also identification of classical modal parameters, the state space representation allows relatively easy recognition of physical matrices provided that the number of outputs $l$ tested is the same as the number of identified modes $n$. This ability to solve the inverse problem is of great interest in the control domain, particularly for discrete structures with concentrated masses and separate elements. Few preliminary numerical results have confirmed the principle. Unfortunately, the main theoretical condition, recalled just above ($l$=$n$), is difficult to apply practically.
In order to overcome this initial limitation, several methods have thus been developed which combine two at a time "sub-state space" models (e.g. partial models including only a few inputs/outputs and a few modes) of the same structure. Finally, the result is an enlarged space-state model including all inputs, outputs and modes coming from these different models. Several data records [1] can thus be associated in order to yield an extended dynamic representation of the structure.
*The theoretical background is then briefly summarized, and the paper presents numerical and experimental results acquired with this identification scheme.*

[1] ...collected sucessively at different points.

*First, to initially obtain good identifications, the algorithm - named N4sid$_r$ - derived from subspace methods is used. Secondly, the modal and physical parameters are computed when a sufficiently enlarged space-state model is reached. Physical matrices identified from corrupted data are compared with true matrices. A real variable mass system is also studied to assess the dynamic tracking of variable factors.*

## 2 Theoretical background

In the identification field, frequency domain techniques are often the best known and the most frequently used. Yet many time domain identification methods have been described over the past few years. They are able to identify dynamic parameters and often produce models more accurately.
More recently, a class of subspace-state methods have been developed and shown their superiority in relation to the classical time domain approaches [2]. Instead of classically pre-estimating a large-sized block Hankel matrix made up of Markov parameters, subspace-state algorithms estimate the model directly from the input/output data without requiring such accuracy-consuming steps. Moreover, they make use of numerically stable methods such as RQ decomposition and singular value decomposition (SVD) [3].
The subspace identification algorithm taken into account in this paper is the robust version of N4sid (N4sid$_r$), which stands for "Numerical algorithm for Subspace State Space System Identification" of P. Van Overschee and B. De Moor [4], [5].

### 2.1 State space representation

It provides a discrete time, linear, time-invariant state space model which is mathematically described by the following set of equations[2]:

$$\begin{cases} x_{k+1} = Ax_k + Bu_k + w_k \\ y_k = Cx_k + Du_k + w_k \end{cases} \text{and } E\left[\begin{Bmatrix} w_p \\ \nu_p \end{Bmatrix} \cdot \begin{Bmatrix} w_q^T & v_q^T \end{Bmatrix}\right] = \begin{bmatrix} Q & S \\ S^T & R \end{bmatrix} \cdot \delta_{pq} \tag{1}$$

where:

- $x_k \in \Re^{2n}$ is the state vector of the process which contains the numerical values of $2n$ states [3], $u_k \in \Re^m$ and $v_k \in \Re^l$ are respectively the measurements of the $m$ inputs (forces) and $l$ outputs (usually accelerations).
- $A \in \Re^{2n \times 2n}$ , $B \in \Re^{2n \times m}$ and $C \in \Re^{l \times 2n}$ are respectively the state matrix, the input and the output matrix. $D \in \Re^{l \times m}$ is the direct feedthrough

[2] $E$ is the expected value operator and $\delta_{pq}$ the Kronecker delta.
[3] Consequently, $n$ is the number of modes.

term[4]. $A$ characterizes the complete dynamics of the system: the $n$ frequencies $\omega_i$ and $n$ damping ratios $\varepsilon_i$ are easily determined from its complex eigenvalue pairs $\lambda_i$
$(i = 1, \cdots, 2n)$ with $\lambda_i = \pm(\varepsilon_i \times \omega_i) + i \times \omega_i\sqrt{(1-\varepsilon_i^2)}$.

$w_k$ and $v_k$ are umeasurable vector signals; they are supposedly zero mean, stationary, white noise vector sequences. $Q$, $S$ and $R$ are their associated covariance matrices.

### 2.2 State space identification problem

Given $N$ measurement of the input and output of the unknown real system, the multi input/output identification problem as described by (1) involves estimating:

- the order $n$ of the unknown system.
- the quadruple state space matrices within a similarity transformation $T$, i.e. $(TAT^{-1}, TB, CT^{-1}, D)$.

The key elements of subspace methods are using geometric tools and manipulation of subspaces spanned by rows and columns. Irrespective of an specific subspace algorithm, the identification process is twofold [3]. The first step performs both state sequence $X_s$ and extended observability matrix $\Gamma_s$ directly from the given input/output data, where:

$$X_s = [x_s \quad x_{s+1} \quad . \quad x_{s+r-1}]_{(2n,r)} \qquad \Gamma_s = \begin{bmatrix} C \\ CA \\ CA^2 \\ CA^{s-1} \end{bmatrix}_{(ls,2n)} \tag{2}$$

In a second step, the state space representation up to a similarity transformation is obtained. Distinctions between the different subspace algorithms are ultimately due to different numerical implementations.

### 2.3 Physical parameters identification

In order to subsequently identify $m$, $c$ and $k$ (mass, damping and stiffness matrix respectively ), the main problem involves retrieving the "physical" form of the estimated state space model usually expressed in continuous time $([\,]_c)$ as:

$$A_c = \begin{bmatrix} 0_{n\times n} & I_n \\ -m^{-1}k & -m^{-1}c \end{bmatrix} \quad B_c = \begin{bmatrix} 0_{n\times n} \\ m^{-1} \end{bmatrix} \quad C_c = [\, I_n \; 0_{n\times n}\,] \quad D_c = [0] \tag{3}$$

[4] $D$ is most often 0 in continuous time.

To solve this problem, it has been shown in [6] that there is a unique similarity transformation $P$, defined as $P = \begin{bmatrix} C_c \\ C_c A_c \end{bmatrix}$, that reaches this "physical" space.

Note that $m$, $c$ and $k$ can be determined only if the number of output $l$ equals the number of modes $n$. $P$ must be invertible and then square; a condition that is seldom satisfied under real tests.

### 2.4 Re-association methods

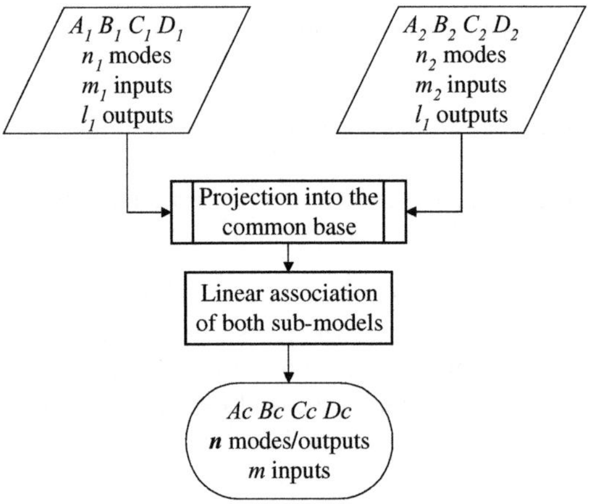


**Fig. 1.** Re-association principle

Methods that combine two at a time sub-state space representations (Figure 1) have been developed in order to increase the number of outputs but also the number of inputs and the number of modes of a reduced model.
A detailed description of three related methods can be found in [7]. The key step lies in the projection properties into a common base [8] of each sub-state space representation allowing subsequently linear association of input, output and modes.

## 3 Simulation results

In this section, a constant 4 degrees of freedom system is considered, defined as:

$$m = diag\left(\left[300\ 220\ 300\ 310\right]\right)$$

$$c = 10^2 \begin{bmatrix} 23 & -15 & 0 & 0 \\ -15 & 26 & 11 & 0 \\ 0 & -11 & 13 & -2 \\ 0 & 0 & -2 & 12 \end{bmatrix} \qquad k = 10^4 \begin{bmatrix} 32 & -16 & 0 & 0 \\ -16 & 36 & -20 & 0 \\ 0 & -20 & 35 & 15 \\ 0 & 0 & -15 & 29 \end{bmatrix} \tag{4}$$

In order to validate the efficiency of the physical parameters autonomous [5] identification principle, we first excite the studied system with 1 white noise input, and 4 outputs are used. Given that the number of modes (4) equals the number of outputs, no combining methods are necessary here.

1200 data samples are used with the sampling frequency set at 30Hz. The number of data block rows $s$ is set at 12. 50 runs are systematically performed and the true system order (namely 4) is specified for each one. The data are then averaged.

For zero noise level, the identification results agree perfectly with the true physical matrices. For 1% and 10% noise level, the Figure 2 shows the bands covered by the evaluated relative estimation errors [6] when the four inputs are selected successively.

Examination of these results highlights the fact that the stiffness matrix is the easiest to identify with the mass matrix. On the other hand, increasing the noise level is detrimental to the overall consistency; the damping matrix remains the most sensitive.

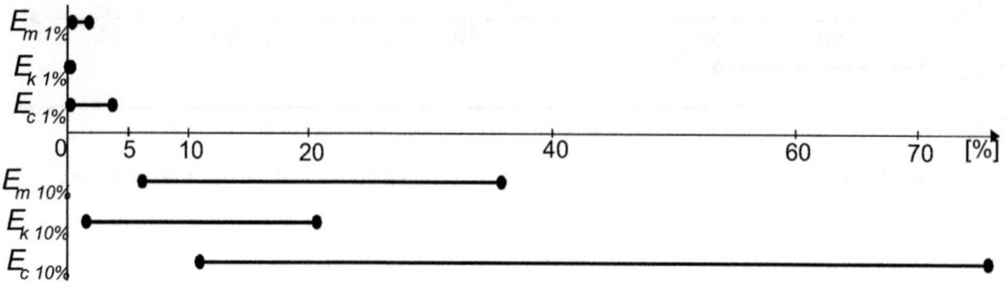


**Fig. 2.** Relative estimation errors of $m$,$c$ & $k$ versus selected input

Finally, Table 1 gives the estimated matrices for 10% noise and demonstrates a good identification process.

[5] Nothing of $m$, $c$ or $k$ is known.

[6] $E_m = 100\frac{|m-\hat{m}|}{|m|}$ where $\hat{m}$ is the true mass matrix. Alike for $c$ and $k$.

| $m = diag([302.56 \quad 222.97 \quad 317.42 \quad 290.85])$ | |
|---|---|
| $c =$ | $k =$ |
| $10^2 \begin{bmatrix} 22.81 & -14.738 & -0.162 & 0.05 \\ -14.373 & 25.223 & -10.679 & 0.107 \\ -0.072 & -11.094 & 13.054 & -2.053 \\ 0.313 & -0.48 & -1.782 & 12.273 \end{bmatrix}$ | $10^4 \begin{bmatrix} 32.026 & -15.94 & -0.106 & -0.071 \\ -15.94 & 35.784 & -19.988 & -0.109 \\ 0.036 & -19.988 & 34.911 & -14.947 \\ 0.255 & -0.444 & -14.947 & 29.443 \end{bmatrix}$ |

**Table 1.** Estimated matrices – 10% noise level

In a second step, with the same 4 d.o.f. system and the same numerical conditions, the capacities of the re-association methods are evaluated. Many inputs/outputs sequences are tested (with at least here one input and two outputs per identification) with the aim of producing an enlarge state-space representation made of 4 outputs. In order to identify the physical matrices, we assume the mass matrix $m$ is entirely known.

For zero noise level, the results are perfect and are not shown. For 1% and 10% noise level, Figure 3 summarizes the evaluated level errors. Comparatively, they are almost equivalent to previous simulations performed without sub-models association highlighting the efficiency of such combining techniques. But lower results can be expected if $m$ is autonomously identified rather than assumed to be known.

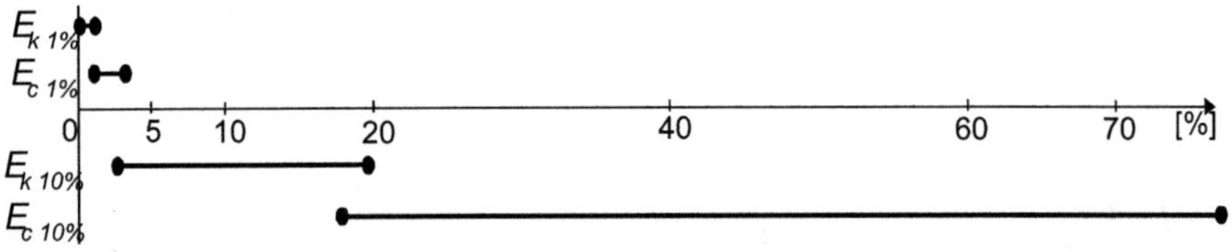


**Fig. 3.** Relative estimation errors of $c$ & $k$ with sub-models association

## 4 Experimental results

This section describes the results of tests which were conducted on a simple cantilevered beam. This structure was excited consecutively at different locations using only one random noise input. 2 or 4 accelerometers were used for measuring the dynamic response. Two analyses were carried out on this structure.

↪ The first experimental sequence involved evaluating the capacity of the whole identification scheme to track noticeable alterations in the physique of the system. We chose to modify the mass distribution of an equivalent

2 degrees of freedom structure by successively overloading each node with a 1.2 kg mass. The analysis was performed on a 0-100 Hz frequency range. The sampling frequency was set at 1000 Hz and 2000 data points were collected during the test. The number of data block rows $s$ was preset at 10 to effectively identify 10 modes (one input only). Nevertheless, after the reduction process (unstable and out of range modes are removed), only 2 modes were clearly selected.

Inspection of the identified mass matrices (Table 2) clearly highlighted the capacity to track weight variations despite a slight overestimation probably due to the comparatively high mass of the electromagnetic exciter.

| no overloading | 1.2 kg on node 1 | 1.2 kg on node 2 |
|---|---|---|
| $\begin{bmatrix} 3.80 & . \\ . & 1.25 \end{bmatrix}$ | $\begin{bmatrix} 5.94 & . \\ . & 1.51 \end{bmatrix}$ | $\begin{bmatrix} 4.28 & . \\ . & 2.85 \end{bmatrix}$ |

**Table 2.** Identified mass matrices under test

As for the stiffness matrix, its different estimations were logically stable (less than 5.5% variation). On the other hand, the damping matrix was more affected; its high sensitivity has already been shown with numerical simulations.

| | no overloading | 1.2 kg on node 1 | 1.2 kg on node 2 |
|---|---|---|---|
| $k$ | $10^5 \begin{bmatrix} 6.91 & -2.62 \\ -2.62 & 1.13 \end{bmatrix}$ | $10^5 \begin{bmatrix} 7.22 & -2.73 \\ -2.73 & 1.19 \end{bmatrix}$ | $10^5 \begin{bmatrix} 6.59 & -2.49 \\ -2.49 & 1.08 \end{bmatrix}$ |
| $c$ | $\begin{bmatrix} 72.27 & 166.45 \\ -137.43 & -24.06 \end{bmatrix}$ | $\begin{bmatrix} 113.79 & 183.19 \\ -153.82 & -26.17 \end{bmatrix}$ | $\begin{bmatrix} 126.63 & 316.60 \\ -147.73 & -77.26 \end{bmatrix}$ |

**Table 3.** Identified stiffness and damping matrices under test

$\hookrightarrow$ The second field tests were aimed at experimentally validating the capacities of the re-association methods. In order to obtain an equivalent 4 d.o.f. model having 4 outputs, each sub-identification involved 3 measurement outputs and one input (either at the middle or end of the beam). 2000 data points were gathered on a 0-500 Hz frequency range with a 2000 Hz sampling frequency (just 1 second!). The same four modes were systematically selected among 10 identified ($s = 10$), thus avoiding the need to combine modes during these tests. Simultaneously, direct identification (without any

sub-models association) using whole outputs was performed to obtain a comparative base.
Regardless of the test, all physical matrices were entirely identified without any preliminary knowledge of them. As shown in Figure 4, the reconstructed outputs calculated with the combined state space model or with the direct identification were very close to the real data. These traces illustrate the high efficiency of the N4sid$_r$ algorithm and the low effects of such re-association processes on each other. Good subsequent autonomous identifications of the physical matrices could thus be expected. The Table 4 sums up the values of $m$, $c$ and $k$ obtained by the direct or the combining method respectively.
The comparison gives similar results. However, note that parameters that are supposed to be zero in the stiffness and damping matrices were not correctly identified.

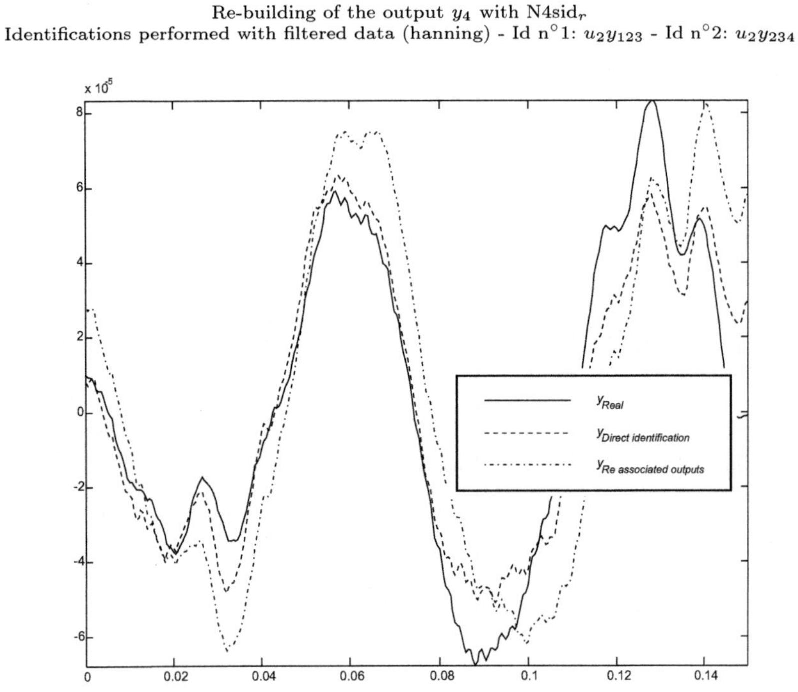


**Fig. 4.** Comparison between reconstructed output and real output

Direct identification

| $m$ | $k \times 10^7$ | $c \times 10^3$ |
|---|---|---|
| $\begin{bmatrix} 3.56 & & & \\ & 4.60 & & \\ & & 3.19 & \\ & & & 0.90 \end{bmatrix}$ | $\begin{bmatrix} 1.12 & -0.92 & 0.35 & -0.08 \\ -0.92 & 1.31 & -0.79 & 0.24 \\ 0.35 & -0.79 & 0.71 & -0.27 \\ -0.11 & 0.27 & -0.27 & 0.11 \end{bmatrix}$ | $\begin{bmatrix} 0.03 & 1.09 & -0.45 & 0.45 \\ -1.04 & -0.34 & 0.07 & -0.89 \\ 0.85 & -0.53 & 0.67 & 0.61 \\ -0.43 & 0.50 & -0.53 & -0.11 \end{bmatrix}$ |

Identification with combining method

| $m$ | $k \times 10^7$ | $c \times 10^3$ |
|---|---|---|
| $\begin{bmatrix} 3.69 & & & \\ & 4.77 & & \\ & & 3.30 & \\ & & & 0.93 \end{bmatrix}$ | $\begin{bmatrix} 1.15 & -0.95 & 0.36 & -0.09 \\ -0.95 & 1.35 & -0.81 & 0.24 \\ 0.36 & -0.81 & 0.73 & -0.28 \\ -0.11 & 0.28 & -0.28 & 0.12 \end{bmatrix}$ | $\begin{bmatrix} 0.10 & 1.34 & -0.01 & 0.71 \\ -1.52 & -0.57 & -0.42 & -1.18 \\ 0.93 & -0.42 & 0.98 & 0.78 \\ -0.46 & 0.46 & -0.65 & -0.17 \end{bmatrix}$ |

**Table 4.** Identified physical matrices with and without re-associated outputs

## 5 Conclusion

In order to perform high level dynamic and autonomous identification of physical parameters of structures, an algorithm from the class of subspace state methods was successfully evaluated. Subsequently, the principles of original methods which are able to combine several sub-state space models of the same system have been succinctly outlined and numerically validated.

Then these different identification processes were tested on a simple real structure and showed their high robustness, particularly concerning mass and stiffness matrices.
These encouraging results allow to extend this identification scheme to larger structures.

## References

1. R. Motro, "Formes et Forces dans les Systèmes Constructifs, Cas des Systèmes Réticulés Spatiaux Autocontraints", Tome I et II, Thèse, Université Montpellier II, 2 Juin 1983.
2. M. Abdelghani, "Identification Temporelle des Structures: Approche des Algorithmes Sous-Espace dans l'Espace Etat", Thèse [95MON2185], Université Montpellier II, 5 December 1995.
3. P. Van Overschee, "Subspace Identification, Theory - Implementation - Application", Thèse [UDC: 519.71], Katholieke Universiteit Leuven, February 1995.
4. P. Van Overschee and B. De Moor, "N4SID: Subspace Algorithms for the Identification of Combined Deterministic-Stochastic Systems",Automatica, Vol.30, N°1, 1994, 75-93.

5. P. Van Overschee and B. De Moor, "N4SID: Subspace Algorithms for the Stochastic Identification Problem",$30^{th}$ IEEE Conference on Decision and Control, Brighton, UK, 1991, 1321-1326.
6. C.-D. Yang and F.-B. Yeh, "Identification, Reduction and Refinement of Model Parameters by the Eigensystem Realization Algorithm", Journal of Guidance, Vol.13, N°6, November-December 1990, 1051-1059.
7. F. Le Rohellec, "Identification des Matrices Physiques dans l'Espace Etat - Recombinaison de Sous-Représentations", Mémoire de Thèse, Université Montpellier II, 10 juillet 1998.
8. P. Borne, G. Dauphin-Tanguy, J.P. Richard, F. Rotella et I. Zambettakis, "Modélisation et Identification des Processus - tome 1", Méthodes et pratiques de l'ingénieur, Editions Technip, 1992.

# Application of the Wavelet-element Method to Linear Random Materials

Frédéric Lebon[1], Serge Dumont[2]

[1] Laboratoire de Mécanique et Génie Civil,
cc 048, Université Montpellier II,
F-34095 Montpellier Cedex 5, France.
[2] Faculté de Mathématiques et d'Informatique,
Université de Picardie,
F-80039 Amiens, France

**Abstract.** This paper is devoted to the statistical study of the effective linear properties of random materials, that is to say microstructures are random lattices given by a stochastic process. The local numerical procedure associated to homogenization techniques is based on a wavelet-element method. The numerical results are compared with classical theories. A new approach is obtained in order to determine these effective properties if the details of the microstructure are not well-known.

## 1 Introduction

The purpose of this paper is to determine and to compute the effective properties of linear random heterogeneous media. While natural geomaterials are heterogeneous, they are regarded as homogeneous with effective properties which manipulate the overall behavior. Grounds, for example, are usually modeled as a body consisting of homogeneous materials with such effective properties. Behavior of these grounds cannot be predicted if the effective properties are poorly determined. The necessity to incorporate more detailled informations on the microstructure is clearly recognized. In the last years, a large part of literature has presented measurements, empirical and theoretical techniques in order to determine these effective properties. Clearly, the numerical cost to attack this problem directly for a given material sample is prohibitive. In former papers, the authors have presented numerical scheme associated to homogenization techniques in order to take into account the details of the microstructure [4–6,8,11]. These algorithms are based on wavelet basis and permit to predict the effective properties from the knowledge of the microstructure. Thus, it is possible to optimally design linear composite materials from real images of the sample.
In most cases, unfortunately, the details of the microstructure are not absolutely known, and leads us to estimate the effective properties from statistical studies. In a famous paper [14], Torquato has presented results concerned with the case where a partial statistical information, as correlation functions, is given. Improved bounds on effectives properties of two-phase

random heterogeneous media are obtained. In this paper, we suppose that only the volume fraction of each phase is known, and we intend to find the averaged effective properties of the composite. In addition, we suppose that the microstructure is given pixel by pixel i.e. the characteristic shape of the microstructure components is a small square and its characteristic lenght is the lenght of the pixel.
In section 1, we present the notations and the mechanical problem. Section 2 is devoted to the numerical algorithm. In section 3, we present a statistical study for a AlSiC composite. The results obtained are compared with Hashin and Shtrickman bounds [9] and with the self-consistent schema [10,13]. At the end of the paper, perspectives to this work are presented.

## 2 The mechanical problem

We consider a two phase isotropic elastic composite and we intend to study the behaviour of this heterogeneous media. A step of homogenization consists in determining the effective properties which characterize the equivalent homogeneous medium. Let us consider a plane periodic composite $\Omega$ (macro-scale, variable $x$) and a rectangular periodicity cell $Y$ (micro-scale, variable $y$). We assume that the bonds at all interfaces of the composite are perfect. The elasticity coefficients $C^{\varepsilon}_{ijkl}$ are supposed to be periodic on the period $Y$. In elastostatics, the problem $(P_{\varepsilon})$ is written

*Problem*$(P_{\varepsilon})$
Find $u^{\varepsilon} \in V$ such that $a_{\varepsilon}(u^{\varepsilon}, v) = L(v), \forall \text{ v} \in V$,

where $a_{\varepsilon}(u, v) = \int\limits_{\Omega} C^{\varepsilon}_{ijkl} e_{kl}(u) e_{ij}(v) dx$,

$e_{kl}(u) = \frac{1}{2}(u_{k,l} + u_{l,k})$,

$L(v) = \int\limits_{\Omega} fv dx + \int\limits_{\partial_1 \Omega} F v dl$

and $V = \{v \in (H^1(\Omega))^2, v = 0 \text{ } on \text{ } \partial_2 \Omega\}$.

The solution of the previous problem $u^{\varepsilon}$ tends to $u$ in $V$, the space of admissible global displacements, u is the solution of problem $P$ [1]

*Problem*$(P)$
Find $u \in V$ such that $a(u, v) = L(v), \forall \text{ v} \in V$,

where a(u,v) $= \int\limits_{\Omega} A_{ijkl} e_{kl}(u) e_{ij}(v) dx$

and $A_{ijkl} = \frac{1}{meas(Y)} \int\limits_{Y} (C_{ijkl} + C_{ijpq} e_{pq}(u^{kl})) dy$.

The determination of the displacements $u^{ij}$ needs to solve three local problems on Y (Problem $P_Y$) in the space $H = (H^1_{per}(Y))^2$, the set of admissible local displacements:

$Problem(P_Y)$
$E^{ij}$ be given, find $u^{ij} \in H$ such that $a_Y(u^{ij}, v) = l(v) \; \forall v \in H$,

where $a_Y(u,v) = \int\limits_Y C_{ijkl}(y) e_{ij}(u) e_{kl}(v) dy$

and $l(v) = -\int\limits_Y C_{ijkl}(y) E^{ij} e_{kl}(v) dy.$

## 3 Numerical procedure

In this section, we focus on the solution of problem $P_Y$. This problem is solved by a wavelet-element method [15] associated to Daubechies wavelets [3]. This method is based on the standard form of the discretized operator [2]. Details concerning this method can be found in previous papers of the authors [5,8,?]. Let $\Lambda_j = [0, 2^j - 1]$, the local displacement $u = (u_1, u_2)$ is approached in the space $V_{j_{max}}$ as

$$u_i(y) = \sum_{k \in \Lambda^2_{j_0}} u^i_{j_0 k} \Psi^0_{j_0 k}(y) + \sum_{l=1}^{l=3} \sum_{j=j_0}^{j=j_{max}} \sum_{k \in \Lambda^2_j} u^i_{jk} \Psi^l_{jk}(y),$$

$$\Psi^0_{jk}(y) = \phi_{jk_1}(y_1) \phi^l_{jk_2}(y_2), \Psi^1_{jk}(y) = \psi_{jk_1}(y_1) \psi^l_{jk_2}(y_2),$$

$$\Psi^2_{jk}(y) = \psi_{jk_1}(y_1) \phi^l_{jk_2}(y_2), \Psi^1_{jk}(y) = \phi_{jk_1}(y_1) \phi^l_{jk_2}(y_2),$$

where $\phi_{jk}(x) = 2^{j/2} \phi(2^j x - l)$ and $\psi_{jk}(x) = 2^{j/2} \psi(2^j x - l)$. $\phi$ is the scale function and $\psi$ is the associated wavelet. Problem $(P_Y)$ is discretized using this approximation, well-conditioned stiffness matrices are obtained [15,2].

## 4 Numerical results

In this part of the paper, numerical tests are presented. They are concerned by a AlSiC composite. The influence of two parameters is studied: the volume fraction of each component (Aluminium and Silicium) and the distribution of each material in the microstructure. For each volume fraction, we proceed to about 5000 draws (random lattices) and we study from a statistical point of view the results obtained. In the following the fourth order tensor $A$ is represented as a matrix of order 3. For a volume fraction of 70 % the variations of each elasticity coefficient are presented in Fig. 1 for about 5000 microstructures. At first sight, the fluctuations seem very large. In the other hand, one

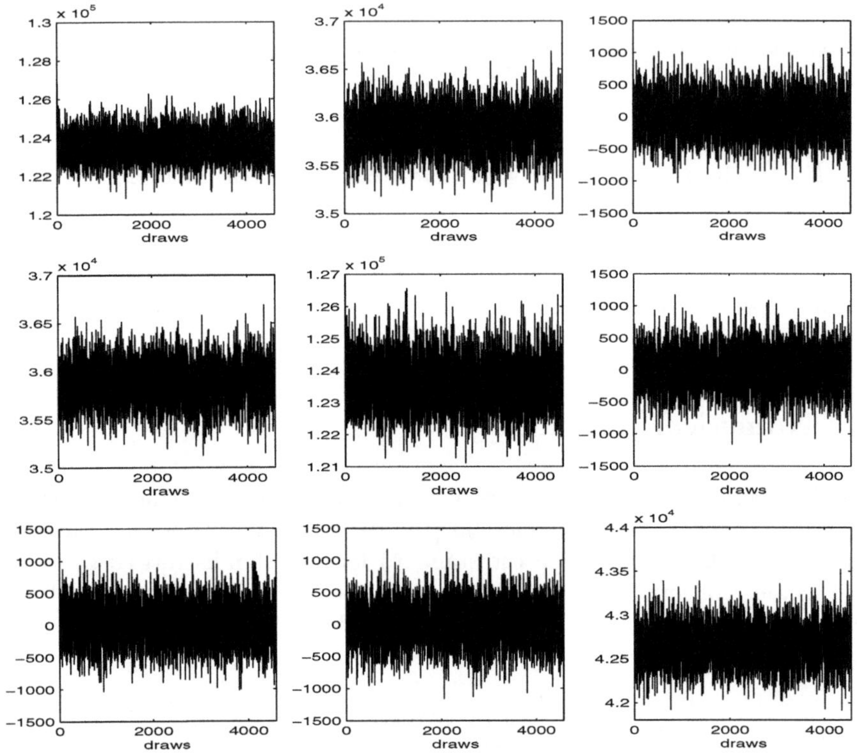


**Fig. 1.** Homogenized plane elasticity tensor (MPa) versus draws (volume fraction 70 %), line n and column m correspond to $A_{nm}$

can present the histogram of each component of the elasticity tensor. We observe a Gaussian distribution of the values (Fig. 2).

The averaged values of each coefficient of the elasticity tensor are given in Fig. 3. We observe the convergence of the coefficients to an mean value along the draws. The values obtained in Fig. 3 show that the numerical limit is an **isotropic material**. In particular, we have $A_{13} = A_{23} = 0$ and $A_{11} = A_{22}$. Thus, it is possible to compute the bulk and shear moduli of the limit material (Averaged bulk and shear moduli).

In the following we denote by L-HS and U-HS the Lower and Upper Hashin and Shtrickman bounds [9]. We denote by WAV the Averaged Value obtain by the Wavelet-element method. Fig. 6 shows that the relative behavior of the value is splitted in three phases: in the first one, for volume fraction lower than 0.25, the LHS and the WAV values are very closed. The second phase corresponds to volume fractions between 0.25 and 0.95. In this phase, the WAV value is intermediate between the two HS bounds. The third phase, for a volume fraction larger than 0.95, shows WAV values closed to the U-HS bound.

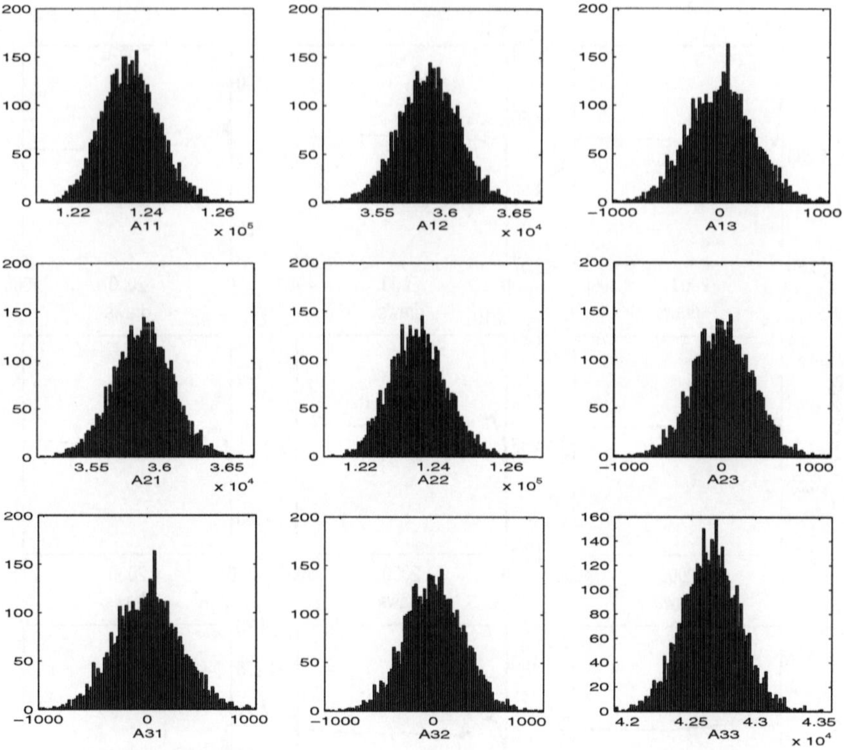


**Fig. 2.** Histogram of plane elasticity tensor (volume fraction 70 %)

We compare the results with the self-consistant schema for the shear modulus and for volume fractions lower than 0.5. We observe in Fig. 7 that for the lower value of the volume fraction (lower than 0.1) the L-HS value, the self-consistant schema (denoted SCS) and the WAV value are quite similar. For larger values from 0.1 to 0.25 the values L-HS and WAV are equal and the SCS value is lower than the two other one. For volume fractions larger than 0.25 and lower than 0.4, the L-HS and SCS values are closed , and lower than the WAV value.

## 5 Concluding remarks and perspectives

Generally, it is impossible to determine exactly the effective properties of random heterogeneous media. In the literature, rigorous statements about these effective properties take the form of bounds. In this paper, we obtain another kind of bounds as the form of an average and a standard deviation. An important result is the numerical convergence of the stochastic process of homogenization to an averaged solution, that is to say that the limit, when the number of draws tends to infinity, of the potential energy corresponding

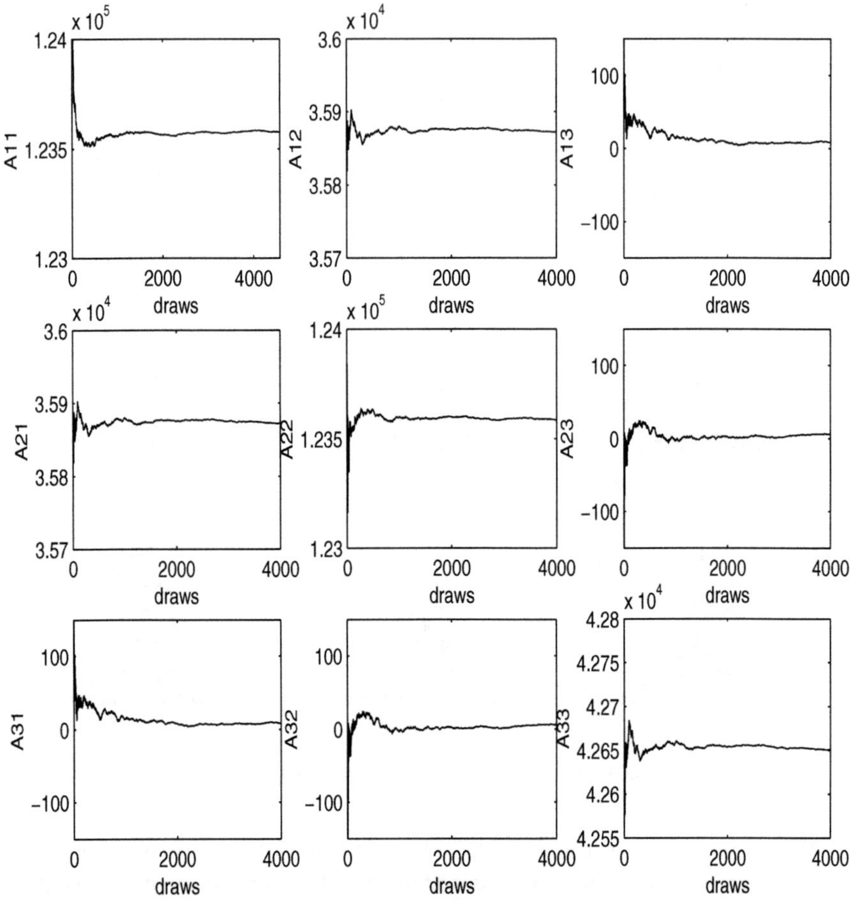


**Fig. 3.** Averaged homogenized elasticity tensor versus draws (volume fraction 70 %)

to a 'random lattice' is the potential energy of an isotropic material. This material has effective properties depending on the volume fraction of each component. For lower volume fraction, the material has properties closed to the lower Hashin and Shtrickman bounds and to the self-consistant schema. A 'numerical theorem' is obtained. Similar results are obtained for different kinds of materials (Resin-Glass mixture for example).
Two ways to this work are proposed : the first one is a study of the influence of the shape of the lattices. We propose to study the same problem with different morphologies : n-pixels in one direction, association of n-pixels in a neibourough of a point, ... Another perspective is to study the effective properties for more complex behavior: piezoelectricity couplings, thermome-

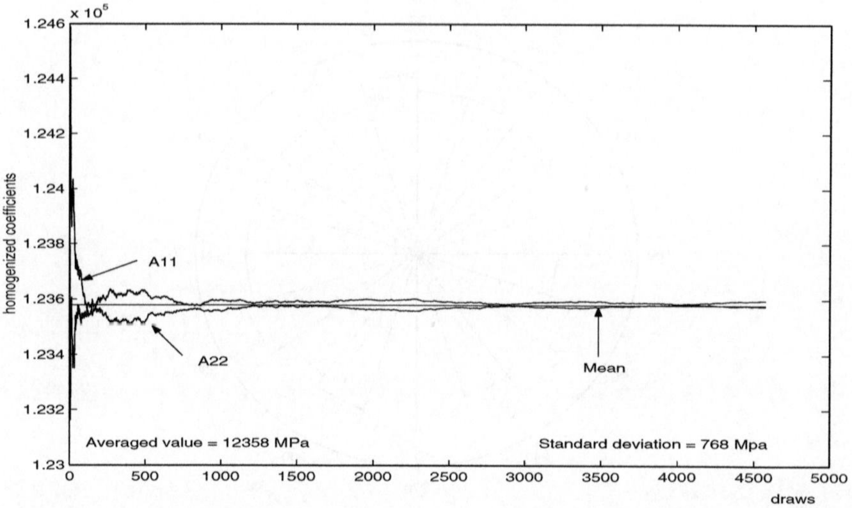


**Fig. 4.** Averaged homogenized elasticity coefficients A11 and A22 (volume fraction 70 %). If we study in details the variations of the averaged coefficients, we observe that the mean is obtained for a small number of draws and that the standard deviation is small compared to the coefficients. In the case presented in Fig. 4, i.e. for a volume fraction equal to 70 %, the mean is equal to 12358 Mpa and the standard deviation to 768 Mpa (± %). The values are included in the interval [12083,12682].

chanical couplings, plasticity [12], ... The efficiency of our method permits us to envisage to treat this kind of problems.

## Acknowledgments

The authors gratefully acknowledge R. Peyroux for helpful discussions, lighting comments and self-consistant schema software.

## References

1. Bensoussan A., Lions J.L., Papanicolaou G. (1978) Asymptotic analysis for periodic structures. North-Holland, 1st edition, Amsterdam
2. Beylkin G. (1992) On the representation of operators in bases of compactly supported wavelets. SIAM J. Num. Anal. **29**, 1716–1740
3. Daubechies I. (1988) Orthonormal bases of compactly supported wavelets. Com. Pure Appl. Math. **41**, 909–998
4. Dumont S., Lebon F. (1996) Representation of plane elastostatics operators in Daubechies wavelets. Comp. and Struct. **60**, 561–569
5. Dumont S., Lebon F. (1996) Wavelet-Galerkin method for heterogeneous media. Comp. and Struct. **61**, 55–65

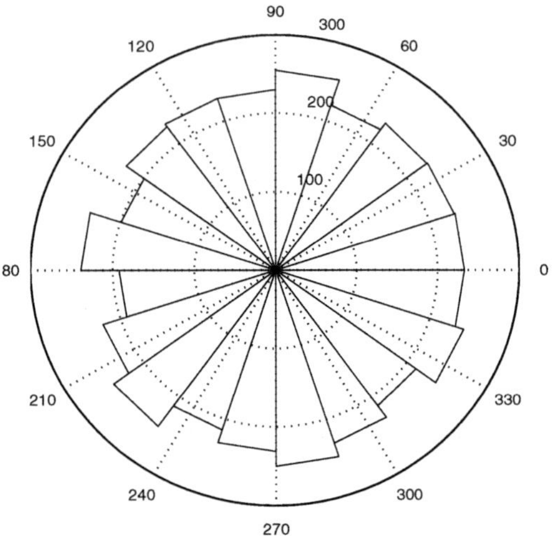


**Fig. 5.** Histogram in polar coordinate system (volume fraction 70 %. The histogram in polar coordinate system shows the good distribution of the results according to the average value.)

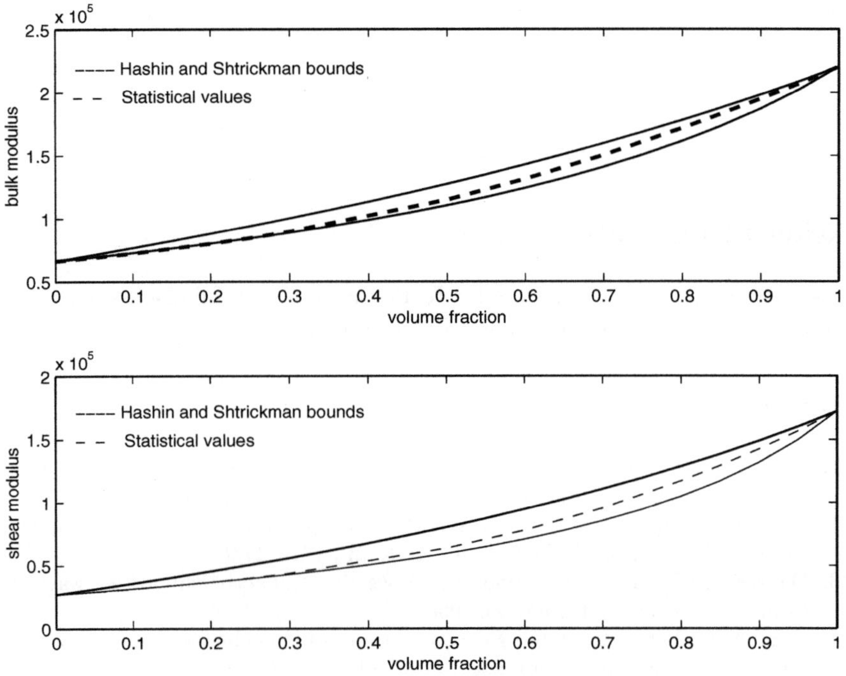


**Fig. 6.** Bulk and shear moduli and Hashin-Shtrickmann bounds

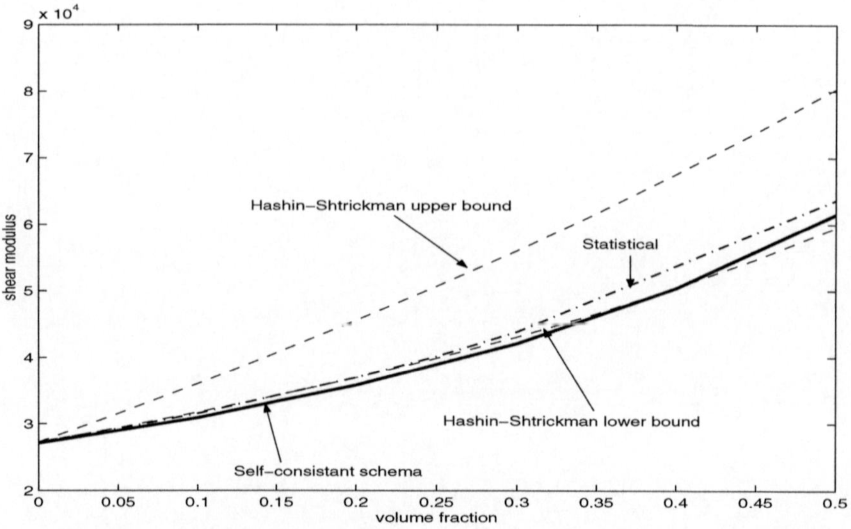


**Fig. 7.** Shear modulus, Hashin-Shtrickmann bounds and self-consistant schema


6. Dumont S., Lebon F. (1999) Wavelet-galerkin method for plane elasticity. Comput. and Appl. Math. **8**, 127–142
7. Dumont S., Lebon F., Nguyen T.T.V. (2001) Wavelet's and PDE's: on two applications in solids mechanics. Ciencas Matematicas **19**, 21–30
8. Dumont S., Lebon F., Ould Khaoua A. (2000) A numerical tool for periodic heterogeneous media. Application to interface in Al/Sic composites. J. of Appl. Mech. **67**, 214–217
9. Hashin Z., Strickman S. (1963) A variational approach to the theory of the elastic behavior of multiphase materials. J. of Mech. Phys. Solids **11**, 127–140
10. Hill R. (1965) A self-consistant mechanics of composite materials. J. of Mech. Phys. Solids **13**, 213–222
11. Lebon F. (2000) Small parameter, homogenization and wavelets. Ciencas Matematicas **18**, 11–30
12. Lebon F., Rodriguez R., Mesejo A. (2002) Homogenization and wavelet-Galerkin method for a nonlinear one-dimensional problem. Int. J. of Engn. Sci. (to appear)
13. Peyroux R. (1990) Caractéristiques thermoélastiques de matériaux composites à fibres courtes, Thesis, Université Montpellier 2
14. Torquato S. (1991) Random heterogeneous media: Microstructure and improved bounds on effective properties. Appl. Mech. Rev., **44**, 37–76
15. Wells R.O., Zhou X. Wavelet solutions for the Dirichlet problem. Numerische Mathematik **70**, 379–396

# Towards Tensegrity Systems Design

Mohammed Nadjib Kazi Aoual, Jérôme Quirant, René Motro

Laboratoire de Mécanique et Génie Civil,
cc 048, Université Montpellier II,
F-34095 Montpellier Cedex 5, France

## Abstract

Tensegrity systems are innovative systems in the field of Civil Engineering. Architects have been spurred on by their light weight and their apparent transparency. "A tensegrity system is system in a stable self equilibrate state comprising discontinuous compressed components inside a continuum set of tensioned components".

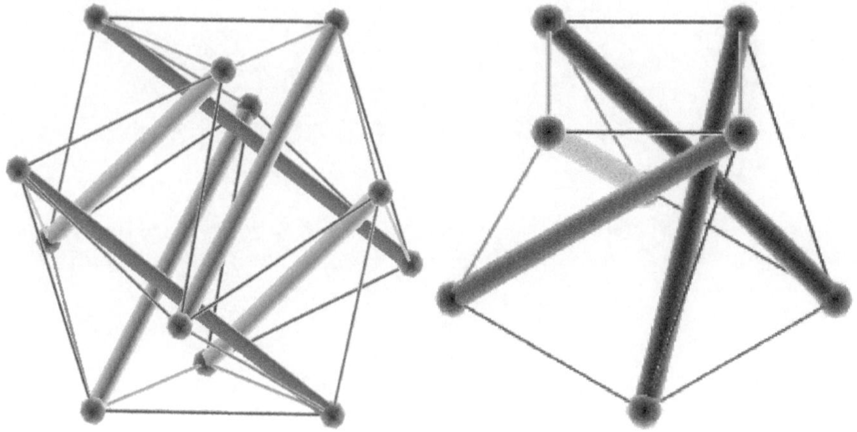

**Fig. 1.** Two tensegrity systems : Expanded octahedron and half-cuboctahedron

This communication leads towards tensegrity systems design. As a first step, behaviour study shows that more than classical parameters design (mechanical characteristics and cross section area), specific ones exist and should be taken in to account (selfstress level, rigidities ratio and selfstress distribution).

Seeing that tensegrity systems are generally steel structures, the design code is the EUROCODE 3 in which the selfstress loads are not taken in account.

A selfstress load can be considered as a dead one which can be resistant load because the rigidity of tensegrity systems is closely dependant on the selfstress level and as an acting load because each element stability is related to the selfstress level.

Consequently, two partial security coefficients should be affected to the selfstress value on load combinations :

$$\gamma_{M,S} = 1.2 \text{ for the acting selfstress case}$$
$$\gamma_{M,S} = 0.8 \text{ for the resistant selfstress case}$$

All these considerations lead to a design procedure conform to EUROCODES. This design procedure is applied to a double layer grid which is a result of an assembly of 36 half-cuboctahedrons.

# Density Methods Applied to Form Finding of Initially Stressed Systems

René Motro, Bernard Maurin and Nicolas Vassart

Laboratoire de Mécanique et Génie Civil,
cc 048, Université Montpellier II,
F-34095 Montpellier Cedex 5, France

## 1 Introduction

Some contemporary structures like cable nets, membranes and tensegrity systems are initially stressed systems and require form finding processes. The final shape has to fulfil equilibrium conditions. Corresponding relations are not linear and initially a so-called "forcedensity method" has been developed by Linkwitz and Scheck (1971) in order to linearise the problem. This paper describes its utilisation, and an innovative form-finding process the Surface Stress Density Method, based upon the use of isotropic stress tensors, which allows the calculation of broad range of structures and appears as an extension of the existing Force Density Method. Several illustrative examples point out the efficiency of the procedures and also their adaptation to the requirements of either architects or engineers.

## 2 Force density method

This method is used when the system is modelled by an assembly of straight links: cable nets, membranes in case of assimilation to cable nets and tensegrity systems.

### 2.1 Equilibrium relation

**Notations** The following notations and definitions are used:

$n$, number of nodes, $b$ number of links

$l_j^{lib}$ , manufacture length of element "$j$", "free" length

$l_j^0$ , Length of element "$j$" in reference configuration (assembled and not loaded)

$l_j$ , Length of element "$j$" in current configuration (assembled and loaded)

Vectors of node coordinates (global reference system)

$\{x\} = \{x_1, ..., x_i, ..., x_n\}^t$, $\{y\} = \{y_1, ..., y_i, ..., y_n\}^t$, $\{z\} = \{z_1, ..., z_i, ..., z_n\}^t$

Vector of external applied actions $\{f\} = \{..., f_{ix}, f_{iy}, f_{iz}, ...\}^t$
External action on node "$i$", $\{f_i\} = \{f_{ix}, f_{iy}, f_{iz}\}^t$
Force density coefficient $q_j = \frac{T_j}{l_j^0}$ with $T_j$, internal effort in element "$j$" , $T_j > 0$, traction, $T_j < 0$, compression.

**Equilibrium equation** Equilibrium equation for node "$i$" with force density coefficient is given by:

$$\begin{aligned} &\sum_{j \ni i} (x_i - x_h) \cdot q_j = f_{ix} \\ &\sum_{j \ni i} (y_i - y_h) . q_j = f_{iy} \\ &\sum_{j \ni i} (z_i - z_h) . q_j = f_{iz} \end{aligned} \tag{1}$$

For the whole system:

$$[A] \cdot \{q\} = \{f\} \tag{2}$$

With $[A]$, equilibrium matrix defined by components of members. This matrix has "$3n$" lines corresponding to the "$n$" nodes, and "$b$" columns, one by element. The number of lines is then reduced to $N$, $N$ being the number of degrees of freedom.

It is convenient to build this matrix with the connectivity matrix $[C]$, which traduces the relational structure of the system. $[C]$ is a $b$ lines, $n$ columns matrix. General term is $C_{ji}$ ; for nodes "$i$" and "$j$" with no link, $C_{ji}$ is equal to zero. When nodes are linked, then:

$$C_{ji} = -1 \quad \text{or} \quad C_{ji} = 1 \tag{3}$$

Negative value for $i < j$ in the numbering of nodes, and positive value for the opposite case. Each line of equilibrium matrix, corresponding to node "$i$" and $X$-direction is given by:

$$[A]_{ix,\bullet} = [C]^t_{\bullet,i} \cdot [X] \tag{4}$$

In this expression $[X]$ is a diagonal matrix comprising the components of the $b$ members along $X$ direction. Similar expressions are derived for $Y$ and $Z$-directions. The whole matrix of equilibrium can be written line by line and organised sequentially according to $X$, $Y$ and $Z$ directions.

The writing of equilibrium equation of node "$i$" along $X$-direction in terms of force densities is given by:

$$[C]^t_{\bullet,i}\,[X]\,\{q\} = \{f_{ix}\} \tag{5}$$

According to diagonal matrix properties, we may express:

$$[X]\,\{q\} = [Q]\,\{X\} \tag{6}$$

With $[Q]$, diagonal matrix of force densities.
And since:

$$[X] = [C]\,\{x\} \tag{7}$$

Equilibrium equation for all nodes with degree of freedom along $X$ direction, takes the following form:

$$[C]^t\,[Q]\,[C]\,\{x\} = \{f_x\} \tag{8}$$

Values associated with nodes are split in two parts to introduce boundary conditions (for pre stressed systems with fixed nodes). This leads to a partition of the connectivity matrix. The first part is built with terms related to free nodes (subscript "$l$", and subscript "$lx$" for $x$-direction), second one with those which qualify fixed nodes along the considered direction (subscript "$f$" is mentioned for these values, and "$f_x$" for $X$-direction). If, in equations (8), we consider only the $n_{lx}$ equilibrium equations associated with $n_{lx}$ free nodes, we have:

$$[C_{lx}]^t[Q][C_{lx}]x_l = f_{lx} - [C_{lx}]^t[Q][C_{fx}]x_f \tag{9}$$

We define $[D_x]$ as "connection matrix" containing force density coefficients of nodes which are free along the direction $X$, as follows:

$$[D_x] = [C_{lx}]^t[Q][C_{lx}] \tag{10}$$

And if we note:

$$[D_{fx}] = [C_{lx}]^t[Q][C_{fx}] \tag{11}$$

We reach the matrix writing of equilibrium with force density method along $X$-direction:

$$[D_x]\{x_l\} = \{f_{lx}\} - [D_{fx}]\{x_f\}, \tag{12}$$

and similar expressions along $Y$ and $Z$-directions.

It is convenient to notice that in case of completely fixed or free nodes, connection matrices are identical for $X$, $Y$ and $Z$.

Solving system (12) for each direction leads to a geometry satisfying equilibrium equation with chosen force density coefficients.

## 2.2 Applications

**Pre and self stressed systems** For initially stressed systems, values of force density coefficients are chosen by the designer. These values are stored in matrix $[Q^0]$, and consequently matrices $[D^0_x]$ and $[D^0_{xf}]$ are defined

Two cases can be examined:

- pre stressed systems: equilibrium equation $[D^0_x].\{x_l\} = -[D^0_{xf}].\{x_f\}$
- self stressed systems: equilibrium equation $[D^0_x].\{x_l\} = \{0\}$

**Cable nets and membranes** Force density method has been developed for cable nets and membranes with an analogy depending on cable net modeling for membranes. Many examples are described in literature. The first studies on tensile membranes were carried out by physical modelling and significant results ensued from F. Otto's works (1973) on soap films. However, these methods are cumbersome and may not provide sufficient accuracy or restrict the variety of possible shapes to minimal area surfaces. Form-finding is today performed with computer-aided numerical procedures. We may firstly distinguish the mechanical approaches based upon a discrete representation of the domain by use of cable networks. This consideration has led to the Force Density Method (Linkwitz and Scheck, 1971). We developed a specific software called Architectural Membranes Design illustrated below according to the different steps of the method

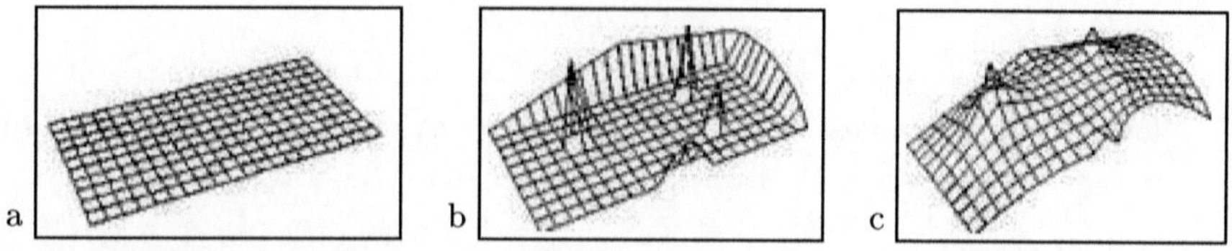


**Fig. 1.** Formfinding with AMD.
a) initial mesh, b) boundary conditions, c) first shape of Form finding with AMD

**Tensegrity systems** Tensegrity systems are systems in a stable equilibrium state comprising a discontinuous set of compressed components inside a continuum of tensioned components. It is necessary to emphasise on the fact that matrices for these self stressed systems, are always singular, since for any column or row, the sum of terms is always equal to zero.

$$rank([D_x^0]) = rank([D_y^0]) = rank([D_z^0]) < n \tag{13}$$

Consequently equilibrium equations admit an infinity of solutions, since all characteristic determinants vanish when the system to be solved is homogeneous. In case of non specific self stress coefficients, the rank of matrices is generally equal to $n-1$. Solutions are then parametered by only one redundant coordinate, but all the other nodes are confounded with this redundant node. In order to have self stressed geometries, which are not restricted to one point (or one straight line), it is necessary to reduce the connection matrix rank to $n-2$ (respectively $n-3$).When this rank is equal to $n-3$, solutions are parametered by three redundant coordinates. The resulting self stressed forms are then planar at best. For planar reticulated systems it is sufficient, but for spatial systems it is necessary to further reduce the rank by one. When the connection matrix rank is reduced to $n-4$, the four redundant nodes which parameter the solutions are then sufficient to generate spatial reticulated self stressed systems.

**Example** We call "triplex" every tensegrity system comprising six nodes, three struts and nine cables, such as every node is connected to one strut and three cables. In order to find irregular triplexes, we know that it is necessary to reduce the rank of the self stress coefficient connection matrix until $n-4$. Matrices being of dimension 6, the rank will be equal to 2. Required self stress coefficients have to be different from zero and to satisfy $q^0 > 0$ for cables and $q^0 < 0$ for struts. All self stress coefficient combinations which lead to a rank equal or less than two can be taken in account.

If, for instance, we choose identical coefficients for members pertaining to a same set of members, that is:

lower triangle cables: $$q_1^0 = q_2^0 = q_3^0 = q_i^0 \tag{14}$$

bracing cables: $$q_4^0 = q_5^0 = q_6^0 = q^0 \tag{15}$$

upper triangle cables: $$q_7^0 = q_8^0 = q_9^0 = q_s^0 \tag{16}$$

struts: $$q_{10}^0 = q_{11}^0 = q_{12}^0 = -q^0 \tag{17}$$

We find after an analytical study (based on Gaussian elimination) that the following relationship has to be satisfied in order to reduce the rank to two:

$$3.q_i^0 q_s^0 - (q^0)^2 = 0 \tag{18}$$

Consequently, if $q_i^0 = q_s^0 = 1$, then $q^0$ can be derived:

$$q_i^0 = q_s^0 = 1 \Rightarrow q^0 = \sqrt{3} \tag{19}$$

For every direction, there remain only two independent equilibrium equations. There are then four redundant nodes, which can be located anywhere. Some examples are given on the following figures (in plane views)

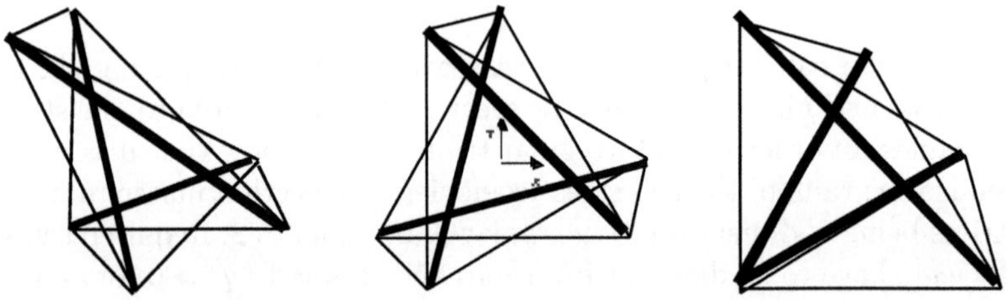

**Fig. 2.** Different triplex resulting from form finding. Plane views

# 3 Surface stress density method

## 3.1 Introduction

The stress distribution in a fabric remains an unknown factor even if reference to tensions in cable are done. Most generally, the question is related to the equivalence between a tensile cable net and a pre stressed membrane. Developments point out the inacurracy of this equivalence and therefore the inadequacy of such approach which may lead to significant mistakes. Thus, several methods based on a continuous representation of the fabric with membrane surfacic elements has been proposed. We may quote processes using the Dynamic Relaxation (Barnes, 1975) or Finite Elements methods under large displacement context (Haug and Powell, 1972). They are related to geometrically nonlinear analysis and determinate an equilibrated geometry by deforming a trial configuration whose boundary conditions are modified according to designer's specifications. However, several drawbacks may be put forward. Both the final shape and stresses in the membrane are difficult to control and some areas of the structure may end up in compression. Moreover, these techniques require most of time the use of unfriendly softwares for the architect and high performances hardwares.

## 3.2 The Surface Stress Density Method

This approach is based upon a continuous representation of the domain with triangular membrane surface elements. We note that the mechanical requirement of absence of compressive areas may be satisfied if every elementary Cauchy stress tensor is isotropic, which implies in its local axis $(\vec{x}_m \vec{y}_m \vec{z}_m)$, that $\{\sigma_{0iso}^m\}^T = \langle \sigma_0^m \sigma_0^m 0 \rangle$ with $\sigma_0^m > 0$. Therefore, the internal forces exerted by the element at its nodes may be written in the global axis of the structure $(\vec{X}\ \vec{Y}\ \vec{Z})$ by:

$$\vec{f}\, i_j^m = \frac{-ep_m}{2} \sigma_0^m \ell_{bj} \vec{n}_j \tag{20}$$

where $ep_m$ is the fabric thickness and $\vec{n}_j$ a normed vector orthogonal to the opposite side of the node. It comes for node 1:

$$\vec{f}\, i_1^m = \frac{ep_m}{2} \sigma_0^m \frac{\ell_{b1}}{\ell_{h1}} \vec{14} = \frac{ep_m}{4} \ell_{b1}^2 \frac{\sigma_0^m}{s_m} \vec{14} \tag{21}$$

with $s_m = \frac{1}{2} \ell_{b1} \ell_{h1}$ being element area

By considering the ratio $q_{sm} = \frac{\sigma_m^0}{s_m}$ as the *surface stress density* for the element

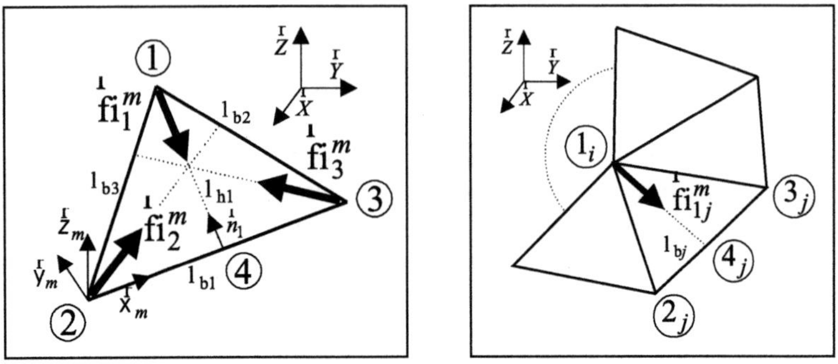

**Fig. 3.** Membrane element and equilibrium of node

$$\overrightarrow{f}i_1^m = \frac{ep_m}{4} q_s \ell_{b1}^2 \overrightarrow{14} \tag{22}$$

When $m_i$ elements are adjacent to the node $i$ (noted $1_i$ on Fig. 3), the total internal force is then

$$\overrightarrow{f}i_j = \sum_{j=1}^{m_i} \overrightarrow{f}i_{1j}^m = \frac{ep_m}{4} \sum_{j=1}^{m_i} q_{sj} \ell_{bj}^2 \overrightarrow{1_i 4_j} \tag{23}$$

With $\{\chi_i\}^T = \langle X_1 Y_1 Z_1 \rangle$, the equilibrium position of node $1_i$ verifies the matrix relationship:

$$\{\chi_i\} = [[Id_3] - [A_i]]^{-1}\{B_i\} \tag{24}$$

This equation is the main feature of the *Surface Stress Density* Method (SSDM)(Maurin and Motro, 1998). It allows to determinate the position of every node in an easy way (the most intricate requirement is a $3 \times 3$ inversion matrix) and to modify it by acting on the $q_{sj}$ values.

Indeed, $[Id_3]$ is the $3 \times 3$ identity matrix and

$$[A_i] = \sum_{j=1}^{m_i} Q_{sj} \begin{bmatrix} X_{3_j2_j}^2 & sym & sym \\ X_{3_j2_j}Y_{3_j2_j} & Y_{3_j2_j}^2 & sym \\ X_{3_j2_j}Z_{3_j2_j} & Y_{3_j2_j}Z_{3_j2_j} & Z_{3_j2_j}^2 \end{bmatrix}$$

$$\{B_i\} = \sum_{j=1}^{m_i} Q_{sj} \left\{ \begin{matrix} Y_{2_j3_j}(X_{3_j}Y_{2_j} - X_{2_j}Y_{3_j}) + Z_{2_j3_j}(X_{3_j}Z_{2_j} - X_{2_j}Z_{3_j}) \\ X_{2_j3_j}(X_{2_j}Y_{3_j} - X_{3_j}Y_{2_j}) + Z_{2_j3_j}(Y_{3_j}Z_{2_j} - Y_{2_j}Z_{3_j}) \\ X_{2_j3_j}(X_{2_j}Z_{3_j} - X_{3_j}Z_{2_j}) + X_{2_j3_j}(Y_{3_j}Z_{2_j} - Y_{2_j}Z_{3_j}) \end{matrix} \right\}$$

with $q_{sj} = Q_{sj}(\sum_{j=1}^{m_i} q_{sj}\ell_{bj}^2)$

**Combining with the Force Density Method (FDM)** The process allows the management of reinforcing cables located at the edges or above the fabric. If $C_i$ cable elements are connected to node $1_i$, the total internal force may be expressed as:

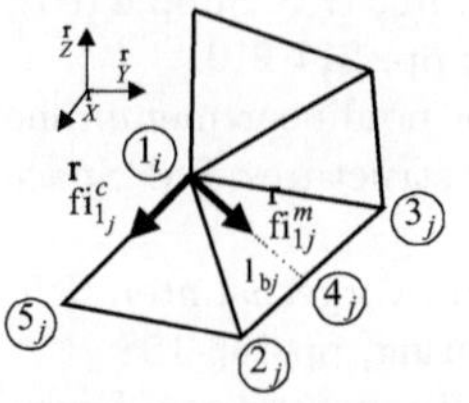


$$\vec{f}i_i = \sum_{j=1}^{m_i} \vec{f}i_{1_j}^m + \sum_{j=1}^{c_i} \vec{f}i_{1_j}^c \tag{25}$$

$$\vec{f}i_i = \frac{ep_m}{4} \sum_{j=1}^{m_i} q_{sj}\ell^2 \overrightarrow{1_i4_j} + \sum_{j=1}^{c_i} q_{\ell j} \overrightarrow{1_i5_j}$$

**Fig. 4.** Membrane with cables

**Applications**

**Fig. 5.** Some examples of SSDM applications for membranes

When cable elements are located on the edges, their curvature may be managed by changing the ratio between surface coefficients $q_{sj}$ and force density coefficients $q_{lj}$ (Fig.6 a to c).

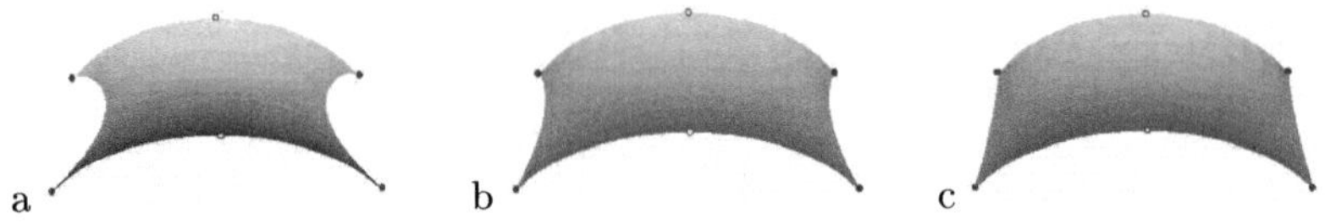

**Fig. 6.** Combination of SSDM and FDM for different values of coefficients a) $q_{sj} = 1$ and $q_{lj} = 6$ b) $q_{sj} = 1$ and $q_{lj} = 25$ c) $q_{sj} = 1$ and $q_{lj} = 100$

# 4 Conclusion

Force density methods appear to be very useful as far as form finding processes are required for initially stressed systems. We contributed to extend the classical force density method to surface stress density method opening also combination possibilities.

# References


1. Barnes M.R.,(1975): Applications of dynamic relaxation to the topological design and analysis of cable, membrane and pneumatic structures. In : W.J. Supple (ed), 2nd International , Conf. on Space Structures, Guildford, pp. 211-219.
2. Haug E., Powell G.H. , (1972): Finite element analysis of nonlinear membrane structures. In : IASS Pacific Symp. Part II on Tension structures and Space Frames, Tokyo and Kyoto, pp 83-92.
3. Linkwitz K. , Schek, H.J., (1971): Einige Bemerkungen von vorgespannten Seilnetzkonstruktionen. In : Ingenieur-Archiv 40, Springer-Verlag, pp 145-158.
4. Maurin B. and Motro R., (1998) :The Surface Stress Density method as a Form-finding tool for tensile membrane. In Engineering Structures Vol 20 (8), pp. 712-719
5. Otto F., (1973) Tensile Structures, Vols 1 and 2. MIT, Cambridge, MA (ed).

# Simple Analytical Models of Tensegrity Structures

Andrea Micheletti

Dipartimento di Ingegneria Civile
Università di Roma "Tor Vergata"
00133 Roma, Italia

**Abstract.** Some simple tensegrity structures are considered, all of which can be given an analytical model. Firstly, a qualitative description of a tensegrity system as a first-order mechanism is given. Next, for a representative three-dimensional structural system, the T3 module [4] in Fig. 1, the condition for the tensegrity configuration and the load-displacement relations are derived; remarkably, the dynamic characterization of the T3 module has general validity. Finally, tower-shaped structures obtained by juxtaposition of T3 modules are briefly considered.

## 1 Introduction

Tensegrity systems were first conceived, in 1948, by the artist Kenneth Snelson and the architect Richard Buckminster Fuller. According to their definition, tensegrity systems are pin-jointed spatial trusses with a large number of tension elements (≡ cables) and a small number of bars, all compressed. Bars are never connected to each other, while cables form a connected set (they have "tensile-integrity", whence the artifact word *tensegrity*). Each joint connects one bar and, at least, three cables (see Fig. 1).

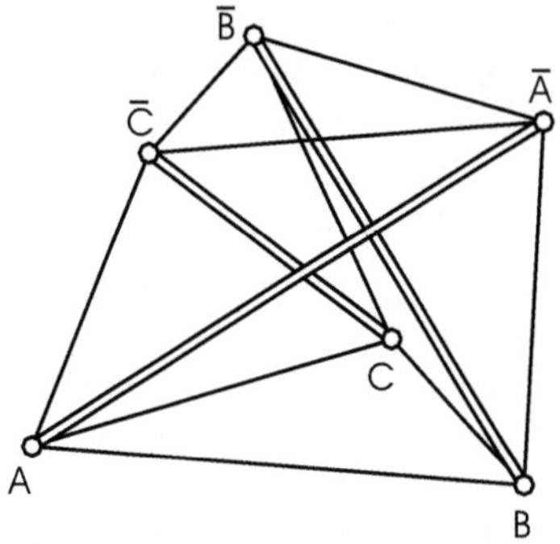


**Fig. 1.** A typical tensegrity, the T3 module. The cable triangles $ABC$ and $\bar{A}\bar{B}\bar{C}$ lay in two parallel planes

Tensegrity are both kinematically and statically indeterminate, i.e., there are mechanisms and self-stress states, but each self-stress state stabilizes and stiffens every mechanism. The indeterminacy condition is related to a condition of minimum or maximum length for one or more elements, the length of all other elements being fixed. This fact becomes evident when we try to build a tensegrity structure by hand [4]. For a system of $n$ elements, we can fix the length of $(n-1)$ elements and then realize the physical connection between them. The partial assembly has no stiffness. The length of the last element is determined when we try to decrease (in case the last element is a cable) or increase (in case it is a bar) the distance between the end nodes. The $(n-1)$ length constraints impose that the distance between the last two joints be, respectively, not less or not more than a certain extremum, a minimum for a cable or a maximum for a bar. If this extremum is attained, the system reaches its *tensegrity configuration* and becomes stable, but it has no stiffness yet: a tensegrity structure becomes stiff only when a prestressed element is inserted between the last two joints.

Tensegrity systems are light-weight structures to be employed as deployable or variable-geometry structures. We can fold the system by removing or lengthening a few cables; we can pass from a geometry to another, through a continuous path of tensegrity configurations, simply by changing the length of two or more elements. These features make tensegrity structures good candidates to realize smart structures, with some elements acting as sensors or actuators. In addition, tensegrity systems have already been used to cover large areas, for example, the roof of the Sun Coast Dome (St. Petersburg, Florida, 1989); another example is the Georgia Dome (Atlanta, Georgia, 1992).

## 2 First-order infinitesimal mechanisms

The configuration of a tensegrity structure is described by the *structural matrix*, that is, either the *equilibrium matrix*, whose linear action on the external-load vector yields the axial forces in the elements, or its transpose, the *compatibility matrix*, linking the joint displacements to the element deformations. The structural matrix of a tensegrity configuration is rank deficient, a condition depending only on the system's geometry.

The simple planar example in Fig. 2 is useful to describe the main tensegrity properties. If the three joints are not aligned, then the system is determinate, i.e., the structural matrix is square and has full rank. Otherwise, if the three joints are aligned, then the rank of the structural matrix decreases by one and the system possesses a mechanism and a possible state of self-stress. Note that the aligned configuration is the one for which the length of an element is minimal if the length of the other is fixed. Systems like this are often called first-order infinitesimal mechanisms [8]. From a kinematic point of view this means that element deformations $\varepsilon$ are infinitesimal of the second order with respect to joint displacements $d$:

$$\varepsilon = O(d^2)\,. \tag{1}$$

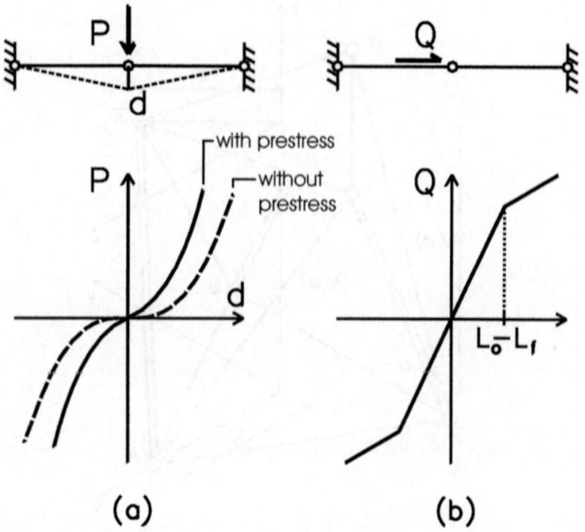


**Fig. 2.** Statics of a first-order mechanism: (**a**) stiffening under an activating load $P$; (**b**) piece-wise linear response under a nonactivating load $Q$ ($L_f$ fabrication length, $L_0$ length at $Q = 0$; $L_0 > L_f$)

The static response depends on the load direction. If the load activates the mechanism, then the load-displacement relation is approximate by a cubic with the inflection point in the origin (Fig. 2(a)); the tangent at the origin gives the initial stiffness, and is proportional to the prestress in the system. The load is balanced by the change in direction of elements, without any first-order change in the axial forces. The cubic embodies the geometric stiffening effect: the system becomes stiffer as the load increases. If the load does not activate the mechanism (Fig. 2(b)), then the response is linear until one of the cables becomes slack.

## 3 Single-module analysis

In this section we analyse the T3 module sketched in Fig. 1. Firstly, we find the geometrical condition under which the system is in the tensegrity configuration. Then we study the static response under two different load conditions. Finally, we lay down the equation of motion and briefly discuss the general dynamic behaviour.

The system is composed by nine cables and three bars. Six cables form the lower and upper horizontal triangles, $ABC$ and $\bar{A}\bar{B}\bar{C}$; three bars connect the joints $A$ and $\bar{A}$, $B$ and $\bar{B}$, $B$ and $\bar{B}$, the last three cables connect the joints $B$ and $\bar{A}$, $C$ and $\bar{B}$, $A$ and $\bar{C}$. An *oppositely oriented* system is obtained by connecting the joints $C$ and $\bar{A}$, $A$ and $\bar{B}$, $B$ and $\bar{C}$ by the last three cables. In order to deal only with cyclic-symmetric configurations, we restrict attention to structures which are left unchanged by a rotation of $2\pi/3$ about the

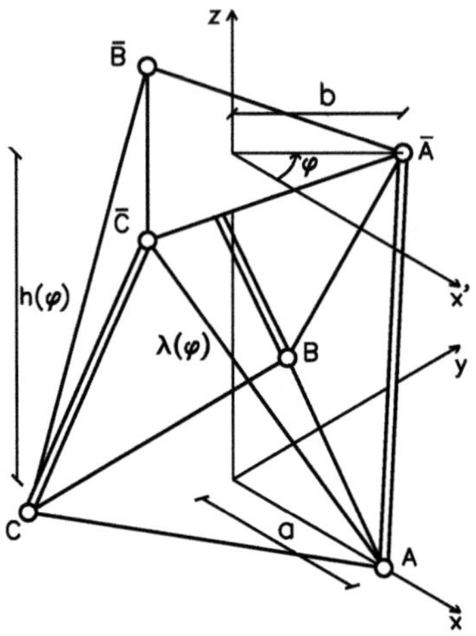


**Fig. 3.** The geometric parameters for the T3 module

axis passing through the centers of the horizontal triangles. Every symmetric geometry is determined by four parameters (Fig. 3): the radii $a$, $b$ of the circles circumscribed to the triangles $ABC$ and $\bar{A}\bar{B}\bar{C}$; the distance $h$ between the planes where the triangles lay; the relative rotation $\varphi$ of the two triangles about the vertical axis. For a general choice of these parameters the structural matrix has full rank, and the tensegrity configuration is determined by finding conditions under which the rank decreases by one. Equivalently, we can consider a model with elastic and prestressed cables $B\bar{A}$, $C\bar{B}$, $A\bar{C}$, while the remaining elements are inextensible. In this way we obtain the tensegrity condition at equilibrium in the absence of external loads. Since elasticity is concentrated, the problem simplifies: in statics, an explicit analytic solution can be found; in dynamics, the response is ruled by an ordinary differential equation.

It can be shown that, due to the length constraint on the anelastic elements, the total energy can be written in the form

$$U_{tot} = U_{el}(\varphi) = \frac{3}{2}k(\lambda(\varphi) - \lambda_f)^2 \,, \tag{2}$$

where $\lambda(\varphi)$ is the common current length of the elastic cables, which have stiffness $k > 0$ and fabrication length $\lambda_f$ (so that $\lambda(\varphi) > \lambda_f$); the minimum is found for $\varphi = 5\pi/6$. In conclusion, the tensegrity condition imposes the value of the relative rotation $\varphi$, but does not depend on the value of $a$, $b$, or $h$.

### 3.1 Statics

To derive the load-displacement relations, it is convenient to count the rotation $\varphi$ from the tensegrity value; therefore, we take the angle $\theta$ to be such that $\varphi = \theta + 5\pi/6$. With this, we obtain:

$$\lambda(\theta) = \lambda(-\theta) , \tag{3}$$

a symmetric expression in $\theta$.

Let us now consider the couple associated with the relative rotation $\theta$, which is given by the first derivative of the energy:

$$C_{T3}(\theta) = U'_{el}(\theta) = 3\sqrt{3}abk\left(1 - \frac{\lambda_f}{\lambda(\theta)}\right)\sin\theta = -C_{T3}(-\theta) , \tag{4}$$

a skew-symmetric function of $\theta$. The initial modulus is given by:

$$C'_{T3}(0) = 3\sqrt{3}abk\left(1 - \frac{\lambda_f}{\lambda_0}\right) , \tag{5}$$

and depends linearly on the initial prestress $k(\lambda_0 - \lambda_f)$.

The second load condition we consider is a vertical thrust associated with the relative change in distance between triangles. This relation is given in parametric form:

$$\begin{cases} F(\theta) = 6\sqrt{3}k\left(1 - \dfrac{\lambda_f}{\lambda(\theta)}\right)h(\theta)\dfrac{\sin\theta}{\sqrt{3}\sin\theta - \cos\theta)} ; \\ \delta(\theta) = h(\theta) - h_0 . \end{cases} \tag{6}$$

The initial modulus is:

$$\left.\frac{dF}{dh}\right|_{h_0} = 12\sqrt{3}\frac{h_0^2}{ab\lambda_0}k(\lambda_0 - \lambda_f) , \tag{7}$$

again linear in the prestress.

### 3.2 Dynamics

The dynamic response is characterized by a low-frequency mode, due to the small stiffness of the structure for any force activating the mechanism. Because of geometric stiffening, the frequency depends on the oscillation amplitude, decreasing if the amplitude decreases, until it reaches a minimum value corresponding to the initial stiffness.

We suppose that a mass is placed in each of the upper joints, while lower joints are fixed to the ground; furthermore, we assume that the dissipation is quadratic in $\dot{\lambda}$, the rate of change in length of vertical cables. The equation of motion, a highly nonlinear second-order equation, has the following form:

$$\mu_1(\theta)\ddot{\theta} + \mu_2(\theta)\dot{\theta}^2 + \gamma(\theta)\dot{\theta} + C_{T3}(\theta) = 0 , \tag{8}$$

where $(\mu_1 \ddot{\theta} + \mu_2 \dot{\theta}^2)$ is the inertial couple, $\gamma \dot{\theta}$ the damping couple, and $C_{T3}$ the elastic couple. Small vibrations are not damped, because the damping coefficient $\gamma$ is essentially quadratic in $\theta$, and hence negligeable. To have marked damping effects, one should envisage dissipation as depending on the relative rotation between elements.

## 4 Serial modules

In this section we consider the complex of two superposed T3 modules shown in Fig. 4(a). The three cables forming the lower triangle of the upper module are removed, while the lower joints of the upper module are connected to the middle points of the upper cables of the lower module (so that each of those cables is effectively split into two distinct elements). The two modules are oppositely oriented. This complex has in general six internal mechanisms, but in the tensegrity configuration it has one more mechanism and one state of self-stress. The geometric parameters necessary to determine every cyclic-

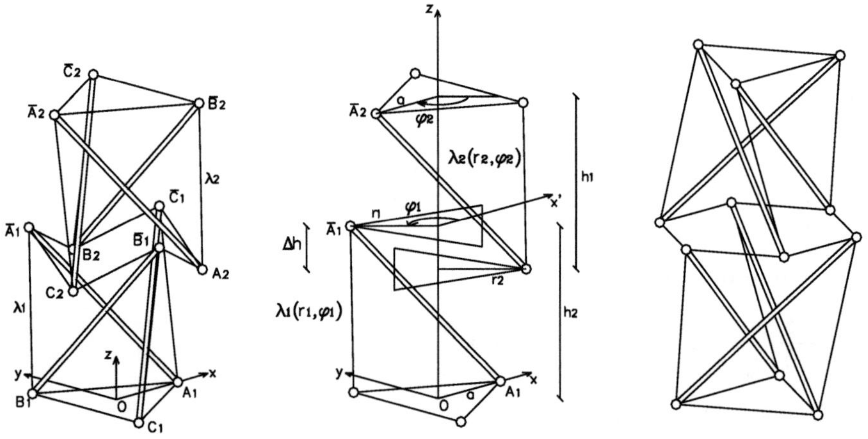


**Fig. 4.** Superposition of two T3 modules: (**a**) generic symmetric configuration; (**b**) geometric parameters; (**c**) tensegrity configuration

symmetric configuration are (Fig. 4(b)): the radius $a$ of the circle circumscribed to each of the two base triangles; the radii $r_1$, $r_2$ of the circumscribed circles of the two middle triangles; the heights $h_1$, $h_2$ of the modules; the two relative rotations $\varphi_1$, $\varphi_2$; the overlap $\Delta h$, that is, the distance between the planes where the middle triangles lay. The lengths of all elements are fixed except for the elastic cables; therefore, a possible choice for the set of independent variables is: $r_1$, $\varphi_1$, $r_2$, $\varphi_2$, $\Delta h$. This set reduce to $r$, $\varphi$, $\Delta h$ if we consider a center of symmetry such that $r_1 = r_2 = r$ and $\varphi_1 = \varphi_2 = \varphi$. The length of the elastic cables, and hence the elastic energy, is a function of $r$ and $\varphi$ only. We also have a constraint $G(r, \Delta h) = 0$, reflecting the fact that the lengths of the new elements are fixed. Consequently, the tensegrity configuration is found by minimizing the energy $U(r, \varphi)$ subject to this constraint. The solution is:

$$\varphi = \frac{5}{6}\pi\,, \quad r = l\,, \quad \Delta h = 0\,, \tag{9}$$

where $l$ denotes the common length of the middle cables. In the tensegrity configuration, the six new middle cables form a planar hexagon (Fig. 4(c)); we see that the identity of the T3 module is preserved.

The only load condition we consider is when two equal and opposite end couples induce the relative rotation $\theta$ of the two base triangles. We find that

$$C(\theta) = C_{T3}\left(\frac{\theta}{2}\right)\,, \tag{10}$$

where the mapping $C_{T3}$ for the single-module case is given by (4).

## 5 Two-level Snelson tower

In this section we describe the simplest version of a tensegrity tower of the type constructed by Snelson. This assembly is obtained by adding six more cables to the previous system; each of these cables connects a base-triangle joint (that is, the end node of a bar belonging to one module) to a middle joint (that is, the end node of a bar belonging to the other module). With reference to Fig. 5, the connected joints are: $A_1$ and $B_2$, $B_1$ and $C_2$, $C_1$ and $A_2$; $\bar{A}_1$ and $\bar{C}_2$, $\bar{B}_1$ and $\bar{A}_2$, $\bar{C}_1$ and $\bar{B}_2$. With the additional cables the tensegrity system has a single mechanism, stabilized by the self-stress state. This system is more difficult to study because of the interdependency of modules due to the new cables. We choose a different approach, and search for conditions under which the structural matrix is rank deficient. In view of the existing symmetry, both cyclic and central, it is sufficient to consider the equilibrium of two joints only: the end-base joint and the middle joint. We write the equilibrium equations projected onto the normal, tangent and vertical directions of a cylindrical reference system. This leads to a square linear system, whose matrix must have null determinant in the tensegrity configuration. This condition is satisfied if the geometric parameters are linked by the relation:

$$\frac{h}{\Delta h} = \frac{1}{2\sin\theta}\left(\frac{1}{2}\frac{a}{r} + \sin\theta + \sqrt{\frac{1}{4}\frac{a^2}{r^2} + 3\frac{a}{r}\sin\theta - 3\sin^2\theta}\right)\,, \tag{11}$$

a condition involving all the geometric parameters describing the structure.

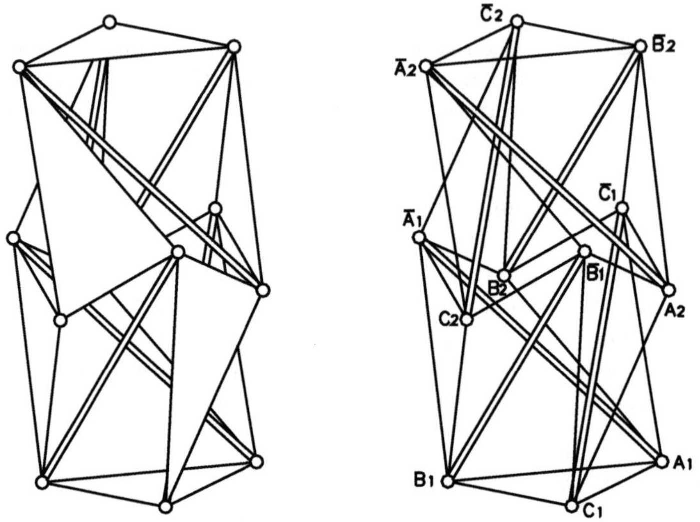


**Fig. 5.** Two level Snelson tower

## References

1. Calladine C. R. (1978) Buckminster Fuller's "Tensegrity" Structures and Clerk Maxwell's Rules for the Construction of Stiff Frames. Int J Solids Structures 14:161–172
2. Kebiche K., Kazi-Aoual M. N., Motro R. (1999) Goemetrical Non-linear Analysis of Tensegrity Systems. Engineering Structures 21:864–876
3. Motro R. (1999) Tensegrity Systems. In: Deployable Structures, CISM Course in Udine, Italy, July 5–9
4. Oppenheim I. J., Williams W. O. (1997) Tensegrity Prisms as Adaptive Structures. Adaptive Structures and Material Systems ASME 54:113–120
5. Oppenheim I. J., Williams W. O. (1998) Geometric Effects in an Elastic Tensegrity Structure. J Elasticity, to appear
6. Pellegrino S., Calladine C. R. (1986) Matrix Analysis of Statically and Kinematically Indetrminate Frameworks. Int J Solids Structures 22:409–428
7. Pellegrino S. (1990) Analysis of Prestressed Mechanisms. Int J Solids Structures 26:1329–1350
8. Pellegrino S., Calladine C. R. (1991) First Order Infinitesimal Mechanisms. Int J Solids Structures 27:505–515
9. Sultan C., Skelton R. E. (1997) Integrated Design of Controllable Tensegrity Structures. Adaptive Structures and Material Systems ASME 54:27–35

# Modelling and Simulation of Long-Span Bridges under Aerodynamic Loads

Franco Maceri, Giuseppe Vairo

Dipartimento di Ingegneria Civile
Università di Roma "Tor Vergata"
00133 Roma, Italia

**Abstract.** This paper deals with the dynamic stability of long-span bridges under non-stationary aerodynamic loads. Some generalizations of classical models are presented in order to check the critical conditions both in the case of flutter and divergence for long-span bridges with particular reference to the cable-stayed scheme. Furthermore, a numerical model based on a finite volume formulation of the flow problem around the girder cross-section is developed, able to simulate the steady and non-steady wind load conditions on the bridge. Good agreement with wind tunnel test results is found for the Normandy Bridge design cross-section models.

## 1 Introduction

The modern bridges aerodynamic design started with the investigations on the aeroelastic instability phenomenon that led to the collapse the first Tacoma Narrows Bridge.

Owing to their high flexibility, long-span bridges are in fact often found to be very sensitive to wind effects and their major wind-related problems are associated with deflections produced by oscillatory instabilities or by response due to the random action of wind gusts.

Undesirable moderately large vibratory motions may occur at lower wind velocities, due to the vortex shedding in the bridge's wake. Moreover, at much higher wind velocities, truly unstable divergent self-excited oscillations can occur. These oscillations are termed *flutter*. Finally, the forces engendered by gusting at very high speeds can lead to random *buffeting* oscillations that are of considerable concern in modern flexible bridges.

The wind-induced deflections lead to non-stationary interactive wind-structure (aeroelastic) effects. In other words, while the structure deflects under wind load, a deflection induces possibly positive feed-back that leads to a modification of the acting load, truly dependent on the girder's cross-section shape.

Furthermore, even if the bridge does not oscillate in the approaching flow, the divergence instability may occur as a consequence of the wind-induced eulerian stiffness singularity for the global wind-structure system.

The bridge aerodynamic analysis and design is traditionally based on experimental wind tunnel tests. However, their cost and realization time may be very large and hence, prohibitive for parametric studies.

Starting from the dynamic modelling aspects of long-span bridges, with particular reference to cable-stayed ones, this paper presents a numerical tool useful to evaluate the static and dynamic wind loads acting on a bridge deck and able to allow an easier acquisition of the aerodynamic data needed in bridge design.

## 2 Bridge decks under cross wind: stability concerns

In general, in proximity of a bridge site the stationary wind flow runs parallel to the sea or to a river surface. On the other hand, the wind velocity component normal to the bridge span gives rise to the largest effects and therefore in the analysis the wind stream may be considered transversally acting on the bridge.

As confirmed by the section model experimental technique, due to the large span-length/girder-width ratio of most bridges, the flow across the girder may be considered, in first approximation, two-dimensional. Then, the wind load analysis can be performed on a characteristic undeformable cross-section orthogonal to the span. Lift ($L$), drag ($D$) and twisting moment ($M$) represent the unit aerodynamic loads on the section and they are defined as:

$$L = \frac{1}{2}\rho U_o^2 B C_L, \qquad D = \frac{1}{2}\rho U_o^2 B C_D, \qquad M = \frac{1}{2}\rho U_o^2 B^2 C_M \tag{1}$$

where $C_L, C_D, C_M$ are dimensionless coefficients, $B$ is the cross-section width, $\rho$ the air density, $U_o$ the stationary cross-wind velocity. These forces can be equivalently introduced either in wind axes reference or in profile axes one.

Since the bridge cross-section profiles have generally *bluff- body* geometry, the unit actions in Eqs. (1) are *time-varying* about the steady value due to signature effects (local turbulence) ascribable to local flow separation, reattachment, etc. close the structure. Furthermore, the previous dimensionless coefficients are also dependent on the angle of incidence of the approaching flow.

As it is known, when a deck structure is moving in the wind flow, interactive or aeroelastic forces are generated. If the quasi-steady theory is assumed to be valid, *i.e.* if the non-linear problem in its incipient small-motion stage is considered, the unit aeroelastic wind loads acting on the bridge cross-section, when oscillating in wind flow, are completely described by the classical Scanlan formulation [1–4]:

$$D_{ae}(z,t) = \frac{1}{2}\rho {U_o}^2 BK \left[ P_1^* \frac{\dot{u}(z,t)}{U_o} + P_2^* \frac{B\dot{\theta}(z,t)}{U_o} + KP_3^* \theta(z,t) \right.$$
$$\left. + KP_4^* \frac{u(z,t)}{B} + P_5^* \frac{\dot{v}(z,t)}{U_o} + KP_6^* \frac{v(z,t)}{B} \right] \tag{2}$$

$$L_{ae}(z,t) = \frac{1}{2}\rho U_o^2 BK \left[ H_1^* \frac{\dot{v}(z,t)}{U_o} + H_2^* \frac{B\dot{\theta}(z,t)}{U_o} + KH_3^* \theta(z,t) \right.$$
$$\left. + KH_4^* \frac{v(z,t)}{B} + H_5^* \frac{\dot{u}(z,t)}{U_o} + KH_6^* \frac{u(z,t)}{B} \right] \tag{3}$$

$$M_{ae}(z,t) = \frac{1}{2}\rho U_o^2 B^2 K \left[ A_1^* \frac{\dot{v}(z,t)}{U_o} + A_2^* \frac{B\dot{\theta}(z,t)}{U_o} + KA_3^* \theta(z,t) \right.$$
$$\left. + KA_4^* \frac{v(z,t)}{B} + A_5^* \frac{\dot{u}(z,t)}{U_o} + KA_6^* \frac{u(z,t)}{B} \right] \tag{4}$$

where $u$, $v$ and $\theta$ are the cross-section in-plane displacement components evaluated from the static bridge equilibrium configuration (Fig. 1); $\omega$ is the circular frequency of girder oscillation; $K = B\omega/U_o$ is the reduced frequency; $P_i^*$, $H_i^*$, $A_i^*$ ($i = 1, \dots, 6$) are the dimensionless aerodynamic (or flutter) derivatives, introduced by Scanlan [1] as functions of $K$ and depending on the cross-section geometry as well as on the static torsional rotation $\theta_o = \theta_o(z)$:

$$P_i^* = P_i^*(K, \theta_o(z)), \qquad H_i^* = H_i^*(K, \theta_o(z)), \qquad A_i^* = A_i^*(K, \theta_o(z)) \tag{5}$$

being $z$ the spanwise co-ordinate.

As a matter of fact, if the steady aerodynamic twisting moment is not vanishing at zero angle of attack, or if an eccentric live load condition is acting on the structure, static torsional rotation along the bridge span length is produced. This occurrence can change of a large amount the non-steady aerodynamic loads also under small steady torsional rotations. As a consequence, relevant effects in bridge stability response may be produced. Hence,

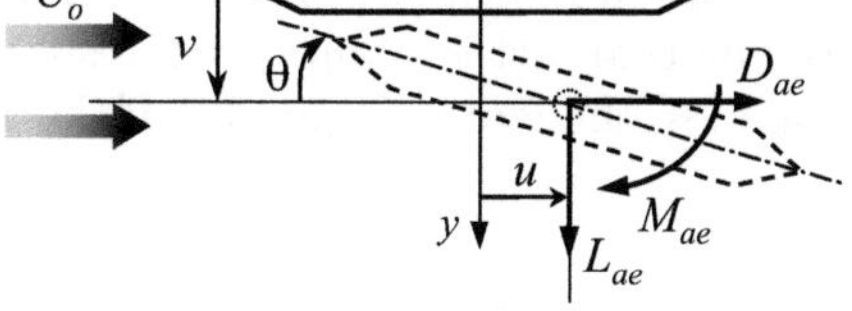


**Fig. 1.** Deck section displacements and aeroelastic forces: notations.

in these cases, the knowledge of the aerodynamic derivatives as functions of the static torsional rotation along the bridge span length may be essential for the design and for a more accurate analysis of the dynamical behaviour of long-span bridges under wind loads.

If the generalized compatible displacement vector at the time $t$ for the section at the $z$ co-ordinate is indicated as $\mathbf{s}(z,t) = [u(z,t), v(z,t), \theta(z,t)]^T$ , the functional relation between $\mathbf{s}(z,t)$ and the distributed generalized forces can be symbolically expressed by the equilibrium equation:

$$\mathbf{Q}_A + \mathbf{Q}_I + \mathbf{Q}_D = \mathcal{S}(\mathbf{s}) \tag{6}$$

where $\mathcal{S}$ is a structural operator and $\mathbf{Q}_A$, $\mathbf{Q}_I$ and $\mathbf{Q}_D$ are the generalized aerodynamic, inertial, and external disturbance forces, respectively. The disturbance force $\mathbf{Q}_D$ is usually assumed to be given, whereas the aerodynamic and inertial forces depend in some manner on the displacement and motion of the system. If a quasi-steady aerodynamic loading is assumed, the operator equation of the aeroelastic problem may be written as:

$$\mathcal{A}(\mathbf{s}, \dot{\mathbf{s}}) + \mathcal{I}(\ddot{\mathbf{s}}) + \mathbf{Q}_D = \mathcal{S}(\mathbf{s}) \tag{7}$$

where $\mathcal{A}$ and $\mathcal{I}$ are an aerodynamic and inertial operator, respectively.

As indicated in [4] and applied in [5–7], assuming the bridge moving in a linearly elastic way from the static configuration, the full three-dimensional wind-induced bridge response can be regarded as a superposition of an adequate number $N$ of natural vibration modes. Consequently, the section displacements can be written as:

$$\frac{u(z,t)}{B} = \sum_{j=1}^{N} u_j(z)\xi_j(t); \ \frac{v(z,t)}{B} = \sum_{j=1}^{N} v_j(z)\xi_j(t); \ \theta(z,t) = \sum_{j=1}^{N} \theta_j(z)\xi_j(t) \tag{8}$$

where $u_j$, $v_j$, $\theta_j$ are the dimensionless spanwise modal (shape) forms and $\xi_j$ is the $j$th generalized modal co-ordinate.

Then, the motion equations equivalent to the Eq. (7) can be obtained through an energy approach considering for each mode the corresponding Lagrange's equation of motion. Hence, disregarding the disturbance force and taking into account the mutual orthogonality with respect to mass weighting of the bridge natural modes, the motion equation of the $j$th mode results [4]:

$$I_j[\ddot{\xi}_j + 2\zeta_j\omega_j\dot{\xi}_j + \omega_j^2\xi_j] = \int_{span} [Lv_jB + Du_jB + M\theta_j]dz \tag{9}$$

where $I_j$, $\zeta_j$, $\omega_j$ are the generalized mass inertia, the damping ratio and the natural circular frequency of mode $j$, respectively. Furthermore, if the vortex-shedding effects and buffeting-aeroelastic interactive ones are disregarded and if the structural reference configuration is assumed to be coincident with the bridge aerostatic configuration under the action of aerostatic stationary wind loads, the unit wind actions $L$, $D$ and $M$ may be completely described by the amount of aeroelastic and buffeting effects.

Since the Eqs. (2-4) hold for sinusoidal motion [1], it can be observed that the bridge motion equation in the frequency domain can be obtained by assuming time complex harmonic motion for the $j$th mode, *i.e.* $\xi_j = \overline{\xi}_j e^{i\omega t}$, where $i$ is the imaginary unit, $\omega$ the real circular frequency and $\overline{\xi}_j$ the complex amplitude of the motion. Moreover, it can be proved that the buffeting excitation terms play a negligible role in stability concerns [4]. Hence, assuming a constant girder cross-section along the bridge span and taking into account the flutter derivatives variability with the $z$ co-ordinate, the $j$th equation of motion can be written as:

$$\begin{aligned}
&\left[-K^2 + 2i\zeta_j K K_j + K_j^2\right]\overline{\xi}_j \\
&= \left(\frac{\rho B^4 K^2 \ell}{2I_j}\right)\sum_{n=1}^{N}\overline{\xi}_n \Big[(iH_1^* + H_4^*)_{v_n v_j} + (iH_2^* + H_3^*)_{\theta_n v_j} \\
&+ (iH_5^* + H_6^*)_{u_n v_j} + (iA_1^* + A_4^*)_{v_n \theta_j} + (iA_2^* + A_3^*)_{\theta_n \theta_j} \\
&+ (iA_5^* + A_6^*)_{u_n \theta_j} + (iP_1^* + P_4^*)_{u_n u_j} + (iP_2^* + P_3^*)_{\theta_n u_j} \\
&+ (iP_5^* + P_6^*)_{v_n u_j}\Big]
\end{aligned} \tag{10}$$

where $\ell$ is the bridge span length, $(Q)_{r_n q_j} = \int_{span} Q(z) r_n(z) q_j(z) dz/\ell$ with $r_j, q_j = u_j, v_j$ or $\theta_j$ and $K_j = B\omega_j/U_o$.

From the Eq. (10), the divergence condition for the $j$th mode is obtained by imposing the balance between the aerodynamic stiffness and the structural one, when circular frequency $\omega$ of girder oscillation approaching zero. As a consequence the corresponding critical wind condition results:

$$\begin{aligned}
U_{o,div}^{(j)} = \lim_{\omega \to 0} \Bigg\{ \sqrt{\frac{I_j \omega_j^2}{\rho B^2 K^2 \ell}} \Big[ (H_3^* + A_4^*)_{\theta_j v_j} + (H_4^*)_{v_j v_j} + (H_6^* + P_6^*)_{u_j v_j} \\
+ (A_3^*)_{\theta_j \theta_j} + (A_6^* + P_3^*)_{u_j \theta_j} + (P_4^*)_{u_j u_j} \Big]^{-1/2} \Bigg\}.
\end{aligned} \tag{11}$$

On the other hand, the single-mode flutter instability criterion is derived by imposing the damping unbalance condition for the $j$th mode:

$$\big[(P_1^*)_{u_ju_j} + (P_5^* + H_5^*)_{u_jv_j} + (P_2^* + A_5^*)_{u_j\theta_j} + (H_1^*)_{v_jv_j} \\ +(H_2^* + A_1^*)_{v_j\theta_j} + (A_2^*)_{\theta_j\theta_j}\big] \geq \frac{4I_j\zeta_j}{\rho B^4 \ell}\frac{K_j}{K} \tag{12}$$

where, for an incipient flutter, the ratio between the natural reduced frequency and the flutter one $K_c$ results from:

$$\frac{K_j^2}{K_c^2} = \frac{\omega_j^2}{\omega_c^2} = 1 + \frac{\rho B^4 \ell}{2I_j}\left[(P_4^*)_{u_ju_j} + (P_6^* + H_6^*)_{u_jv_j} + (P_3^* + A_6^*)_{u_j\theta_j} \right. \\ \left. +(H_4^*)_{v_jv_j} + (H_3^* + A_4^*)_{v_j\theta_j} + (A_3^*)_{\theta_j\theta_j}\right]. \tag{13}$$

From knowledge of the modal forms, the mechanical properties and the flutter derivatives as functions of $K$ and of the static torsional rotation, the stability tendency of any structural mode and the corresponding critical flutter wind speed, *i.e.* the wind velocity corresponding to self-sustained harmonic oscillations, can be evaluated through the criterion (12).

The single-mode approach is enough to evaluate the critical flutter condition for typical bluff-body bridge decks, because the energy transferred from the flow comes mainly into a single mode instead of being spread into more than one. However, when the bridge cross-section is properly designed in an aerodynamic sense, the occurrence of one-degree-of-freedom flutter can be often excluded, even if flutter with two degree-of-freedom or more can arise.

Writing the Eq. (9) for each value of $j$ to be considered, the multi-mode flutter condition may be investigated. In detail, the following homogeneous problem results in the frequency domain:

$$\mathbb{E}(K)\overline{\boldsymbol{\xi}} = 0 \tag{14}$$

being $(\overline{\boldsymbol{\xi}})_j = \overline{\xi}_j$ and

$$E_{jn}(K) = -K^2\delta_{jn} + iKA_{jn}(K) + B_{jn}(K) \tag{15}$$

where $\delta_{jn}$ represents the Kronecker symbol and

$$A_{jn}(K) = 2\zeta_jK_j\delta_{jn} - \frac{\rho B^4 K\ell}{2I_j}\big[(H_1^*)_{v_nv_j} + (H_2^*)_{\theta_nv_j} + (H_5^*)_{u_nv_j} \\ + (A_1^*)_{v_n\theta_j} + (A_2^*)_{\theta_n\theta_j} + (A_5^*)_{u_n\theta_j} + (P_1^*)_{u_nu_j} + (P_2^*)_{\theta_nu_j} + (P_5^*)_{v_nu_j}\big] \tag{16}$$

$$B_{jn}(K) = K_j^2 \delta_{jn} - \frac{\rho B^4 K^2 \ell}{2 I_j} \left[ (H_3^*)_{\theta_n v_j} + (H_4^*)_{v_n v_j} + (H_6^*)_{u_n v_j} \right. \tag{17}$$
$$\left. + (A_3^*)_{\theta_n \theta_j} + (A_4^*)_{v_n \theta_j} + (A_6^*)_{u_n \theta_j} + (P_3^*)_{\theta_n u_j} + (P_4^*)_{u_n u_j} + (P_6^*)_{v_n u_j} \right] .$$

The existence of steady-state oscillations with amplitude $\overline{\boldsymbol{\xi}}$ is possible only if both the real and imaginary parts of the determinant of $\mathbb{E}$ vanish. Two real equations in $K$ and $\omega$ are so obtained, whose solution, if it exists, leads to evaluate the critical wind speed $U_{oc} = B\omega_c/K_c$, where $\omega_c$ denotes the circular frequency of the critical flutter mode, to which several natural modes contribute, possibly out of phase but synchronized one another due to aeroelastic forces. The flutter mode $\overline{\boldsymbol{\xi}}$ corresponds to the eigenvector of the problem (14) with $K = K_c$ and $\omega = \omega_c$.

The problem, completely analogous in its general form to the classic aircraft flutter problem, has received a variety of treatments. In what follows, with reference to cable-stayed bridges, it is presented a simple approach to analyze the two degree-of-freedom flutter conditions.

## 3 Aerodynamic instabilities of long-span cable-stayed bridges

A fan-shaped scheme for cable-stayed bridges is considered (Fig. 2). Its physical behaviour is very similar to a large truss structure where the main state of stress is given by axial forces in the stays and in the girder, while girder's bending and torsion are of a local nature. A beam-like girder is assumed hung to the tops of two towers by means of two stay curtains and girder's ends are simply supported, without longitudinal constraints. $H$-shaped or $A$-shaped piles (Fig. 3) are commonly used. With reference to the usual erection procedures the girder is also assumed free from bending moments under dead load $g$. Moreover, girder cross-section is assumed constant along the span length and the bridge is symmetric with respect to both the axial vertical plane and the mid-span cross-plane.

In modern long-span cable-stayed bridges, stays spacing $\Delta$ is very small compared to central span length $L_c$, and therefore a continuous stays distri-

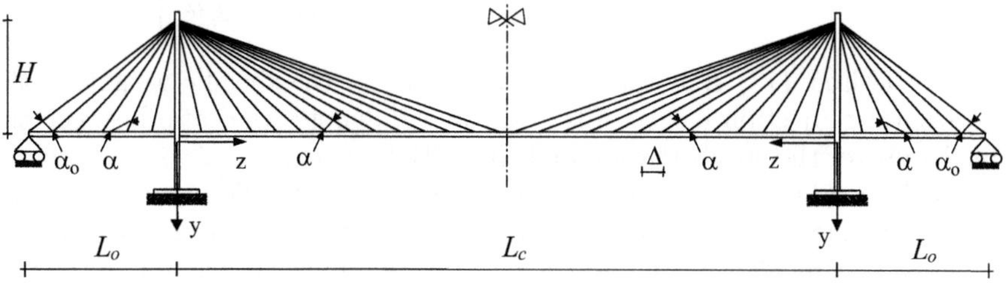


**Fig. 2.** Fan-shaped cable-stayed bridge scheme: notations.

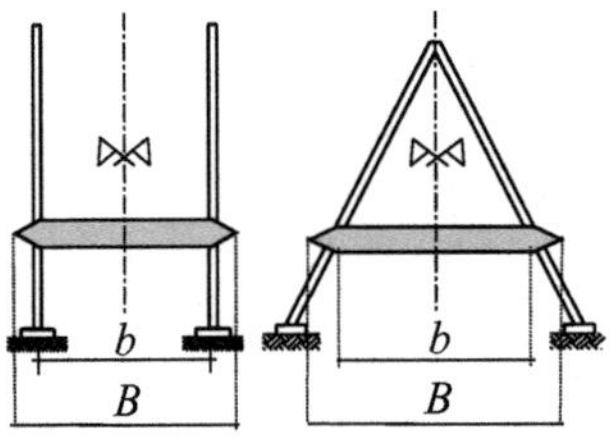


**Fig. 3.** *H*- and *A*-shaped tower schemes.

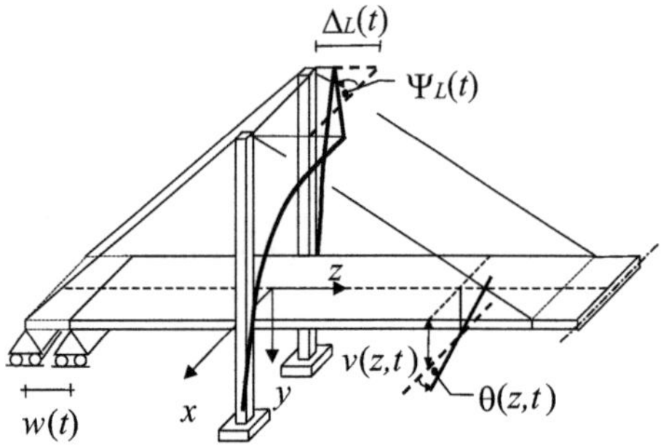


**Fig. 4.** Displacement parameters.

bution along the deck may be reasonably taken to describe the stays-girder interaction. Hence, the continuous model proposed in [8–12] can be used to characterize the non-linear dynamics of the structure. Neglecting the towers and girder's axial extensions, the Euler-Bernoulli bending theory and the De Saint Venant torsion theory are applied to describe the constitutive behaviour of the girder. Assuming a two-degrees of freedom model for the in-plane cross-section motion, the bridge deformation can be described by means of the following displacement parameters (Fig. 4): the girder vertical deflection $v(z,t)$; the girder torsional rotation $\theta(z,t)$; the girder longitudinal displacement $w(t)$; the $z$-directed displacements $\Delta_L(t)$, $\Delta_R(t)$ and the torsional rotations around the vertical axis $\Psi_L(t)$, $\Psi_S(t)$ of the towers tops (these latter rotations disappear in the case of $A$-shaped towers [8]).

As it is known, the behaviour of cable-stayed bridges is non-linear due to the intrinsic non-linearities of the stays. The non-linear elastic response of a stay is modelized through the usual Dischinger's theory [8]. Therefore the non-linear constitutive equation $\Delta\sigma = E^*\Delta\varepsilon$ is assumed to hold, where the fictitious elasticity modulus $E^*$ can be written through the tangent theory ($E_t^*$) or the secant theory ($E_s^*$):

$$E_t^* = E\left[1 + \frac{\gamma_c^2 l_o^2 E}{12\sigma_o^3}\right]^{-1} \qquad E_s^* = E\left[1 + \frac{\gamma_c^2 l_o^2 E}{12\sigma_o^3}\frac{1+\tilde{\beta}}{2\tilde{\beta}^2}\right]^{-1}. \tag{18}$$

In the Eqs. (18) $E$ is the Young modulus, $\gamma_c$ the specific weight, $l_o$ the horizontal projection length of the stay and $\tilde{\beta}$ denotes the ratio between the final value of the stay's tension and the initial one $\sigma_o$.

As far as the dynamic equilibrium is concerned, the bridge aerostatic configuration under the action of aerostatic stationary wind loads is considered practically coincident to the bridge configuration under the action of the vertical dead loads. Moreover, it is assumed that the non-stationary oscillations will occur in a small neighbourhood of the initial configuration. As shown in

[10], considering $H$-shaped towers, for the case of a symmetrical motion with respect to mid-span, the dynamic equilibrium equations under wind load can be written in dimensionless form as:

$$\frac{\tilde{\varepsilon}^4}{4}\frac{\partial^4 V(\zeta,t)}{\partial\zeta^4} + \Phi(\zeta)V(\zeta,t) - \zeta\Phi(\zeta)U(t) + M\ddot{V}(\zeta,t) = \frac{\sigma_g}{Eg}L_{ae}(\zeta,t) \quad (19)$$

$$-(\vartheta+\chi)U(t) + \int_L \zeta\Phi(\zeta)V(\zeta,t)d\zeta = 0 \quad (20)$$

$$\tau^2\frac{\partial^2\theta(\zeta,t)}{\partial\zeta^2} - \Phi(\zeta)\theta + \zeta\Phi(\zeta)\Psi(t) - \hat{I}_o\ddot{\theta}(\zeta,t) = -\frac{H\sigma_g}{Egb^2}M_{ae}(\zeta,t) \quad (21)$$

$$-(\vartheta+\chi)\Psi(t) + \int_L \zeta\Phi(\zeta)\theta(\zeta,t)d\zeta = 0 \quad (22)$$

being $w=0$, $U = \Delta_L/H = -\Delta_R/H$, $\Psi = \Psi_L = -\Psi_R$.

The following dimensionless quantities are involved in Eqs. (19-22) [10]:

$$\zeta = \frac{z}{H}, \quad V = \frac{v}{H}, \quad \frac{\tilde{\varepsilon}^4}{4} = \frac{I\sigma_g}{H^3 g}, \quad \tau^2 = \frac{C_t\sigma_g}{Eb^2Hg}, \quad \chi = \frac{k^T\sigma_g}{Eg} \quad (23)$$

$$\vartheta = \int_L \frac{\cos^2\alpha(\zeta)}{1+a\zeta^2}d\zeta + \chi_o, \quad a = \frac{\gamma_c^2 EH^2}{12\sigma_g^3}, \quad M = \frac{\tilde{\mu}H\sigma_g}{Eg}, \quad \hat{I}_o = \frac{I_oH\sigma_g}{Egb^2} \quad (24)$$

$$\chi_o = 2\frac{E_o^* A_o}{E}\frac{\sigma_{go}}{Hg}\sin\alpha_o\cos^2\alpha_o, \qquad \Phi(\zeta) = \left[(1+a\zeta^2)(1+\zeta^2)\right]^{-1} \quad (25)$$

where $\tilde{\mu}$ is the girder's mass per unit length, $I$, $I_o$, $C_t$ are the flexural inertia, the polar mass moment of inertia and the torsional stiffness factor of girder cross-section, respectively. Moreover, $\sigma_g$ is the stress produced in the stays by dead load, $k^T$ is the tower flexural stiffness, $\alpha_o$ is the slope of anchor stays, whose stress, fictitious tangent Dischinger's modulus and cross-section area are $\sigma_{go}$, $E_o^*$ and $A_o$, respectively.

Due to the motion symmetry hypothesis, Eqs. (19) and (21) describe the flexural and torsional equilibrium of the girder, while Eqs. (20) and (22) give the translational and rotational equilibrium of the left ($L$) tower's top.

In practical cases the parameters $\tilde{\varepsilon}$ and $\tau$ are small ($\tilde{\varepsilon} \le 0.3$, $\tau \le 0.1$). This fact allows using a *truss assumption*, that is disregarding terms in $\tilde{\varepsilon}$ and in $\tau$ with respect to the others in the equilibrium equations. By using this assumption and Eqs. (3-4), the Eqs. (19) and (21) may be written as:

$$+\Phi V - \zeta\Phi U = -M\ddot{V} + Q\left[HH_1^*\dot{V} + BH_2^*\dot{\theta} + KU_oH_3^*\theta + KHU_oH_4^*\frac{V}{B}\right] \tag{26}$$

$$-\Phi\theta + \zeta\Phi\Psi = \hat{I}_o\ddot{\theta} - Q\frac{HB}{b^2}\left[HA_1^*\dot{V} + BA_2^*\dot{\theta} + KU_oA_3^*\theta + KHU_oA_4^*\frac{V}{B}\right] \tag{27}$$

where $Q = \rho U_o K B \sigma_g/(2Eg)$.

In order to characterize the torsional divergence and the one degree-of-freedom torsional flutter conditions, it is assumed a zero degree of freedom $v$. As a consequence, Eqs. (27) and (22) represent the significant motion equations. Assuming time complex harmonic rotational motion for the girder and for the towers top with the same frequency, *i.e.* $\theta(\zeta,t) = \overline{\theta}(\zeta)e^{i\omega t}$ , $\psi(t) = \overline{\psi}e^{i\omega t}$, and solving the Eq. (27) into the Eq. (22) with the simplified framework of the truss assumption [11], the following motion equation in the frequency domain is obtained:

$$I_o(\omega_\theta^2 - \omega^2) - \frac{1}{2}\rho U_o B^2[U_o K^2\overline{A}_3^* + i\omega BK\overline{A}_2^*] = 0 \tag{28}$$

where $\overline{A}_i^*(K) = \int_L A_i^*(\zeta,K)\zeta^2 d\zeta/v$ represents the $i$th generalized flutter derivatives, $v = \int_L \zeta^2 d\zeta$ and $\omega_\theta$ is the torsional free oscillation circular frequency in still air of the bridge, already given in [9]: $\omega_\theta^2 = (\chi + \chi_o)/(vI_o)$.

In this way, the torsional divergence condition is obtained similarly to the Eq. (11), by imposing that the aerodynamic stiffness balances the structural one when circular frequency $\omega$ approaching zero:

$$U_{o,div}^{(\theta)} = \lim_{\omega\to 0}\frac{B\omega_\theta}{\sqrt{\frac{\beta}{2}\gamma K^2\overline{A}_3^*}} \tag{29}$$

being $\beta = \frac{\rho B^2}{\tilde{\mu}}$ and $\gamma = \frac{\tilde{\mu}B^2}{I_o}$ the dimensionless flexural and torsional inertial parameters of the bridge, respectively.

Furthermore, the one degree-of-freedom torsional flutter critical conditions are obtained imposing that the global damping vanish, *i.e.* $\overline{A}_2^* = 0$. As a consequence, the critical value of the reduced frequency $K_{\theta c}$ is obtained. By using the Eq. (28), the correspondent critical circular frequency of motion results: $\omega_{\theta c} = \omega_\theta\left[1 + \frac{\beta}{2}\gamma\overline{A}_3^*(K_{\theta c})\right]^{-1/2}$.

Analogously, assuming a zero degree-of-freedom $\theta$ the flexural flutter critical conditions may be obtained. In detail, imposing that the global damping vanish, *i.e.* $\overline{H}_1^* = 0$, the critical reduced frequency $K_{vc}$ may be characterized. Hence, the critical circular frequency of motion results: $\omega_{vc} = \omega_v \left[1 + \frac{\beta}{2}\overline{H}_4^*(K_{vc})\right]^{-1/2}$, where $\omega_v$ denotes the flexural free oscillation circular frequency in still air [12]: $\omega_v^2 = (\chi + \chi_o)/(vM)$.

The two degree-of-freedom flutter instability conditions and the corresponding critical wind speed value can be obtained by putting the stationary solution of the Eqs. (20), (22), (26),(27) in the form:

$$V(\zeta,t) = \overline{V}(\zeta)e^{i\omega t}, \quad U(t) = \overline{U}e^{i\omega t}, \quad \theta(\zeta,t) = \overline{\theta}(\zeta)e^{i\omega t}, \quad \Psi(t) = \overline{\Psi}e^{i\omega t} \tag{30}$$

where $\omega$ is the flutter circular frequency. Introducing in Eqs. (20) and (22) the values of $\overline{V}$ and $\overline{\theta}$ obtained from Eqs. (26) and (27), a linear homogeneous system in the unknowns $\overline{U}$ and $\overline{\Psi}$ is derived. Imposing equal to zero the determinant of this system, after some algebra and in the simplified framework of the truss assumption [12], the following frequency equation yields:

$$\begin{aligned}
&\Omega^4 - \Omega^2 \left[(1+\varphi^2) + \frac{\beta^2}{4}\gamma\Omega^2(\overline{H}_1^*\overline{A}_2^* - \overline{H}_2^*\overline{A}_1^*) - \frac{\beta}{2}\gamma\Omega^2\overline{A}_3^* - \frac{\beta}{2}\Omega^2\overline{H}_4^*\right] \\
&\quad + \left[\varphi^2(1 - \frac{\beta}{2}\Omega^2\overline{H}_4^*) + \frac{\beta^2}{4}\gamma\Omega^4(\overline{H}_4^*\overline{A}_3^* - \overline{H}_3^*\overline{A}_4^*) - \frac{\beta}{2}\gamma\Omega^2\overline{A}_3^*\right] \\
&\quad + i\frac{\beta}{2}\left\{\Omega^3\left[\Omega(\overline{H}_1^* + \gamma\overline{A}_2^*)\right] - \Omega\left[\frac{\beta}{2}\gamma\Omega^3(\overline{H}_3^*\overline{A}_1^* - \overline{H}_1^*\overline{A}_3^* + \overline{H}_2^*\overline{A}_4^* - \overline{H}_4^*\overline{A}_2^*)\right.\right. \\
&\qquad \left.\left. + \Omega(\gamma\overline{A}_2^* + \varphi^2\overline{H}_1^*)\right]\right\} = 0
\end{aligned} \tag{31}$$

where $\Omega = \frac{\omega}{\omega_v}$ and $\varphi = \frac{\omega_\theta}{\omega_v}$ are the dimensionless parameters of the oscillating motion.

The imaginary part of Eq. (31) leads to the significant root:

$$\Omega^2 = \frac{\varphi^2\overline{H}_1^* + \gamma\overline{A}_2^*}{(\overline{H}_1^* + \gamma\overline{A}_2^*) + \frac{\beta}{2}\gamma(\overline{H}_1^*\overline{A}_3^* - \overline{H}_3^*\overline{A}_1^* + \overline{H}_4^*\overline{A}_2^* - \overline{H}_2^*\overline{A}_4^*)} \tag{32}$$

or in equivalent way

$$\varphi^2 = \frac{\Omega^2\left[(\overline{H}_1^* + \gamma\overline{A}_2^*) + \frac{\beta}{2}\gamma(\overline{H}_1^*\overline{A}_3^* - \overline{H}_3^*\overline{A}_1^* + \overline{H}_4^*\overline{A}_2^* - \overline{H}_2^*\overline{A}_4^*)\right] - \gamma\overline{A}_2^*}{\overline{H}_1^*}. \tag{33}$$

Substitution of Eq. (33) into the real part of Eq. (31) gives the equation in $\Omega^2$ from which its critical value can be obtained:

$$\Omega_c^2 = \left[4\overline{A}_2^* - \beta\left(\Xi_3 - \overline{H}_4^*\overline{A}_2^*\right) \pm \beta\sqrt{(\overline{H}_3^*\overline{A}_1^* + \overline{H}_2^*\overline{A}_4^*)^2 - 4\overline{H}_1^*\overline{A}_2^*\Xi_1}\right] / \left[4\overline{A}_2^* - 2\beta\left(\Xi_3 - \overline{H}_4^*\overline{A}_2^*\right) + \beta^2\left(\overline{H}_1^*\Xi_1 - \overline{H}_4^*\Xi_3\right)\right] \tag{34}$$

being $\Xi_1 = \overline{H}_1^*\overline{A}_2^* - \overline{H}_2^*\overline{A}_1^* + \overline{H}_3^*\overline{A}_4^*$; $\Xi_2 = -\Xi_1 + \overline{H}_4^*\overline{A}_3^*$ and $\Xi_3 = \overline{H}_2^*\overline{A}_4^* - \overline{H}_4^*\overline{A}_2^* + \overline{H}_3^*\overline{A}_1^*$.

In this way, the correspondent critical wind speed value can be easily obtained for a two degree-of-freedom flutter.

It can be observed that, disregarding the terms in $\beta^2$, taking $A_4^* = H_4^* = 0$ and assuming the aerodynamic derivatives independent from the static torsional rotation, *i.e.* constant along the bridge span length, Eq. (31) reduces to the simplified frequency condition given in [10].

## 4 Numerical model for wind loads simulation

It is worth to observe that the analytical approach shown in section 2 and, for the particular cable-stayed scheme in section 3, enables the designer to evaluate the critical values of wind speed on the basis of a deep engineering understanding of the bridge's overall behaviour. The mechanical parameters governing the physical behaviour are put in evidence and therefore design concepts and procedures can follow. But the simple analytical criteria expressed by the algebraic equations stated above need, for an effective application, a computation of the flutter derivatives and this point motivates the development of a numerical tool.

In what follows, in order to evaluate the steady and non-steady wind loads on a bridge cross-section, it is proposed a numerical model employing an arbitrary Lagrangian Eulerian (A.L.E.) formulation of the flow problem. In other words, an arbitrary reference domain is introduced as a third domain additionally to the commonly used material and spatial domains [13].

The kinematics of a compressible, viscous fluid flow, with fluid's constitutive law of an ideal Newtonian gas in a fluid domain $\Omega_f$ about a rigid body $\Omega_b$, is governed by the mass, momentum and energy balance equations. In A.L.E. formulation they are written as:

$$\frac{\partial \rho}{\partial t} + c_i\frac{\partial \rho}{\partial x_i} + \rho\frac{\partial u_i}{\partial x_i} = 0 \text{in } \Omega_f \tag{35}$$

$$\frac{\partial(\rho u_i)}{\partial t} + c_j \frac{\partial(\rho u_i)}{\partial x_j} + \rho u_i \frac{\partial u_k}{\partial x_k} + \frac{\partial}{\partial x_i}\left(p + \frac{2}{3}\rho k\right) = \frac{\partial S_{ij}}{\partial x_j} \text{in } \Omega_f \tag{36}$$

$$\frac{\partial(\rho e)}{\partial t} + c_i \frac{\partial(\rho e)}{\partial x_i} + (\rho e + p)\frac{\partial u_i}{\partial x_i} = S^{\ell}_{ij}\frac{\partial u_i}{\partial x_j} - \frac{\partial}{\partial x_i}(\Lambda \frac{\partial T}{\partial x_i}) + \rho\varepsilon \text{in } \Omega_f \tag{37}$$

where $x_i$ are the spatial co-ordinates, $p$ is the fluid pressure, $T$ is the fluid temperature, $\Lambda$ is the air's thermal conductivity, $e$ is the specific internal energy, $S_{ij} = S^{\ell}_{ij} + S^{T}_{ij} = \mu\left(\frac{\partial u_i}{\partial x_j} + \frac{\partial u_j}{\partial x_i}\right) - \frac{2}{3}\mu\frac{\partial u_k}{\partial x_k}\delta_{ij}$ is the Newtonian viscous stress tensor with Stokes hypothesis, $\mu = \mu_\ell + \mu_T$ is the total (laminar and turbulent) fluid viscosity and $c_i$ is the A.L.E. convection velocity vector, given as difference between the material velocity $u_i$ and the reference domain velocity. It can be observed that if the reference domain is assumed to be coincident with the computational grid it appears evident the possibility to treat the body movement into the fluid domain.

The whole set of equations is defined on the time interval $(0, t^*)$ and on the bounded domain $\Omega_f$ with boundary $\partial\Omega_f$, split into its complementary subsets $\Gamma$ and $\Gamma_w$ (Fig. 5). The former, $\Gamma = \Gamma_u \cup \Gamma_s$, is the external boundary subset split into the Dirichlet boundary $\Gamma_u$ and Neumann boundary $\Gamma_s$. The latter, $\Gamma_w$, is the fluid-structure interface. The boundary and initial conditions for Eqs. (35-37) are imposed according to:

$$u_i = U_{oi} \quad \text{on } \Gamma_u \times (0, t^*); \qquad -pn_i + S_{ij}n_j = s_i \quad \text{on } \Gamma_s \times (0, t^*); \tag{38}$$

$$u_i = u_{oi} \quad \text{in } \Gamma_f \text{ for } t = 0; \qquad u_i = u_{bi} \quad \text{on } \Gamma_w \times (0, t^*); \tag{39}$$

where $n_i$ is the outward unit normal vector for $\Gamma_s$ and $u_{bi}$ is the rigid velocity of $\Omega_b$.

Furthermore, energy dissipation due to turbulence effects is represented through a two-equation $k-\varepsilon$ RNG (Renormalization Group) model. In detail, turbulent kinetic energy $k$ and its dissipation rate $\varepsilon$ are governed by the following A.L.E. partial differential equations:

$$\frac{\partial(\rho k)}{\partial t} + c_i \frac{\partial(\rho k)}{\partial x_i} + \frac{5}{3}\rho k \frac{\partial u_i}{\partial x_i} = S^{T}_{ij}\frac{\partial u_i}{\partial x_j} + \frac{\partial}{\partial x_i}\left(\frac{\mu}{\mathrm{Pr}_k}\frac{\partial k}{\partial x_i}\right) - \rho\varepsilon \quad \text{in } \Omega_f \tag{40}$$

$$\frac{\partial(\rho\varepsilon)}{\partial t} + c_i \frac{\partial(\rho\varepsilon)}{\partial x_i} + (1 + a_{\varepsilon 13})\rho\varepsilon\frac{\partial u_i}{\partial x_i} = \frac{\varepsilon}{k}[(a_{\varepsilon 1} - \lambda^*)S^{T}_{ij}\frac{\partial u_i}{\partial x_j} - a_{\varepsilon 2}\rho\varepsilon] + \frac{\partial}{\partial x_i}\left(\frac{\mu}{\mathrm{Pr}_\varepsilon}\frac{\partial \varepsilon}{\partial x_i}\right) \quad \text{in } \Omega_f \tag{41}$$

where closure coefficients and relative conditions are reported in [14–16].

In the simplified framework of a two-dimensional like approaching wind flow, the discretization of the whole set of equations is based on a two-dimensional transient finite volume approach and a weak form of the governing equations is [14]:

$$\int_{V(t)} \frac{\partial \mathbf{W}}{\partial t} dV + \int_{S(t)} [F(\mathbf{W}) - N(\mathbf{W})] \cdot \mathbf{n} dS = 0 \tag{42}$$

where $\mathbf{W} = (\rho, \rho\mathbf{u}, \rho e, \rho k, \rho\varepsilon)^T$ is the conserved quantities numerical vector, $F$ and $N$ are the advection and diffusion operator, respectively, $V(t)$ is the control (two-dimensional) *volume*, $S(t)$ its boundary (one-dimensional) *surface* with outward unit normal vector $\mathbf{n}$.

The transient solution is marched out in a sequence of finite time increments. The calculation for each time step is divided into two phases. A material phase is performed first, in which the mesh moves with the fluid particles. In this phase the changes in velocity and other thermodynamic quantities due to the diffusion effects are calculated. In the second phase, the advection phase, transport of mass, internal energy, momentum and turbulent quantities across cell boundaries are computed. This may be thought as re-mapping the displaced mesh at material phase back to its original or arbitrary position. In other words, the positions of the fluid nodes may be arbitrarily specified functions of time, thereby allowing a Lagrangian, Eulerian or mixed description.

Let $\Omega_h = \cup_j Q_j$ a discretization of $\Omega_f$ through quadrangular *volumes* with a structured topological scheme and let $\Omega_h = \cup_i C_i$ an additional partition of $\Omega_f$ by node-centred cells obtained as represented in Fig. 6.

Spatial discretization of the governing equations is obtained by a second order approximation and it is performed by integrating the differential terms over the *volumes* of the cells $Q_j$ or $C_i$. *Volume* integrals of gradient or divergence terms are converted into *surface* area integrals using the divergence theorem. In this way, reducing the derivation order, the velocity field and the thermodynamic quantities may be approximated by piecewise constant functions on the integration domain. In detail, the transient solution at discrete time $t^n$ for the velocity field is approximated with uniform functions

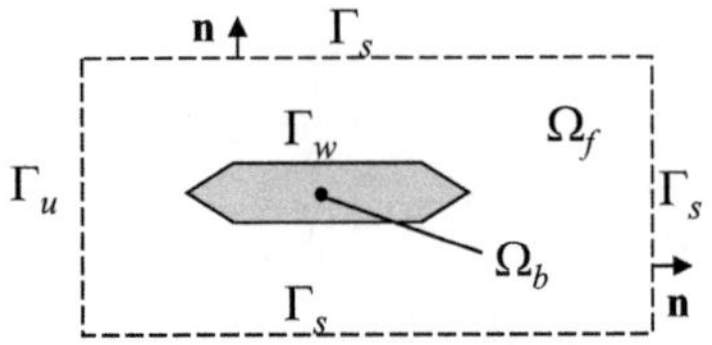


**Fig. 5.** Computational domain.

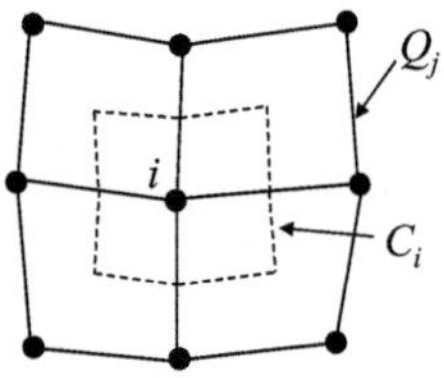


**Fig. 6.** Control *volume* definition.

on each $C_i$ cell. Moreover, the other thermodynamic quantities fields are approximated by uniform functions on the $Q_j$ quadrangular *volumes*. That is, the weak form of the conservation conditions corresponding to the Eq. (36) is formulated assuming the $C_i$ cells as control *volumes*. On the other hand, $Q_j$ cells represent the integration *volumes* for the other conservation laws.

The *volume* integrals of time derivatives are related to the derivatives of the integrals by means of Reynolds' transport theorem. A first order approximation is used for time derivatives, with an implicit temporal difference scheme. In detail, implicit differencing is used for all the diffusion and pressure terms during the material phase. The coupled equations are solved by an implicit method, with individual equations being solved by the conjugate residual method [17].

The convective transport associated with the mesh moving is accomplished in a sub-cycled explicit calculation using a sub-cycle time step, that is a sub-multiple of the main computational time step $\Delta t$. In this way only the sub-cycle time step is restricted by the well known Courant stability condition of explicit methods, whereas the main time step is forced to respect accuracy conditions which arises by imposing a first order approximation in time and limiting the amount of cell distortion that occur due to mesh motion in the material phase.

Hence, advection terms are evaluated by a quasi-second order upwind (QSOU) scheme, based on the assumption that the fluxes in each co-ordinate direction depend only on gradients in that co-ordinate direction.

In detail, if $F_j^\nu$ represents the generic thermodynamic quantity to be convected (or momentum term) defined on the $j$th $Q$-type (or $C$-type) control *volume* and $\nu$ is the convective sub-cycle number relative to the time level $n$, the cell-centered derivatives on $j$th $Q$-type (or $C$-type) cell of $F$ with respect to distance $s$ along the $i$th co-ordinate direction is evaluated as:

$$\left.\frac{\partial F}{\partial s}\right|_j^\nu = \begin{cases} \operatorname{sgn}(\Delta F_j^\nu) \min\left(\frac{|\Delta F_j^\nu|}{|\Delta \mathbf{x}_j|}, \frac{|\Delta F_{j-1}^\nu|}{|\Delta \mathbf{x}_{j-1}|}\right) & \text{if } \Delta F_j^\nu \Delta F_{j-1}^\nu > 0 \\ 0 & \text{if } \Delta F_j^\nu \Delta F_{j-1}^\nu < 0 \end{cases} \qquad (43)$$

where $\Delta F_j^\nu = F_{j+1}^\nu - F_j^\nu$ and $\Delta \mathbf{x}_j = \bar{\mathbf{x}}_{j+1} - \bar{\mathbf{x}}_j$, being $\bar{\mathbf{x}}_j$ the cell-centered locations for the $j$th $Q$-type (or $C$-type) cell, evaluated using the updated vertex spatial locations. The indices $(j+1)$ and $(j-1)$ indicate the computational cells adjacent to $j$th $Q$-type (or $C$-type) cell along the $i$th topological direction.

It can be proved that the QSOU scheme is strongly monotone and thus it does not introduce numerical oscillations in the computed solution, *i.e.* it does not produce undershoots nor overshoots of many higher order schemes [17].

Dirichlet boundary condition is imposed by pseudo-*infinite* control *volumes* on $\Gamma_u$ so that the prescribed velocity is imposed at a great distance from the structural domain $\Omega_b$. In particular, the velocity condition on the inflow mesh boundary is evaluated by considering the approximated one-dimensional momentum conservation equation expressed in spatial form, in which dissipation terms and compressibility effects are disregarded:

$$\frac{\partial u}{\partial t} + \frac{\partial (u^2)}{\partial x} + \frac{1}{\rho}\frac{\partial p}{\partial x} = 0. \tag{44}$$

With reference to Fig. 7, Eq. (44) is integrated over inlet node-centred control *volumes* and discretized imposing the free-field velocity $U_o$ and the atmospheric pressure at a distance $d$ from the inlet mesh boundary greater than 5div $10\delta x$. An explicit approach is used and the advection terms are evaluated by a first-order full upwind scheme.

Furthermore, Neumann condition is performed by the imposition of a zero traction gradient on $\Gamma_s$, *i.e.* by imposing a continuative outflow condition. In other words, the viscous stresses $S_{ij}$ inside the outflow boundary cells are computed and these are iteratively imposed as traction outside the outflow boundary. Moreover, for each outlet boundary control *volume* the external pressure $p_j^{out}$ acting on the Neumann boundary part of the $j$th cell is obtained by a linear interpolation, taking into account the computed boundary cell pressure value $p_j$ and the atmospheric pressure $p_{atm}$ at a distance $d_{atm}$ much greater than the characteristic cell dimension $\delta x$, so that it results $\partial p/\partial \mathbf{n} \approx 0$ and

$$p_j^{out} = \frac{\delta x p_{atm} + 2d_{atm} p_j}{\delta x + 2d_{atm}}. \tag{45}$$

In this way, the oscillatory wake downstream $\Omega_b$ is not numerically reflected from the external boundaries, as shown in [14].

Finally, turbulent law-of-the-wall conditions are adopted on $\Gamma_w$ in order to simulate the boundary layer turbulent effects. In detail, the wall stress $\tau_w = {u^*}^2 \rho_w$ is evaluated by matching to a logarithmic profile, depending on the near-wall flow velocity field, and the flow momentum equilibrium about $\Gamma_w$ is imposed as a function of the local Reynolds number with respect to its critical value $Re_c$ [14–16]:

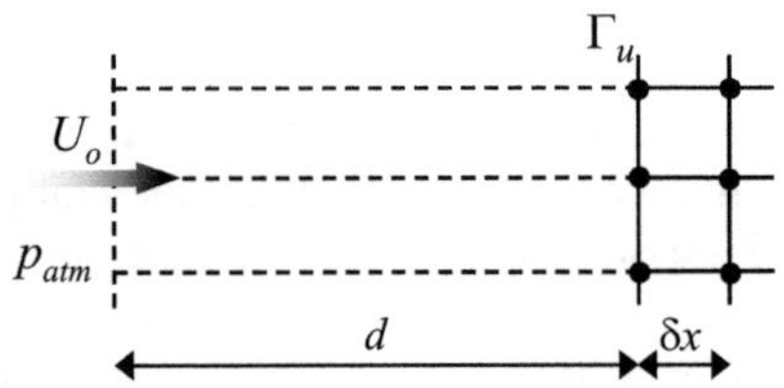


**Fig. 7.** Pseudo-*infinite* control *volume* on the Dirichlet boundary.

$$\frac{u}{u^*} = \begin{cases} Re_c^{1/2} + \frac{7}{8}(\kappa)^{-1} \ln\left(\frac{Re}{Re_c}\right) & Re > Re_c \\ Re^{1/2} & Re < Re_c \end{cases} \tag{46}$$

where $\kappa$ is the well known Kármán constant, related to the turbulence model's closure coefficients [16].

The mesh moving, which enables to consider the arbitrary structural domain movement, is performed with an algorithm able to preserve the convexity of the cells. This condition is necessary in order to ensure a suitable computational accuracy. In detail, the arbitrary displacement conditions on the structure are prescribed as rigid movements of the fluid nodes core close $\Omega_b$, whereas for external fluid nodes a linear decaying law propagates the structural displacement. In this way no topological modification of the grid is requested and it is easy to preserve a fine mesh close $\Omega_b$ [14] (Fig. 8).

To obtain the time-traces of aerodynamic forces acting on $\Omega_b$ the pressure distribution and the wall viscous stresses are integrated on the boundary $\Gamma_w$ for each time step. Moreover, the numerical extraction of the non-stationary aeroelastic actions may be performed by the following considerations.

If drag aeroelastic contribution is disregarded and if a two degree-of-freedom model for the in-plane cross-section motion is assumed, only $L_{ae}$, $M_{ae}$ and eight flutter derivatives ($H_i^*$, $A_i^*$, $i = 1, \ldots, 4$) are involved in Eqs. (2-4). In the spirit of the quasi-steady theory, the displacement functions are assumed to be harmonic in time, *i.e.* $v = \overline{v}e^{i\omega t}$ and $\theta = \overline{\theta}e^{i\omega t}$ ($i$ is the imaginary unit). Consequently, the aeroelastic forces themselves can be considered harmonic with circular frequency $\omega$, but with a phase difference $\tilde{\varphi}$ in comparison with the motion. With these hypotheses and considering the flutter derivatives constant along the bridge span length, the following set of equations can be written [15]:

$$\frac{C_{Lo}Be^{i\tilde{\varphi}_L^{(v)}}}{K^2\overline{v}} = iH_1^* + H_4^* \qquad \frac{C_{Mo}Be^{i\tilde{\varphi}_M^{(v)}}}{K^2\overline{v}} = iA_1^* + A_4^* \tag{47}$$

$$\frac{C_{Lo}e^{i\tilde{\varphi}_L^{(\theta)}}}{K^2\overline{\theta}} = iH_2^* + H_3^* \qquad \frac{C_{Mo}e^{i\tilde{\varphi}_M^{(\theta)}}}{K^2\overline{\theta}} = iA_2^* + A_3^* \tag{48}$$

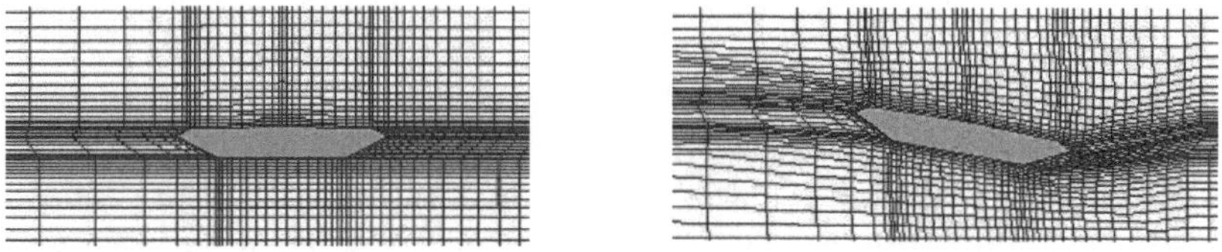

**Fig. 8.** Example of rotational mesh moving.

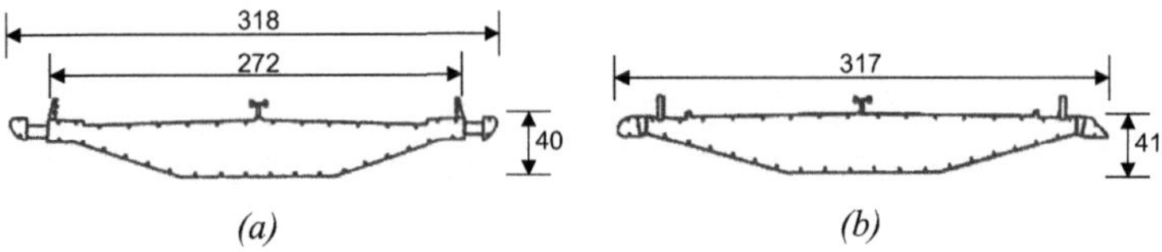


**Fig. 9.** Design section model profiles (scale 1:70, dimension in mm) of the 1987 (*a*) cross-section (*Avant Projet*) and of the 1989 (*b*) cross-section (*Projet Détaillé*) for the Normandy Bridge.

Hence, imposing a forced harmonic motion in vertical or angular degree of freedom and extracting the amplitude $C_{Lo}$, $C_{Mo}$ and the phase shift of the induced dimensionless lift and moment at the same frequency of the imposed motion, the aerodynamic derivatives may be evaluated from the Eqs. (47-48).

The proposed numerical model has been successfully validated by comparison with available data concerning some simple cross-section shapes (circular, square, rectangular, etc.) and good agreement with the steady and non-steady experimental wind tunnel test results has also been found for the Great Belt East and West Bridge cross-section models, as reported in [14,15].

## 5 Numerical results

Numerical simulations are carried out for the design profiles of the 1987 (*Avant-Projet*) and 1989 (*Projet Détaillé*) cross-section models of the Normandy Bridge (Fig. 9).

Scope of these simulations is the assessment of the steady and non-steady unit wind forces acting on the bridge. All simulations are run for a dimensionless time interval $0 < tU_o/B < 35$. Moreover, input models reproduce the large scale geometry of the cross-section models but finer details (railings and median dividers) are omitted.

### 5.1 Simulation of steady aerodynamic forces

With reference to the 1989 design cross-section model of the Normandy Bridge a mesh with about 27000 cells and 36000 nodes has been used and, close $\Gamma_w$, with a ratio $\delta x/B \cong 1.9 \times 10^{-2}$, being $\delta x$ the characteristic mesh dimension.

The Fig. 10 depicts the comparison between the experimental wind tunnel test results measured by the ONERA and the CEBTP on the 1989 profile [18] and the numerical ones for the dimensionless static coefficients defined in the Eqs. (1) and referred to the profile axes reference. Their variability with the angle of incidence is numerically proved varying $\theta$ from $-6$ to 6 degrees. The Reynolds number for the calculations is set to $Re = \rho U_o B/\mu = 1.5\times 10^5$,

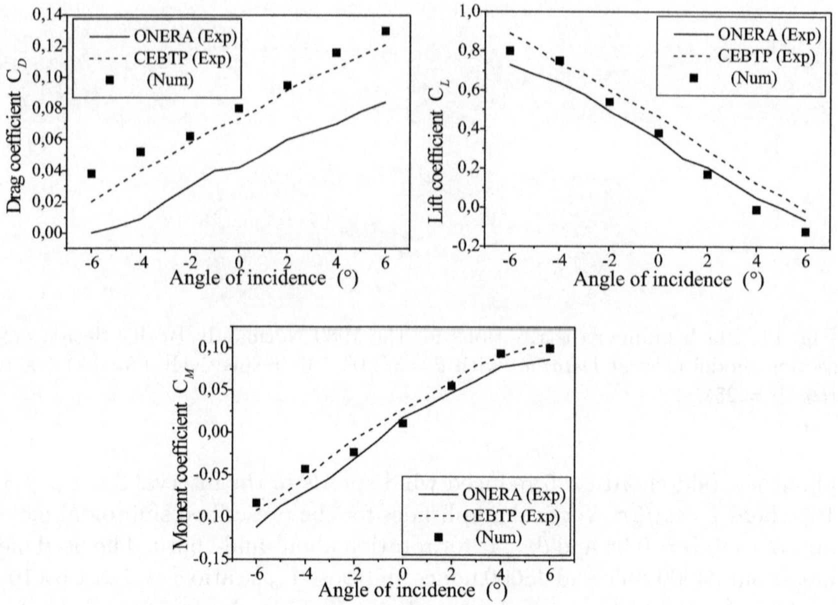


**Fig. 10.** Comparison of the steady aerodynamic coefficients (profile-axes referred) measured by the ONERA and the CEBTP and the numerical ones for the 1989 Normandy Bridge design cross-section model (*Projet Détaillé*).

whereas the inflow turbulence intensity is assumed close to 5%. Comparison with the 1989 ONERA results reveals to be very satisfactory for the lift and moment coefficients, but the drag coefficient appears greater than in ONERA tests, as confirmed by the CEBTP results obtained in turbulent wind [18]. Furthermore, numerically obtained Mach number distributions are represented in Fig. 11 at the dimensionless time $tU_o/B = 25$ for several values of $\theta$.

Finally, it can be observed that the cause of the little discrepancies between the computed and measured values, especially in the $\theta$ negative cases, can be considered as a consequence of the absence in the computed model of finer details, *i.e.* railings and safety barriers. Consequently, with negative $\theta$, the computed flow is more attached to the upper surface than the experimental one in wind tunnel tests.

## 5.2 Simulation of non-steady aeroelastic forces

The non-steady aeroelastic forces can be evaluated through the Eqs. (2-4) by employing the flutter derivatives. With reference to the aerodynamic derivatives extraction procedure explained above, several cross-section forced motion simulations are carried out for the 1987 Normandy Bridge design profile. The simulations involves separate oscillatory vertical and rotational motion

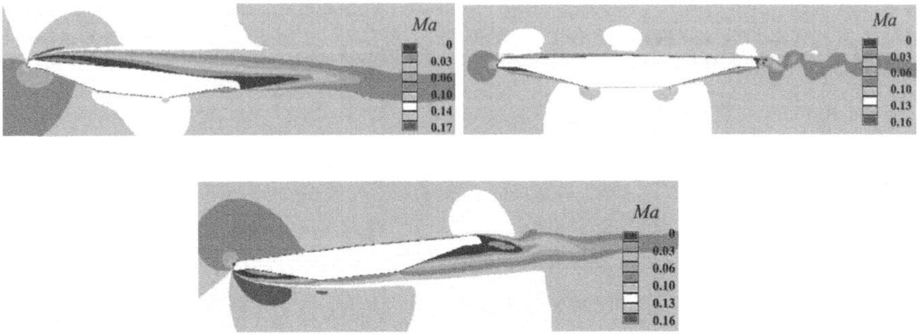


**Fig. 11.** Mach number distributions for the 1989 Normandy Bridge design cross-section model (*Projet Détaillé*) with $\theta = 6°, 0°, -6°$ respectively ($Re = 1.5 \times 10^5$, $tU_o/B = 25$).

about the mid-chord for 5 reduced wind speeds in the interval $2 < U_o/fB < 10$, where $f = \omega/2\pi$. Vertical amplitudes for the prescribed sinusoidal motion are set to $\overline{v}/B = 0.04$ and $\overline{\theta} = 4°$ for rotation about mid-chord. The used mesh has about 34000 cells and 45000 nodes and close $\Gamma_w$ a ratio $\delta x/B \cong 1.6\times10^{-2}$.

In order to characterize the aerodynamic derivatives as functions of the static torsional rotation, three different values of the mean angle of incidence are considered in the computation. As shown in Fig. 12, the flutter derivatives numerically obtained are in good agreement with the correspondent wind tunnel test results and they present an high variability also under small steady torsional rotation. Moreover, tendency to a one degree-of-freedom vertical and torsional flutter appears evident for $\theta_o$ greater than zero.

It is worth to observe that the experimental data, obtained by the ONERA and given in [18], have been transformed to generate the curves of Fig. 12 which are compatible with the expressions (2-4).

The little discrepancies noted between simulated and measured values can be resolved by two plausible causes. Simulated derivatives are obtained directly from the non-steady forces originated for harmonic single degree-of-freedom motions. On the other hand, the wind tunnel test results are generally obtained imposing to the section model a complex movement, combining flexion and torsion and combining different frequencies. Another important aspect is a possible non-linear effect of motion amplitudes, whose experimental test values are not available.

## 6 Concluding remarks

In this paper the generalization of some classical models in order to check the aeroelastic stability to divergence and flutter phenomena for long-span bridges, with particular reference to the cable-stayed scheme, has been presented. This latter scheme is analyzed through a continuous approach based

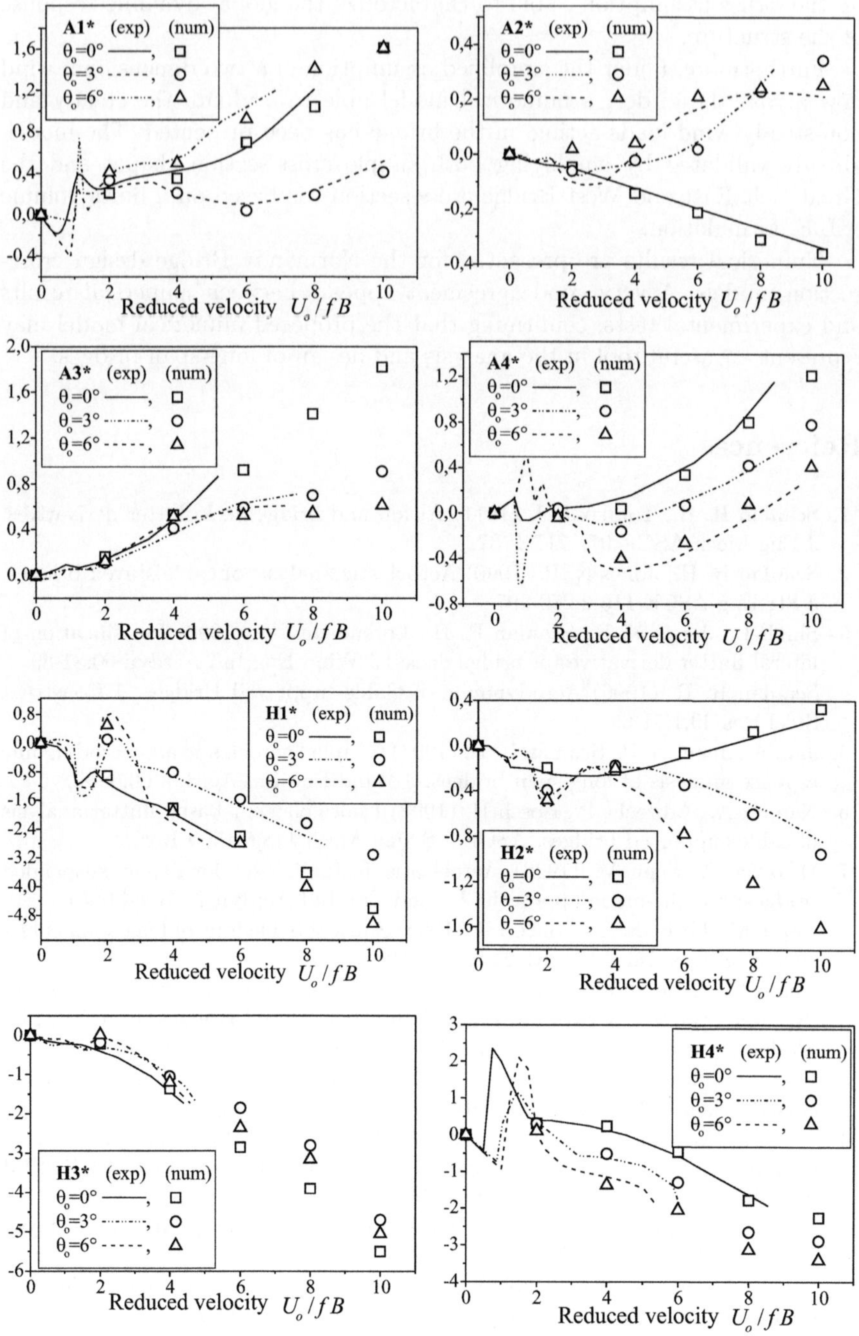


**Fig. 12.** Flutter derivatives for the 1987 Normandy Bridge design cross-section model (*Avant Projet*).

on the 'truss assumption', able to characterize the global dynamic response of the structure.

Furthermore, under the simplified assumption of a two-dimensional wind flow across the girder, a numerical model able to evaluate the steady and non-steady wind loads acting on the bridge has been presented. The model, already validated by employing both simple cross-section shapes and the Great Belt East and West Bridge cross-sections, is based on a finite volume A.L.E. formulation.

Numerical results are presented for the Normandy Bridge design cross-section models. A very good agreement appears between numerical results and experimental tests, confirming that the proposed numerical model may represent an useful tool in the analysis and design of long-span bridges.

## References

1. Scanlan R. H., Tomko J. J. (1971) Airfoil and bridge deck flutter derivatives. J Eng Mech ASCE 97:1717-1737
2. Scanlan R. H., Jones N. P. (1990) Aeroelastic analysis of cable-stayed bridges. J Str Eng ASCE 116,2:279-297
3. Singh L., Jones N. P., Scanlan R. H., Lorendeaux O. (1996) Identification of lateral flutter derivatives of bridge decks. J Wind Eng Ind Aerdyn 60:81-89
4. Scanlan R. H. (1996) Aerodynamic of Cable-supported Bridges. J Construct Steel Res 39,1:51-68
5. Jain A., Jones N. P., Scanlan R. H. (1996) Coupled aeroelastic and aerodynamic esponse analysis of long-span bridges. J Wind Eng Ind Aerdyn 60:69-80
6. Namini A., Albrecht P., Bosch H. (1992) Finite element-based flutter analysis of cable-supported bridges. ASCE J Struct Mech 118,6:1509-1526
7. D'Asdia P., Sepe V. (1998) Aeroelastic instability of long-span suspended bridges: a multi-mode approach. J Wind Eng Ind Aerdyn 74-76:849-857
8. Como M., Grimaldi A., Maceri F. (1985) Statical behaviour of long-span cable-stayed bridges. Int J Sol Str 21,8:831-850
9. Como M., Grimaldi A., Maceri F. (1981) A Continuous Model for the Aerodynamic Stability Analysis of Long-Span Cable-Stayed Bridges. Proc Eur Sim Modelling and Simulation of Large Scale Structural Systems, Capri
10. Bruno D., Leonardi A., Maceri F. (1987) On the non-linear dynamics of cable-stayed bridges. Proc Int Conf Cable-Stayed Bridges, Bangkok
11. Como M., Leonardi A., Maceri F. (1993) Modelling aerostatic and aerodynamic coupling in flutter analysis of long-span cable-stayed bridges. Proc. EUROSIM '92, Simulation Congress. Elsevier, 377-392
12. Bruno D., Leonardi A. (1997) Natural periods for long-span cable-stayed bridges J Bridge Eng 2,3:105-115
13. Hirt W., Amsden A. A., Cook J. L. (1974) An arbitrary Lagrangian-Eulerian computing method for all flow speeds. J Comput Phys,14:227
14. Vairo G., Ammassari L. (2000) Una tecnica di simulazione numerica per il calcolo dei coefficienti aeroelastici delle sezioni da ponte. Proc XXIX Convegno Associazione Italiana Analisi Sollecitazioni:779-788 - Lucca

15. Vairo G. (2000) Wind loads simulation for long-span cable-stayed bridges. Proc Italian Society for Computer Simulation (ISCS) Conference - Lecce
16. Wilcox D. C. (1998) Turbulence Modelling for CFD. 2nd Ed, DCW Ind, California
17. Roache P. J. (1982) Computational Fluid Dynamics. Hermosa Publishers, Albuquerque, New Mexico
18. Virlogeux M. (1992) Wind design and analysis for the Normandy Bridge. In: Aerodynamics of Large Bridges, Balkema, Rotterdam

# Collisions Involving Solids and Fluids

Eric Dimnet[1], Michel Frémond[1], Raul Gormaz[2], Jorge San Martin[2]

[1] Laboratoire Central des Ponts et Chaussées
75732 Paris Cedex, France
[2] Departamento de Ingeniería Matemática
Universidad de Chile
Santiago, Chile

**Abstract.** Predictive theories of instantaneous collisions involving rigid and deformable solids as well as fluids are described. They are based on the description of interior percussions to the system made of the colliding bodies. The system made of all the elements (solids or fluids) that are colliding is a deformable system: its form changes even if it is made of rigid elements! If the duration of a collision is small compared to the duration of the evolution, we assume that the collision is instantaneous; thus the velocities are discontinuous We describe the collision of a point with a fixed plane and the simultaneous collisions of a collection of rigid bodies. The impact of an hammer with a bar is an example of collisions of deformables bodies. Experiments show the adequation of the theory. The collision of fluids and solids is illustrated by the description of the belly flop of a diver in a swimming pool.

## 1 Collisions of solids

### 1.1 Collision of a rigid body with a plane

**Collision of a point with a plane** Let us consider a point, moving above a rigid fixed plane. Its position at time $t$ is $\boldsymbol{x}(t)$. The system made of the point and the plane is deformable because the distance of the point to the plane changes. The deformation velocity of the system point-plane is the velocity of the point with respect to the plane, $\boldsymbol{U}(t) = d\boldsymbol{x}(t)/dt$. In many circumstances the duration of the collisions of the point and of the plane is small compared to the duration of the motion: thus we consider that the collisions are instantaneous. There is the velocity $\boldsymbol{U}^-(t)$ before a collision at time $t$, and the velocity $\boldsymbol{U}^+(t)$ after the collision. A virtual velocity $\boldsymbol{V}(\tau)$ is a bounded variation function.

We specify the interior forces of the system point-plane by defining their virtual work. The virtual work of the interior forces is a linear function of the velocity of deformation which has to be zero for any rigid system motion. In our situation, because one element of the system (the plane) is fixed, the rigid system motions, i.e., the motions which do not change the form of the system, are reduced to the motion with zero velocity.

Consider the actual motion shown on figure 1 and a virtual velocity which is discontinuous at time $t$. The virtual work of the interior forces we choose is

$$W^{int}(t_1,t_2,\boldsymbol{V}) = -\int_{t_1}^{t_2} \boldsymbol{R}^{int}(\tau)\cdot\boldsymbol{V}(\tau)d\tau - \boldsymbol{P}^{int-}(t)\cdot\boldsymbol{V}^-(t) - \boldsymbol{P}^{int+}(t)\cdot\boldsymbol{V}^+(t),$$

where $\boldsymbol{R}^{int}(\tau)$ is the contact force between the point and the plane. Both quantities, $\boldsymbol{P}^{int-}(t)$ and $\boldsymbol{P}^{int+}(t)$ are percussions.

The virtual work of the acceleration forces is

$$W^{acc}(t_1,t_2,\boldsymbol{V}) = \int_{t_1}^{t_2} m\frac{d\boldsymbol{U}(\tau)}{d\tau}\cdot\boldsymbol{V}(\tau)d\tau + m(\boldsymbol{U}^+(t)-\boldsymbol{U}^-(t))\cdot\frac{\boldsymbol{V}^-(t)+\boldsymbol{V}^+(t)}{2}.$$

Let us note that the virtual work of the acceleration forces is such that the actual work is the variation of the kinetic energy between times $t_1$ and $t_2$. We suppose that there are no exterior forces applied to the system. The equations of motion result from the principle of virtual work

$$\forall t_1, \forall t_2, \forall \boldsymbol{V}, W^{acc}(t_1,t_2,\boldsymbol{V}) = W^{int}(t_1,t_2,\boldsymbol{V})$$

They are

$$m\frac{d\boldsymbol{U}}{d\tau} = -\boldsymbol{R}^{int},$$

almost everywhere; and

$$\boldsymbol{0} = \boldsymbol{P}^{int-}(t) - \boldsymbol{P}^{int+}(t),$$

$$m(\boldsymbol{U}^+(t) - \boldsymbol{U}^-(t)) = -\boldsymbol{P}^{int}(t),$$

where $\boldsymbol{P}^{int}$ is defined by $2\boldsymbol{P}^{int}(t) = \boldsymbol{P}^{int-}(t) = \boldsymbol{P}^{int+}(t)$ at any time $t$ (figure 1). The power of the interior forces becomes

$$W^{int}(t_1,t_2,\boldsymbol{V}) = -\int_{t_1}^{t_2} \boldsymbol{R}^{int}(\tau).\boldsymbol{V}(\tau)d\tau - \boldsymbol{P}^{int}(t)\cdot\frac{\boldsymbol{V}^-(t)+\boldsymbol{V}^+(t)}{2}. \tag{1}$$

*Remark 1.* A more sophisticated theory involving the two percussions $\boldsymbol{P}^{int-}(t)$ and $\boldsymbol{P}^{int+}(t)$ is investigated in [10].

From now on we consider only the collisions and do not investigate the smooth motion.

**The rigid body is not a point** A solid with mass $m$, center of mass $G$, mass moment of inertia $I$, collides the plane at a unique point $A$. The velocity of $G$ is $\boldsymbol{U}$, and the rotation velocity of the solid is $\boldsymbol{\Omega}$. The equations of motion at the time of collision result from the principle of virtual work

$$\begin{aligned}
&\forall \boldsymbol{V}, \forall \boldsymbol{\omega},\ m(\boldsymbol{U}^+ - \boldsymbol{U}^-)\cdot\frac{\boldsymbol{V}^+ + \boldsymbol{V}^-}{2} + I(\boldsymbol{\Omega}^+ - \boldsymbol{\Omega}^-)\cdot\frac{\boldsymbol{\omega}^+ + \boldsymbol{\omega}^-}{2}\\
&= -\boldsymbol{P}^{int}\cdot(\frac{\boldsymbol{D}(\boldsymbol{V}^+,\boldsymbol{\omega}^+,A) + \boldsymbol{D}(\boldsymbol{V}^-,\boldsymbol{\omega}^-,A)}{2})\\
&\boldsymbol{D}(\boldsymbol{V},\boldsymbol{\omega},A) = \boldsymbol{V} + \boldsymbol{\omega}\times\boldsymbol{GA}.
\end{aligned} \tag{2}$$

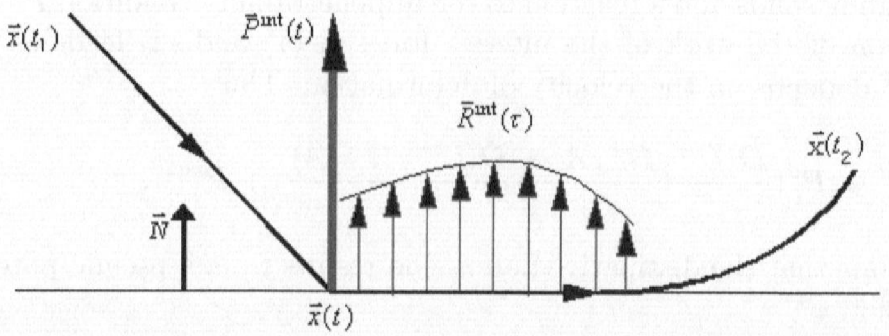


**Fig. 1.** A point with position $\boldsymbol{x}(\tau)$ moves above a plane. It collides with the plane at time $t$. An interior percussion $\boldsymbol{P}^{int}(t)$ results at time $t$, and an interior force $\boldsymbol{R}^{int}(\tau)$ when the point slides on the plane.

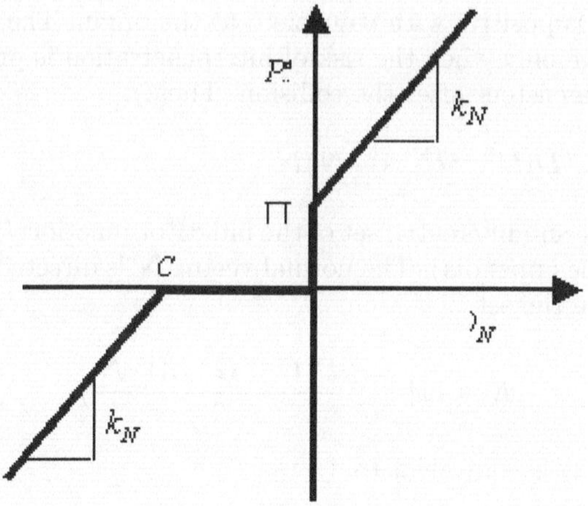


**Fig. 2.** Constitutive law describing collisions with adhesion.

where $\boldsymbol{D}(\boldsymbol{V}, \boldsymbol{\omega}, A)$ is the velocity of deformation of the system at point $A$, i.e. the relative velocity of the point with respect to the plane. They are

$$m(\boldsymbol{U}^+ - \boldsymbol{U}^-) = -\boldsymbol{P}^{int},$$
$$I(\boldsymbol{\Omega}^+ - \boldsymbol{\Omega}^-) = -\boldsymbol{GA} \times \boldsymbol{P}^{int}.$$

The situation where the contact occurs at points, lines or surfaces is investigated in [10].

*Remark 2.* A consequence of the chosen work of the interior forces in (2) is that the interior percussion $\boldsymbol{P}^{int}$ is applied at the collision point $A$.

**Constitutive laws** The interior percussion $\boldsymbol{P}^{int}$ is split between a dissipative percussion $\boldsymbol{P}^d$ which modelises all the dissipative interactions between the colliding solids and a reaction to the impenetrability condition $\boldsymbol{P}^{reac}$. The expression of the work of the interior force in (1) and (2) leads to assume that $\boldsymbol{P}^{d}$ depends on the velocity of deformation. Thus

$$\boldsymbol{P}^{int} = \boldsymbol{P}^{d}(\frac{\boldsymbol{D}(\boldsymbol{U}^{+}, \boldsymbol{\Omega}^{+}, A) + \boldsymbol{D}(\boldsymbol{U}^{-}, \boldsymbol{\Omega}^{-}, A)}{2}) + \boldsymbol{P}^{reac}. \tag{3}$$

We assume that the dissipative percussion results from a pseudo-potential of dissipation $\phi^{d}$

$$\boldsymbol{P}^{d} \in \partial\phi^{d}(\frac{\boldsymbol{D}(\boldsymbol{U}^{+}, \boldsymbol{\Omega}^{+}, A) + \boldsymbol{D}(\boldsymbol{U}^{-}, \boldsymbol{\Omega}^{-}, A)}{2}),$$

where $\partial\phi^{d}$ is the subdifferential set (see the appendix). Let us recall that a pseudo-potential of dissipation, as introduced by J.J. Moreau, is a convex function which is positive with value zero at the origin. The impenetrability reaction is active only when the risk of interpenetration is present, i.e. when the contact is persistent after the collision. Thus

$$\boldsymbol{P}^{reac} \in \partial I_{+}(\boldsymbol{D}(\boldsymbol{U}^{+}, \boldsymbol{\Omega}^{+}, A) \cdot \boldsymbol{N})\boldsymbol{N}. \tag{4}$$

where $\partial I_{+}$ is the subdifferential set of the indicator function $I_{+}$ of the positive numbers (see the appendix). The normal vector $\boldsymbol{N}$ is directed upward (figure 1). Let us define the set

$$K = \{x |\, x \geq \frac{\boldsymbol{D}(\boldsymbol{U}^{-}, \boldsymbol{\Omega}^{-}, A) \cdot \boldsymbol{N}}{2}\}.$$

Then relation (4) is equivalent to

$$\boldsymbol{P}^{reac} \in \partial I_{K}(\frac{\boldsymbol{D}(\boldsymbol{U}^{+}, \boldsymbol{\Omega}^{+}, A) + \boldsymbol{D}(\boldsymbol{U}^{-}, \boldsymbol{\Omega}^{-}, A)}{2} \cdot \boldsymbol{N})\boldsymbol{N},$$

where $I_{K}$ is the indicator function of the set $K$ (see the appendix). Thus by defining the functions of $\boldsymbol{D}$

$$\widetilde{I}_{K}(\boldsymbol{D}) = I_{K}(\boldsymbol{D} \cdot \boldsymbol{N})$$

and

$$\phi = \phi^{d} + \widetilde{I}_{K},$$

the constitutive law (3) becomes

$$\boldsymbol{P}^{int} \in \partial\phi(\frac{\boldsymbol{D}(\boldsymbol{U}^{+}, \boldsymbol{\Omega}^{+}, A) + \boldsymbol{D}(\boldsymbol{U}^{-}, \boldsymbol{\Omega}^{-}, A)}{2}). \tag{5}$$

Let us note that the pseudo-potential $\phi$ depends on the velocity of deformation before the collision via the function $\widetilde{I}_{K}$.

*Examples of constitutive laws depending on a pseudo-potential.* Let us consider the point and plane described in paragraph **1.1.1.** The normal velocity of the point before the collision $\boldsymbol{U}^- \cdot \boldsymbol{N}$ and after the collision $\boldsymbol{U}^+ \cdot \boldsymbol{N}$ are denoted $U_N^-$ and $U_N^+$ . For the sake of simplicity, we only investigate the normal part of the percussion and we assume that the tangential percussions are zero.

A very simple pseudo-potential is a quadratic function $\Phi(\boldsymbol{D}) = k_N(\boldsymbol{D} \cdot \boldsymbol{N})^2$, which corresponds to a linear dissipative percussion which is for the point $\boldsymbol{P}^d = \left\{k_N(U_N^+ + U_N^-)/2\right\} \boldsymbol{N}$. In the example of paragraph **1.1.1** involving the system point-plane, one can easily show [7] that the linear percussion leads to a normal restitution coefficient $e = (k_N - m)/(k_N + m)$ if $k_N > m$ and $e = 0$ if $k_N \leq m$.

Collisions involving adhesion are collisions such as the interior or exterior percussion has to be large enough for separation to occur after. The constitutive law shown in the graph of figure 2 has this property. The two parameters $C$ and $\Pi$ characterise the properties of adhesion. In the example of the system point-plane:

- if the relative normal velocity of the point before the collision is small ($|U_N^-| \leq Ck_N/(k_N - m)$ ) the adhesion forces are strong enough for the point to remain in contact after a collision ($k_N$ is the slope of the linear parts of the law assumed greater than $m$ (otherwise contact of the point is always persistent after a collision));
- if the point is in contact with the plane, an exterior percussion has to be larger than $\Pi$ to separate them [7].

*A constitutive law which does not result from a pseudo-potential: the Coulomb law.* Experiments of collisions in two dimensions of dry rigid solids with a fixed plane (dry angular steel particles colliding at the point $A$ a dry marble plane, whose outward normal unit vector is $\boldsymbol{N}$) (figures 3 and 4) show that the normal behaviour can be modelised by a pseudo-potential, even quite well by a quadratic one (figure 3): $P_N^d = \boldsymbol{P}^d \cdot \boldsymbol{N} \in \partial\Phi_N\left(U_N^-(A) + U_N^+(A)\right)$,where $U_N(A)$ is the normal component of the velocity of the point $A$ belonging to the solid. But as far as the tangential behaviour is concerned, $P_T^d = \boldsymbol{P}^d \cdot \boldsymbol{T}$ ($\boldsymbol{T}$ is the tangent vector in the plane of experiments) can not be considered as a function of the only variable $U_T^-(A) + U_T^+(A)$: it depends also on $P_N^d$ (figure 4).

*Remark 3.* In our experiments, the colliding solids do not maintain contact after the collision so $\boldsymbol{P}^{reac}$ is always zero.

This behaviour is well described by

$$
\begin{aligned}
&\|\boldsymbol{P}_T\| \leq \mu\,|P_N| \;\; \textit{with}\;\; \mu > 0 \;\;\textit{and}\\
&\textit{if}\;\; \|\boldsymbol{P}_T\| < \mu\,|P_N| \;\;\textit{then}\;\; \boldsymbol{D}_T = \boldsymbol{0},\\
&\textit{if}\;\; \|\boldsymbol{P}_T\| = \mu\,|P_N| \;\;\textit{then}\\
&\exists\lambda > 0 \;\;\textit{such as}\;\; \boldsymbol{D}(\boldsymbol{U}^+, \boldsymbol{\Omega}^+, A)_T + \boldsymbol{D}(\boldsymbol{U}^-, \boldsymbol{\Omega}^-, A)_T = -\lambda \boldsymbol{P}_T,
\end{aligned} \tag{6}
$$

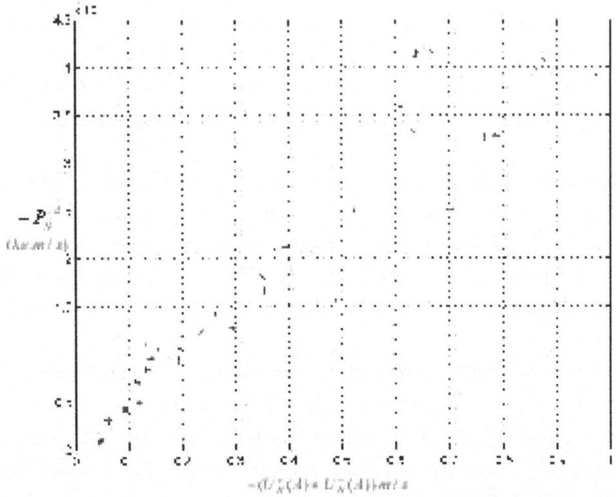

**Fig. 3.** The normal percussion $P_N^d$ versus the normal part of the velocity of deformation, $U_N^+(A) + U_N^-(A)$, for collisions of dry steel particles with a dry marble plane.

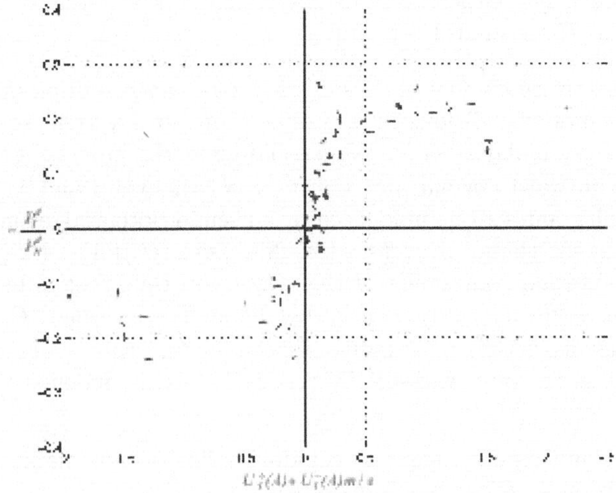

**Fig. 4.** The quotient $-(P_T^d/P_N^d)$ versus the tangential part of the velocity of deformation, $U_T^+(A) + U_T^-(A)$ , for collisions of dry steel particles with a dry marble plane.

noting $\boldsymbol{D}(\boldsymbol{U},\boldsymbol{\Omega},A)_T = \boldsymbol{D}(\boldsymbol{U},\boldsymbol{\Omega},A) - (\boldsymbol{D}(\boldsymbol{U},\boldsymbol{\Omega},A)\cdot\boldsymbol{N})\cdot\boldsymbol{N}$ and $\boldsymbol{P}_T = \boldsymbol{P}^{int} - \boldsymbol{P}^{int}\cdot\boldsymbol{N}$.

This constitutive law, called the Coulomb law, does not derive from a pseudo-potential.

## 1.2 Simultaneous collision of a collection of rigid bodies

Consider $N$ solids with mass $m_i$, center of mass $G_i$ mass moment of inertia tensor $I_i$, colliding together at time $t$. The velocity of the center of mass of solid $_i$ is $\boldsymbol{U}_i$ and its rotation velocity is $\boldsymbol{\Omega}_i$. We assumed that all the contacts are reduced to points. Multiple collisions can occur between solids $i$ and $j$ at points $A_{i,j,k}$. The set $S_{i,j}$ contains the points of contact of the two solids. If the two solids do not collide, this set is empty. The percussion $\boldsymbol{P}^{int}_{i,j,k}$ is applied at the collision point $A_{i,j,k}$. Let the virtual velocities of the centers of mass $G_i$ be $\boldsymbol{V}_i$ and the virtual rotation velocity be $\boldsymbol{\omega}_i$. Let us define the vectors $\hat{V} = (\boldsymbol{V}_i, \boldsymbol{\omega}_i)$, $\hat{U} = (\boldsymbol{U}_i, \boldsymbol{\Omega}_i)$ and the function

$$D_{i,j}(\hat{V}, A_{i,j,k}) = \boldsymbol{V}_i + \boldsymbol{\omega}_i \times G_i A_{i,j,k} - (\boldsymbol{V}_j + \boldsymbol{\omega}_j \times G_j A_{i,j,k}),$$

which gives the velocity of deformation at point $A_{i,j,k}$ (the relative velocity of the point). The principle of virtual work may be written

$$\begin{aligned}
&\forall \hat{V}, \sum_{i=1}^{N} \left\{ m_i(\boldsymbol{U}_i^+ - \boldsymbol{U}_i^-)\cdot(\boldsymbol{V}_i - \frac{\boldsymbol{U}_i^+ + \boldsymbol{U}_i^-}{2}) \right.\\
&\left. + I_i(\boldsymbol{\Omega}_i^+ - \boldsymbol{\Omega}_i^-)\cdot(\boldsymbol{\omega}_i - \frac{\boldsymbol{\Omega}_i^+ + \boldsymbol{\Omega}_i^-}{2}) \right\}\\
&+ \sum_{i=1}^{N-1} \sum_{j=i+1}^{N} \sum_{A_{i,j,k}\in S_{i,j}} \boldsymbol{P}^{int}_{i,j,k} \cdot \left\{ D_{i,j}(\hat{V}, A_{i,j,k}) \right.\\
&\left. - \left( D_{i,j}(\hat{U}^+, A_{i,j,k})) + D_{i,j}(\hat{U}^-, A_{i,j,k}) \right)/2 \right\} = 0.
\end{aligned} \tag{7}$$

By using the constitutive laws (5)

$$\boldsymbol{P}^{int}_{i,j,k} \in \partial\Phi_{i,j,k}\left(\frac{D_{i,j}(\hat{U}^+, A_{i,j,k})) + D_{i,j}(\hat{U}^-, A_{i,j,k})}{2}\right),$$

which take into account the impenetrability condition, the principle (7) becomes

$$\forall \hat{V}, \sum_{i=1}^{N} \left\{ m_i(\boldsymbol{U}_i^+ - \boldsymbol{U}_i^-) \cdot (\boldsymbol{V}_i - \frac{\boldsymbol{U}_i^+ + \boldsymbol{U}_i^-}{2}) \right.$$
$$\left. + I_i(\boldsymbol{\Omega}_i^+ - \boldsymbol{\Omega}_i^-) \cdot (\boldsymbol{\omega}_i - \frac{\boldsymbol{\Omega}_i^+ + \boldsymbol{\Omega}_i^-}{2}) \right\}$$
$$+ \sum_{i=1}^{N-1} \sum_{j=i+1}^{N} \sum_{A_{i,j,k} \in S_{i,j}} \left\{ \Phi_{i,j,k}(D_{i,j}(\hat{V}, A_{i,j,k})) \right.$$
$$\left. - \Phi_{i,j,k}(\Big(D_{i,j}(\hat{U}^+, A_{i,j,k})) + D_{i,j}(\hat{U}^-, A_{i,j,k})\Big) /2) \right\} \geq 0. \tag{8}$$

In order to use law (5), we assume that there exist a normal vector directed from solid $j$ toward solid $i$ with $i < j$.

*Remark 4.* The more general situation where irregular solids collide at angular vertices is studied in [4], [10] and [9].

The function

$$\hat{V} \to \Phi_{i,j,k}(D_{i,j}(\hat{V}, A_{i,j,k})) = \Phi_{i,j,k}\left(\boldsymbol{V}_i + \boldsymbol{\omega}_i \times G_i A_{i,j,k} - (\boldsymbol{V}_j + \boldsymbol{\omega}_j \times G_j A_{i,j,k})\right),$$

is convex, thus the function

$$\hat{V} \to \sum_{i=1}^{N-1} \sum_{j=i+1}^{N} \sum_{A_{i,j,k} \in S_{i,j}} \Phi_{i,j,k}(D_{i,j}(\hat{V}, A_{i,j,k})) = \Phi(\hat{V}),$$

is a pseudo-potential. Let us define the scalar product

$$\left\langle \hat{U}, \hat{V} \right\rangle = \sum_{i=1}^{N} \{ m_i \boldsymbol{U}_i \cdot \boldsymbol{V}_i + I_i \boldsymbol{\Omega}_i \cdot \boldsymbol{\omega}_i \}.$$

Thus relation (8) is

$$\left\langle \hat{U}^+ - \hat{U}^-, \hat{V} - \frac{\hat{U}^+ + \hat{U}^-}{2} \right\rangle + \Phi(\hat{V}) - \Phi(\frac{\hat{U}^+ + \hat{U}^-}{2}) \geq 0,$$

or by having $R^{6N}$ equipped with the scalar product $\langle .,. \rangle$

$$-(\hat{U}^+ - \hat{U}^-) \in \partial\Phi(\frac{\hat{U}^+ + \hat{U}^-}{2}),$$

or by letting

$$X = \frac{\hat{U}^+ + \hat{U}^-}{2},$$
$$2\hat{U}^- \in 2X + \partial\Phi(X). \tag{9}$$

The equation (9), assuming $\hat{U}^-$ is known, has one and only one solution because the operator $X \to 2X + \partial\Phi(X)$ is maximal, surjective and strictly monotone. It is classical that the equation (9) is equivalent to the minimization problem

$$Inf\left\{\frac{1}{2}Y^2 + \Phi(Y) - \left\langle 2\hat{U}^-, Y\right\rangle |Y \in R^{6N}\right\}.$$

*The interior percussions are described by the Coulomb law.* If we set in Coulomb law the normal percussion $P_N$ at its actual value, $X$, the constitutive law results from a pseudo-potential of dissipation. Thus to find $X$ is equivalent to solve the equation

$$X = ArgInf\left\{\frac{1}{2}Y^2 + \Phi(\hat{P}_N(X), Y) - \left\langle 2\hat{U}^-, Y\right\rangle |Y \in R^{6N}\right\},$$

where the vector $\hat{P}_N(X)$ contains the actual (but unknown!) values of the quantities $\mu P_N$ which intervene in the tangential constitutive law (6). One can try to solve this problem by an iterative method

$$X_{n+1} = ArgInf\left\{\frac{1}{2}Y^2 + \Phi(\hat{P}_N(X_n), Y) - \left\langle 2\hat{U}^-, Y\right\rangle |Y \in R^{6N}\right\}.$$

It can be shown that the method converges if the coefficients $\mu_{i,j,k}$ of the laws (6) are not too large [7] .

### 1.3 Collision of a deformable solid and a rigid body

**Instantaneous collisions of deformable solids** When it is no more possible to consider that the colliding bodies are rigid (for example, when they vibrate after the collision), we keep the assumption that the collision is instantaneous: each deformable solid has a velocity field $\boldsymbol{U}^-(\boldsymbol{x}, t)$ before the collision and a velocity field $\boldsymbol{U}^+(\boldsymbol{x}, t)$ after the collision. For the sake of simplicity let us investigate one deformable solid with density $\rho$, which occupies a time $t$ the domain $\Omega(t)$, with boundary $\partial\Omega(t)$. This solid collides a fixed plane at time $t$ (figure 5). The contact surface is $\Gamma_1(t)$. An exterior percussion is applied to the solid on $\Gamma_2(t) \subset \partial\Omega(t)$ ($\Gamma_1(t)$ and $\Gamma_2(t)$ are not reduced to a point).

As already said, we focus only on collisions.

*Equations of motion.* During the short duration of the collision, kinematic incompatibilities produce very large stresses in the solid and very large surface forces on the contact surface with the fixed plane. This assumption of instantaneity of the collision leads us to represent these forces by forces concentrated in a very short period of time, i.e., percussion stresses and contact

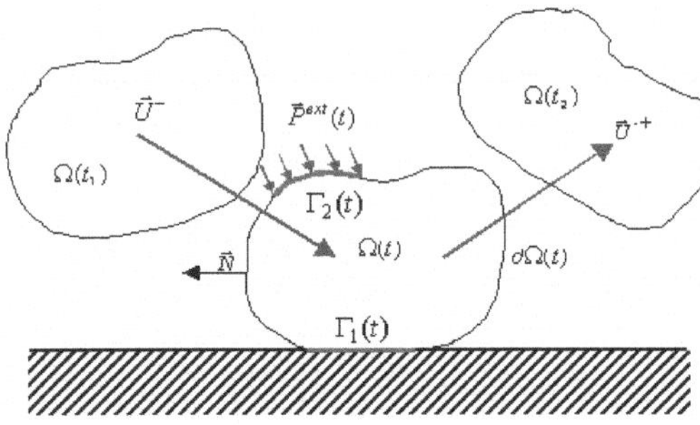


**Fig. 5.** Instantaneous collision of the deformable solid $\Omega$ with a rigid and fixed plane at time $t$.

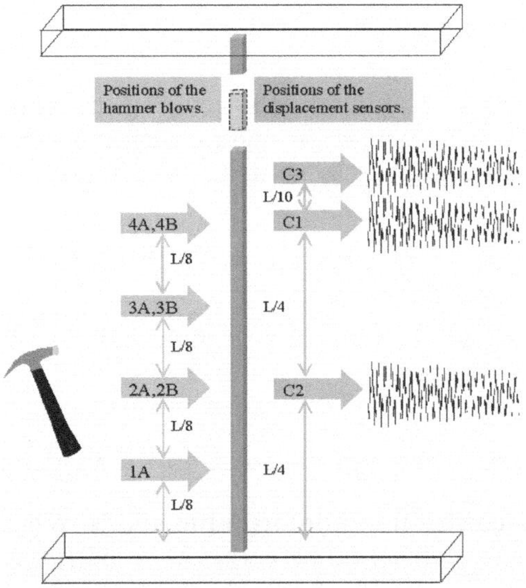


**Fig. 6.** A fixed bar, with length $L$, is blown by a hammer at one of the points $1A$, $2A$, $2B$, $3A$, $3B$, $4A$, or $4B$ and vibrates. For each collision, the displacements of the vibrating bar are recorded at points $C_1$,$C_2$, $C_3$ every $10^{-3}s$ during 10 $s$.

percussions. Thus we choose the work of the interior forces at time $t$ of collision

$$W^{int}(\boldsymbol{V}) = -\int_{\Omega(t)} \Sigma^{int}(\boldsymbol{x},t) : D(\boldsymbol{V}(\boldsymbol{x}))d\Omega - \int_{\Gamma_1(t)} \boldsymbol{P}^{int}(\boldsymbol{x},t)\cdot \boldsymbol{V}(\boldsymbol{x})d\Gamma,$$

where $D(\boldsymbol{V}) = ((V_{i,j} + V_{j,i})/2)$ is the usual strain rate. The brief but very large forces which appear in the collision are represented, as already said, by percussions: the percussion stresses $\Sigma^{int}(\boldsymbol{x},t)$ and the surface percussions $\boldsymbol{P}^{int}(\boldsymbol{x},t)$ on the contact surface $\Gamma_1(t)$. The virtual work of the acceleration forces is

$$W^{acc}(\boldsymbol{V}) = \int_{\Omega(t)} \rho\,[\boldsymbol{U}(\boldsymbol{x},t)] \cdot \boldsymbol{V}(\boldsymbol{x}) d\Omega,$$

where $[\boldsymbol{U}(\boldsymbol{x},t)]$ is the actual discontinuity of velocity. The virtual work of the exterior forces is

$$W^{ext}(\boldsymbol{V}) = \int_{\Gamma_2(t)} \boldsymbol{P}^{ext}(\boldsymbol{x},t) \cdot \boldsymbol{V}(\boldsymbol{x}) d\Omega.$$

The equations of motion at the collision time result from the principle of virtual work

$$\begin{aligned}
&\rho\,[\boldsymbol{U}]\,(\boldsymbol{x},t) = div\Sigma^{int}(\boldsymbol{x},t),\ in\ \Omega(t),\\
&\Sigma^{int}(\boldsymbol{x},t)\boldsymbol{N} = -\boldsymbol{P}^{int}(\boldsymbol{x},t)\ in\ \Gamma_1(t),\\
&\Sigma^{int}(\boldsymbol{x},t)\boldsymbol{N} = \boldsymbol{P}^{ext}(\boldsymbol{x},t)\ in\ \Gamma_2(t),\\
&\Sigma^{int}(\boldsymbol{x},t)\boldsymbol{N} = 0\ in\ \partial\Omega(t)\backslash\left(\Gamma_1(t) \cup \Gamma_2(t)\right),
\end{aligned}$$

where $\boldsymbol{N}$ is the outward normal vector. The constitutive laws can be define by pseudo-potentials of dissipation which involve the impenetrability condition as explained above

$$\Sigma^{int} = \frac{\partial\Phi}{\partial(D(\boldsymbol{U}^+ + \boldsymbol{U}^-))},\ \boldsymbol{P}^{int} = \frac{\partial\Phi_{\Gamma_1}}{\partial(\boldsymbol{U}^+ + \boldsymbol{U}^-)}.$$

Examples of contact percussion with Coulomb law are given in [7]. The most simple pseudo-potential one can use is quadratic $\Phi = (k_v/2)\left(D(\boldsymbol{U}^+ + \boldsymbol{U}^-)\right)^2$, $k_v > 0$. We show in the next section that this simple pseudo-potential is sufficient to describe the impact of a hammer on a deformable bar. Moreover it is possible to identify $k_v$ by experiments.

**Experimental identification of the parameter $k_v$ of the constitutive law** Let us investigate the collision of a hammer with an elastic deformable bar ($\rho = 7800\ kg.m^{-3}$ , $E = 210\ GPa$, $\nu = 0.3$, $\Omega = 4 \times 40 \times 1870\ mm$). The bar is fixed at both ends. The blow of the hammer is an exterior percussion $\boldsymbol{P}^{ext}$ applied to the bar on the contact surface $\Gamma_2$ (the experiments are noted $1A$, $2A$, $2B$, $3A$, $3B$ , $4A$ and $4B$, following the place where the hammer hits the bar (figure 6)).

The quadratic pseudo-potential gives a linear constitutive law

$$\Sigma^{int} = k_v D(\boldsymbol{U}^+ + \boldsymbol{U}^-), k_v > 0. \tag{10}$$

As the bar is at rest before the collision, the velocity $\boldsymbol{U}^+(\boldsymbol{x})$ of the bar after the impact satisfies

$$\begin{aligned}
&k_v \boldsymbol{grad}(div\boldsymbol{U}^+) + k_v \Delta\boldsymbol{U}^+ - \rho\boldsymbol{U}^+ = \boldsymbol{0} \text{ in } \Omega,\\
&\Sigma^{int}\boldsymbol{N} = \boldsymbol{0} \text{ in } \partial\Omega\backslash\left(\Gamma_0 \cup \Gamma_2\right),\\
&\Sigma^{int}\boldsymbol{N} = \boldsymbol{P}^{ext}(\boldsymbol{x}) \text{ in } \Gamma_2,\\
&\boldsymbol{U}^+ = \boldsymbol{0} \text{ in } \Gamma_0.
\end{aligned} \tag{11}$$

The field $\boldsymbol{U}^+(\boldsymbol{x})$ which satisfies this system depends only on $k_v$ for a given $\boldsymbol{P}^{ext}$. To identify $k_v$ we need information on the velocity after the collision, $\boldsymbol{U}^+(\boldsymbol{x})$. Rather than trying to measure it directly, we perform a spectral decomposition of the vibration after the collision and measure the projections of the field $\boldsymbol{U}^+(\boldsymbol{x})$ on the orthogonal eigendisplacements of the bar. Therefore, after the collision, the displacements $\boldsymbol{d}$ of three different points, $\boldsymbol{C}_1, \boldsymbol{C}_2, \boldsymbol{C}_3$ are recorded every $10^{-3}s$.

The collision occurs at time $t = 0$. The assumptions that the collision is instantaneous and that the displacements are continuous functions of time after, imply that the displacement $\boldsymbol{d}(\boldsymbol{y}, t)$ of any point $\boldsymbol{y}$ of the bar is given by

$$\boldsymbol{d}(\boldsymbol{y}, t) = \sum_{s=1}^{\infty} \boldsymbol{d}_s(\boldsymbol{y}) \frac{\sin(\omega_s t)}{\omega_s} \int_{\Omega} \boldsymbol{d}_s(\boldsymbol{z}) \cdot \boldsymbol{U}^+(\boldsymbol{z}) d\Omega,$$

where $(\boldsymbol{d}_s(\boldsymbol{z}), \omega_s)$ is the $s^{th}$ eigenmode of the bar.

On the one side, a spectral analysis of the measured signals $\boldsymbol{d}(\boldsymbol{C}_i, t)$ allows us to get the first three $\alpha_s = \int_{\Omega} \boldsymbol{d}_s(\boldsymbol{z}) \cdot \boldsymbol{U}^+(\boldsymbol{z}) d\Omega$ which are the projection of $\boldsymbol{U}(\boldsymbol{z})$ on the $s^{th}$ eigendisplacement of the bar (the measured eigenfrequencies are 2.8, 10.5 and 23.4 $Hz$). On the other side, the numerical solution to the system (11) is computed for different values of $k_v$ and projected on the three first eigendisplacement of the bar, which are also computed, the values of these projections being called $\beta_s$ (the numerical computations are performed with the program CESAR-LCPC) [13]. The table 1 shows a good agreement between the $\alpha$'s and the $\beta$'s for $k_v = 1{,}2\ 10^7\ Ns^2m^{-2}$. The agreement of the $\alpha_s$ with the $\beta_s$ means that the model and the very simple constitutive law (10) give, at least in this situation, a satisfying description of the collision of the hammer with the bar.

## 2 Collision of solids and fluids

Consider a swimmer diving in a swimming pool. When impacting the water the diver can be horizontal and do a belly flop. In this situation a violent collision occurs with the water. Consider a flat boat in a rough sea: the flat bottom of the ship collides violently with the water. When a solid collides with water very large internal forces appear because of the incompatibility of the velocities of the solid and water: very large stresses inside and very large forces on the contact surface. In many circumstances the duration of the violent contact is very short compared to the duration of the evolution: for instance when skipping stones on the still water of a lake, the duration of the contact of the stone with the water is small compare to its time flight. We decide as a schematisation to consider that the collisions are instantaneous. Thus the velocities are discontinuous functions of the time: for the fluid, there is the velocity $\boldsymbol{U}^-(\boldsymbol{x}, t)$ before a collision at time $t$ and the velocity $\boldsymbol{U}^+(\boldsymbol{x}, t)$ after the collision; for the solid, there is the velocity $\boldsymbol{U}_s^-(\boldsymbol{x}, t) =$

$\boldsymbol{X}^-(t)+\boldsymbol{\Pi}^-(t)\times(\boldsymbol{x}-\boldsymbol{x}_G(t))$ before the collision and the velocity $\boldsymbol{U}_s^+(\boldsymbol{x},t) = \boldsymbol{X}^+(t)+\boldsymbol{\Pi}^+(t)\times(\boldsymbol{x}-\boldsymbol{x}_G(t))$ after the collision. The velocity of the center of mass $\boldsymbol{x}_G(t)$ of the rigid body is $\boldsymbol{X}(t)$ and $\boldsymbol{\Pi}(t)$ is its rotation velocity. Because the collision is instantaneous, the very large internal forces are also concentrated in time: thus they become percussions.

| Collision | Position of measure | $\alpha_1$ | $\alpha_2$ | $\alpha_3$ | $\beta_1$ | $\beta_2$ | $\beta_3$ |
|---|---|---|---|---|---|---|---|
| 1A | C1 | 1.27 | NA | 1.35 | 1.27 | 1.84 | 1.25 |
| | C2 | 1.28 | 1.82 | 1.25 | | | |
| | C3 | 1.25 | 1.54 | 1.32 | | | |
| 2A | C1 | 3.33 | NA | 1.49 | 3.32 | 3.69 | 1.33 |
| | C2 | 3.36 | 3.79 | 2.07 | | | |
| | C3 | 3.27 | 3.32 | 1.43 | | | |
| 2B | C1 | 6.37 | NA | 2.96 | 6.35 | 7.05 | 2.54 |
| | C2 | 6.43 | 7.77 | 2.76 | | | |
| | C3 | 6.26 | 6.83 | 2.30 | | | |
| 3A | C1 | 7.95 | NA | 1.53 | 7.94 | 4.68 | 1.41 |
| | C2 | 8.04 | 5.52 | 1.33 | | | |
| | C3 | 7.82 | 4.43 | 1.60 | | | |
| 3B | C1 | 8.82 | NA | 1.66 | 8.92 | 5.26 | 1.59 |
| | C2 | 8.91 | 5.24 | 1.46 | | | |
| | C3 | 8.67 | 4.88 | 1.75 | | | |
| 4A | C1 | 8.18 | NA | 3.40 | 8.17 | 0.74 | 3.27 |
| | C2 | 8.28 | 0.00 | 3.13 | | | |
| | C3 | 8.04 | 0.00 | 3.41 | | | |
| 4B | C1 | 8.91 | NA | 4.01 | 8.90 | 0.80 | 3.56 |
| | C2 | 9.02 | 0.00 | 3.64 | | | |
| | C3 | 8.76 | 0.00 | 4.03 | | | |

**Table 1.** The recorded projections $\alpha_1$, $\alpha_2$ and $\alpha_3$ and the corresponding computed projections $\beta_1$, $\beta_2$ and $\beta_3$ with $k_v = 1,2\ 10^7\ Ns^2m^{-2}$. When the eigendisplacement is zero at a given point, the corresponding $\alpha$ cannot be measured by analysing the vibration of this point: it is NA (not avaible).

Let consider the system made of the fluid and moving solids. For instance the container of the fluid and a moving obstacle (figure 7). For the sake of simplicity, we suppose there is only one solid (for instance the container). The fluid occupies the domain $\Omega(t)$ and is in contact with the solid on the part $\Gamma_1(t)$ of its boundary $\partial\Omega(t)$. Any time $t$, we consider the virtual velocities $(\boldsymbol{V}(\boldsymbol{x}),\boldsymbol{Y},\boldsymbol{\omega})$ defined on $\Omega(t)$. The mass of the rigid body is $M$ and its mass moment of inertia is $I$. The density of the fluid is $\rho$.

### 2.1 The equations of motion

They result from the principle of virtual work, where the virtual work of the acceleration forces is,

$$W^{acc}(\boldsymbol{V},\boldsymbol{Y},\boldsymbol{\omega}) = \int_\Omega \rho[\boldsymbol{U}(x,t)]\cdot\boldsymbol{V}(\boldsymbol{x})d\Omega + M[\boldsymbol{X}(t)]\cdot\boldsymbol{Y} + I[\boldsymbol{\Pi}(t)]\cdot\boldsymbol{\omega},$$

The virtual work of the interior forces we choose, is a linear function of the virtual strain rate, $D(\boldsymbol{V}) = (1/2)(V_{i,j} + V_{j,i})$, and of the difference of the velocities on the contact surface, $\boldsymbol{V} - \boldsymbol{V}_s$. The percussion stress $\Sigma^{int}(\boldsymbol{x}, t)$ and the interaction percussion $\boldsymbol{P}^{int}(\boldsymbol{x}, t)$ are generalized interior forces which appear when collisions occur. They are, as already said, usual interior forces concentrated in a very short period of time. The virtual work of the interior forces we choose, is

$$W^{int}(\boldsymbol{V}, \boldsymbol{Y}, \boldsymbol{\pi}) = -\int_{\Omega} \Sigma^{int}(\boldsymbol{x}, t) : D(\boldsymbol{V}) d\Omega$$
$$-\int_{\Gamma_1} \boldsymbol{P}^{int}(\boldsymbol{x}, t) \cdot (\boldsymbol{V}(\boldsymbol{x}) - \boldsymbol{V}_s(\boldsymbol{x}))\, d\Gamma$$

The equations of motion are

$$\rho\,[\boldsymbol{U}(\boldsymbol{x}, t)] = div\Sigma^{int}(\boldsymbol{x}, t) \text{ for } \boldsymbol{x} \in \Omega(t),$$
$$\Sigma^{int}(\boldsymbol{x}, t)\boldsymbol{N} = -\boldsymbol{P}^{int}(\boldsymbol{x}, t) \text{ for } \boldsymbol{x} \in \Gamma_1(t),$$
$$\Sigma^{int}(\boldsymbol{x}, t)\boldsymbol{N} = 0 \text{ for } \boldsymbol{x} \in \partial\Omega(t)\backslash\Gamma_1(t),$$
$$M[\boldsymbol{X}(t)] = \int_{\Gamma_1(t)} \boldsymbol{P}^{int}(\boldsymbol{x}, t) d\Gamma,$$
$$I[\boldsymbol{\Pi}(t)] = \int_{\Gamma_1(t)} (\boldsymbol{x} - \boldsymbol{x}_G(t)) \times \boldsymbol{P}^{int}(\boldsymbol{x}, t) d\Gamma.$$

Constitutive laws are needed for $(\Sigma^{int}, \boldsymbol{P}^{int})$.

## 2.2 The constitutive laws

The expression of $W^{int}$ suggests to assume $\Sigma^{int}$ to depend on $D(\boldsymbol{U}^+ + \boldsymbol{U}^-)$. We choose the simple constitutive law:

$$\Sigma^{int} = 2\gamma D(\boldsymbol{U}^+ + \boldsymbol{U}^-) - pI,\ \mathrm{div}(\boldsymbol{U}^+ + \boldsymbol{U}^-) = 0,\ \text{with } \gamma > 0.$$

The incompressibility condition $div(\boldsymbol{U}^+ + \boldsymbol{U}^-) = 0$ or $\mathrm{div}\boldsymbol{U}^+ = 0$ because $\mathrm{div}\boldsymbol{U}^- = 0$, results in the percussion pressure $p$. The interaction percussion $\boldsymbol{P}^{int}$ between the solid and the fluid has to take into account their impenetrability. Thus the percussion is split between a reaction $\boldsymbol{P}^{reac}$ to the impenetrability constraint and a dissipative percussion $\boldsymbol{P}^d$ which modelises all the dissipative interactions between the fluid and the solid,

$$\boldsymbol{P}^{int} = \boldsymbol{P}^d + \boldsymbol{P}^{reac}.$$

The expression of $W^{int}$ suggests to assume $\boldsymbol{P}^d$ to depend on $(\boldsymbol{U}^+ - \boldsymbol{U}_s^+ + \boldsymbol{U}^- - \boldsymbol{U}_s^-)$. Thus we choose again a simple constitutive law

$$\boldsymbol{P}^d(\boldsymbol{x}, t) = K(\boldsymbol{U}^+(\boldsymbol{x}, t) - \boldsymbol{U}_s^+(\boldsymbol{x}, t) + \boldsymbol{U}^-(\boldsymbol{x}, t) - \boldsymbol{U}_s^-(\boldsymbol{x}, t)),\ \text{with } K > 0.$$

Of course the physical parameter $K$ depends on the nature of the solid and fluid colliding. The impenetrability reaction is active only when the risk of interpenetration is present, i.e. when the contact is persistent after the collision. Thus

$$\boldsymbol{P}^{reac} \in \partial I_-(\boldsymbol{U}^+(\boldsymbol{x},t) - \boldsymbol{U}_s^+(\boldsymbol{x},t) \cdot \boldsymbol{N})\boldsymbol{N}, \tag{12}$$

where $\partial I_-$ is the sub-differential set of the indicator function $I_-$ of the set of the negative numbers $R^-$ (see the appendix).

## 2.3 The diver problem

For the sake of simplicity a 2D problem is investigated. It is assumed the swimming pool to be a fixed rectangle. The horizontal swimmer $\Gamma_1$ dives at the middle of the swimming pool (figure 7). It is assumed the thickness of the diver is zero. Its velocities when hitting the water are, $\boldsymbol{X}^- = (0, X^-), \boldsymbol{\Pi}^- = \boldsymbol{0}$.

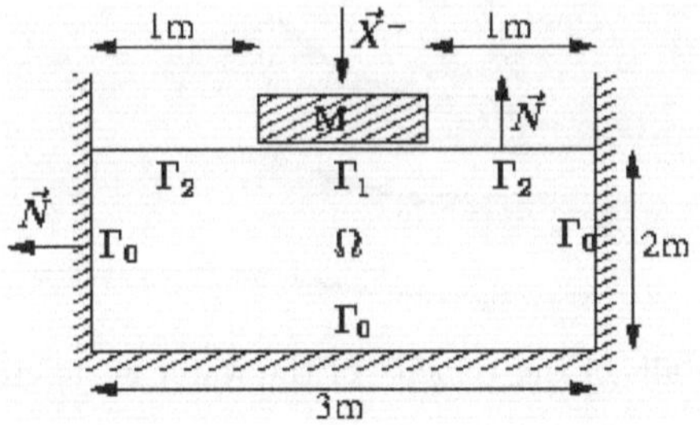


**Fig. 7.** The diver and the swimming pool. The normal velocity of the diver before the collision, $X^-$, is negative.

The diver with mass $M = 100\ Kg$ falls vertically from a height of $1m$. Its vertical velocity is $U^- = \sqrt{2g} = 4.47m/s$. The density of the water is $\rho = 1000Kg/m^3$. The parameter defining the collision constitutive laws are

$$K = 1N/m^2,\ \gamma = 0.25N/m.$$

The equations may be solved by the classical numerical methods for variational inequalities. The water splashes up on the two sides of the diver as shown on the figure 8. The velocity of the diver after the collision is the velocity of the water ($-0.75\ ms^{-1}$), in agreement with every day experiments! Details are shown on figure 9: the velocity is drawn at different depth of the pool. The percussion pressure is shown on the figure 10. The average percussion stress $\Sigma^{int}\boldsymbol{N}$ under the diver is $-373\ Nm^2s^{-1}$ and the average percussion pressure is $371\ Nm^2s^{-1}$.

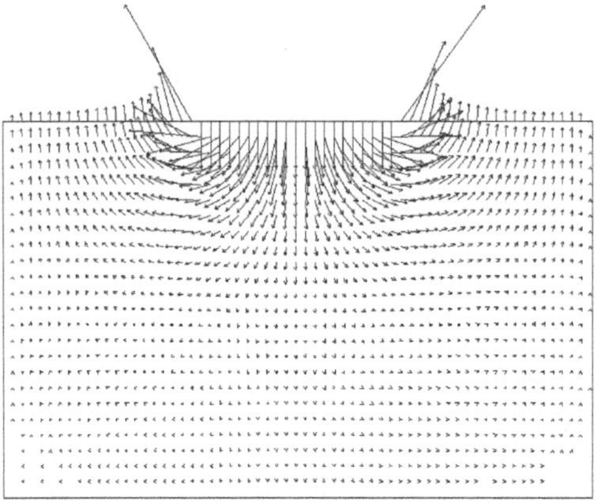

**Fig. 8.** The velocity of the water which splashes up on the two sides of the diver.

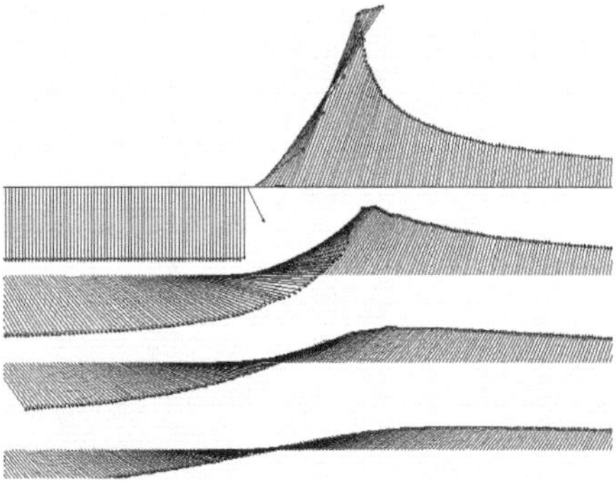

**Fig. 9.** Details of the velocity of the water at different depths.

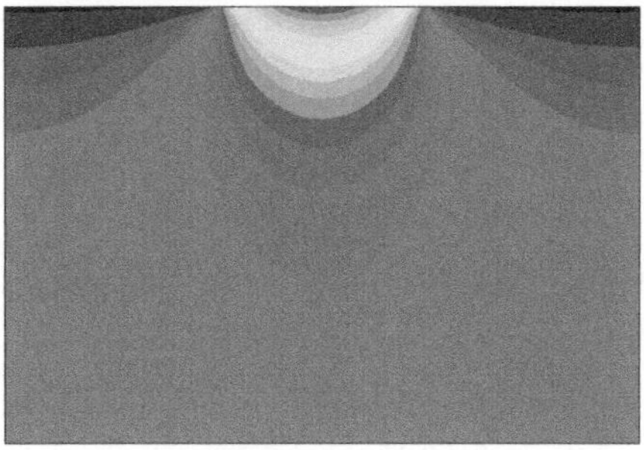

**Fig. 10.** The pressure in the water. Red is a high pressure and blue a low pressure.

## References

1. S. Appol, 2000, Mise au point d'une méthode d'identification pour l'étude des chocs de solides déformables, Projet de fin d'études, Institut Galilée, Université Paris Nord, Villetaneuse.
2. C. Cholet, 1998 a, Chocs de solides rigides, thèse de l'Université Pierre et Marie Curie, Paris.
3. C. Cholet, 1998 b, Collisions d'un point et d'un plan, C. R. Acad. Sci., I, 328, pp 455-458, Paris.
4. E. Dimnet, M. Frémond, 1999, Chocs de solides rigides, 4e Colloque national de calcul des structures, Giens.
5. E. Dimnet, M. Frémond, 1999, Choc instantané d'un solide déformable sur un support rigide, 14e congrès français de mécanique, Toulouse.
6. E. Dimnet, M. Frémond, 2000, Instantaneous collisions of solids, European Congress on Computational Methods in Applied Sciences and Engineering, Barcelona.
7. E. Dimnet, 2001, Chocs de solides et intégration des équations d'évolution des systèmes multi-solides, thèse de l'Ecole Nationale des Ponts et Chaussées, Paris.
8. M. Frémond, 1995 a, Rigid bodies collisions, Physics letters A, 204, 33-41.
9. M. Frémond, 2000, Collision of a wedge with a plane, Computational and Applied Mathematics, Vol 19, N°2, pp 1-10.
10. M. Frémond, 2001, Non-smooth Thermomechanics, Springer-Verlag, Heidelberg.
11. P. Germain, 1973, Mécanique des milieux continus, Masson, Paris.
12. J. J. Moreau, 1966, Fonctionnelles convexes, Séminaire sur les équations aux dérivées partielles, Collége de France, Paris.
13. F. Platel, 2000, Identification d'une loi de choc instantané pour un solide déformable en acier, Rapport de stage scientifique de l'Ecole nationale des Ponts et Chaussées, Paris.

# 3 Appendix

Let us consider the set $R \cup \{+\infty\}$ where the addition and the multiplication by strickly positive numbers are completed by $+\infty + b = +\infty$ and $a(+\infty) = +\infty$ $(a > 0)$. A function from $R^n$ into $R \cup \{+\infty\}$ is convex if and only if

$$\forall x, y \in \mathbf{R}^n, \forall \theta \in \left]0, 1\right[, f(\theta x + (1-\theta)y) \leq \theta f(x) + (1-\theta) f(y).$$

A convex function in not always differentiable but a generalised derivative may be defined: an element $z \in R^n$ is a subgradient of the convex function $f$ at point $x$ if and only if

$$\forall y \in \mathbf{R}^n, f(y) \geq f(x) + (y - x) \cdot z, \tag{13}$$

the set of the sub-gradients is denoted $\partial f(x)$. The indicator function $I_+$ of the set of the positive numbers $R^+$ is defined by $I_+(x) = 0 \; if \; x \geq 0$ and $I_+(x) =$

$+\infty$ *if* $x < 0$. Its subdifferential set is shown in figure 11. The indicator function, $I_-$, of the set of the negative numbers $R^-$ is defined in the same way. The indicator function $I_K$ of the interval $K = [(\boldsymbol{D}(\boldsymbol{U}^-, \boldsymbol{\Omega}^-, A) \cdot \boldsymbol{N})/2, \infty]$ is defined by $I_K(x) = 0$ *if* $x \in K$ and $I_K(x) = +\infty$ *if* $x \notin K$. Their subdifferential sets are shown in figure 11.

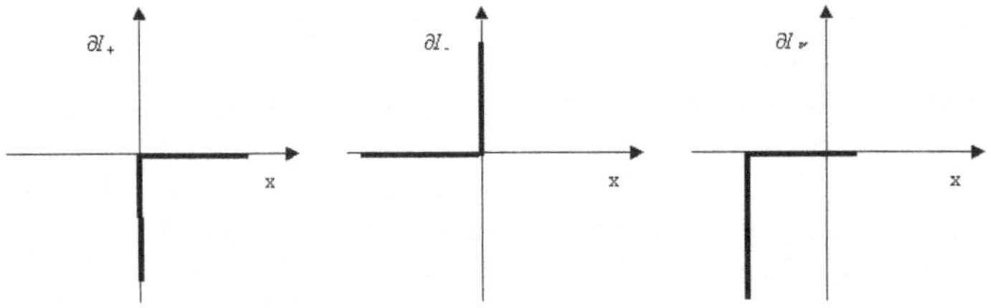


**Fig. 11.** The subdifferential set $\partial I_+$ of the indicator function of the set of the positive numbers $R^+$: $\partial I_+(0) = R^-$, $\partial I_+(x) = \{0\}$ for $x > 0$ and $\partial I_+(x) = \emptyset$ for $x < 0$. The subdifferential set $\partial I_-$ of the indicator function $I_-$ of the set of the negative numbers $R^-$. The subdifferential set $\partial I_K$ of the indicator function $I_K$ of the set $K$.

The following theorem is useful for reading the constitutive laws

**Theorem 1.** *Let a convex function* $f \neq +\infty$. *If this function is subdifferentiable at point* $x$, *i.e. if* $\partial f(x) \neq \emptyset$, *it is finite at this point:* $f(x) < +\infty$.

This theorem applies in the numerous constitutive laws: $P^{reac} \in \partial I_{\hat{K}}(D)$ (with $\hat{K} = R^+$, $R^-$ *or* $K$): because $\partial I_{\hat{K}}(D)$ is not empty it results that $I_{\hat{K}}(D) < +\infty$ which implies $I_{\hat{K}}(D) = 0$ and $D \in \hat{K}$. Thus the constitutive law $P^{reac} \in \partial I_{\hat{K}}(D)$ implies that the constraint on the velocity $D$ is satisfied. It gives also the value of the reaction to the constraint.

Printing: Strauss GmbH, Mörlenbach
Binding: Schäffer, Grünstadt

GPSR Compliance

*The European Union's (EU) General Product Safety Regulation (GPSR) is a set of rules that requires consumer products to be safe and our obligations to ensure this.*

*If you have any concerns about our products, you can contact us on ProductSafety@springernature.com*

In case Publisher is established outside the EU, the EU authorized representative is:

Springer Nature Customer Service Center GmbH
Europaplatz 3
69115 Heidelberg, Germany

**Batch number: 10755312**

Printed by Printforce, the Netherlands